Lecture Notes in Mathematics 1577

Nobuaki Obata

White Noise Calculus and Fock Space

Springer-Verlag
Berlin Heidelberg New York
London Paris Tokyo
Hong Kong Barcelona
Budapest

Author

Nobuaki Obata
Department of Mathematics
School of Science
Nagoya University
Nagoya, 464-01, Japan

Mathematics Subject Classification (1991): 46F25, 46E50, 47A70, 47B38, 47D30, 47D40, 60H99, 60J65

ISBN 3-540-57985-0 Springer-Verlag Berlin Heidelberg New York
ISBN 0-387-57985-0 Springer-Verlag New York Berlin Heidelberg

CIP-Data applied for

SPIN: 10130019 46/3140-543210 - Printed on acid-free paper

Contents

Introduction

The white noise calculus (or analysis) was launched out by Hida [1] in 1975 with his lecture notes on generalized Brownian functionals. This new approach toward an infinite dimensional analysis was deeply motivated by Lévy [1] who considerably developed functional analysis on $L^2(0,1)$ and actually analysis of Brownian functionals. The root of white noise calculus is to switch a functional of Brownian motion $f(B(t); t \in \mathbb{R})$ with one of white noise $\phi(\dot{B}(t); t \in \mathbb{R})$, where $\dot{B}(t)$ is a time derivative of a Brownian motion $B(t)$. Although each Brownian path $B(t)$ is not smooth enough, $\dot{B}(t)$ is thought of as a generalized stochastic process and ϕ is realized as a generalized white noise functional in our language. We may thereby regard $\{\dot{B}(t)\}$ as a collection of infinitely many independent random variables and hence a coordinate system of an infinite dimensional space.

The mathematical framework of the white noise calculus is based upon an infinite dimensional analogue of the Schwartz distribution theory, where the role of the Lebesgue measure on $\mathbb{R}^n$ is played by the Gaussian measure μ on the dual of a certain nuclear space E. In the classical case where $\dot{B}(t)$ is formulated, we take $E = \mathcal{S}(\mathbb{R})$ and the Gaussian measure μ on E^* defined by the characteristic functional:

$$\exp\left(-\frac{|\xi|^2}{2}\right) = \int_{E^*} e^{i\langle x, \xi\rangle} \mu(dx), \qquad \xi \in E,$$

where $|\xi|$ is the usual L^2-norm of ξ. Then the Hilbert space $(L^2) = L^2(E^*, \mu)$ is canonically isomorphic to the (Boson) Fock space over $L^2(\mathbb{R})$ through the Wiener-Itô-Segal isomorphism and links the test and generalized functionals. Namely, in a specific way (called *standard construction*) we construct a nuclear Fréchet space (E) densely and continuously imbedded in (L^2), and by duality we obtain a Gelfand triple:

$$(E) \subset (L^2) = L^2(E^*, \mu) \subset (E)^*.$$

An element in (E) is a *test white noise functional* and hence an element in $(E)^*$ is a *generalized white noise functional.* The above picture is easily understood as a direct analogy of $\mathcal{S}(\mathbb{R}^n) \subset L^2(\mathbb{R}^n) \subset \mathcal{S}'(\mathbb{R}^n)$ which is a frame of the Schwartz distribution theory. Then, $\dot{B}(t) = x(t)$, $x \in E^*$, gives us a realization of the time derivative of a Brownian motion and, in fact, $x \mapsto x(t)$ becomes a generalized white noise functional for each fixed $t \in \mathbb{R}$.

In our actual discussion we do not restrict ourselves to the case of $E = \mathcal{S}(\mathbb{R})$ and $H = L^2(\mathbb{R})$ but deal with a more general function space on a topological space T. Typically T is a time-parameter space and is often taken to be a more general topological space where quantum field theory may be formulated. Again $\{x(t); t \in T\}$ is considered as a coordinate system of E^* intuitively. In fact, within our framework we may discuss not only functionals in $\{x(t); t \in T\}$ but also operators derived from this coordinate system. The coordinate differential operator $\partial_t = \partial/\partial x(t)$ is well defined as a continuous derivation on (E). We have also multiplication operators by coordinate functions $x(t)$, which are, in fact, operators from (E) into $(E)^*$. Furthermore, ∂_t^* is a continuous linear operator on $(E)^*$. The operators ∂_t and ∂_t^* correspond respectively to an *annihilation operator* and a *creation operator* at a point $t \in T$ and they satisfy the so-called canonical commutation relation in a generalized sense. The above mentioned formulation was consolidated in the basic works of Kubo and Takenaka [1]-[4] and has been widely accepted.

The main purpose of these lecture notes is to develop operator theory on white noise functionals as well as to offer a systematic introduction to white noise calculus. From that point of view it is most remarkable that we are free from smeared creation and annihilation operators. In other words, ∂_t and ∂_t^* are not operator-valued distributions but usual operators for themselves. This leads us to an integral kernel operator:

$$\Xi_{l,m}(\kappa) = \int_{T^{l+m}} \kappa(s_1, \cdots, s_l, t_1, \cdots, t_m)\partial_{s_1}^* \cdots \partial_{s_l}^* \partial_{t_1} \cdots \partial_{t_m} ds_1 \cdots ds_l dt_1 \cdots dt_m,$$

where κ is a *distribution* in $l+m$ variables. The use of distributions as integral kernels allows us to discuss a large class of operators on Fock space. In fact, *every* continuous operator Ξ from (E) into $(E)^*$ admits a unique decomposition into a sum of integral kernel operators:

$$\Xi\phi = \sum_{l,m=0}^{\infty} \Xi_{l,m}(\kappa_{l,m})\phi, \qquad \phi \in (E),$$

where the series converges in $(E)^*$. Moreover, if Ξ is a continuous operator from (E) into itself, the series converges in (E). In the process we investigate precise norm estimates of such operators and obtain a method of reconstructing an operator from its symbol. The above expression is called *Fock expansion* and will play a key role in our discussion.

Although applications of white noise calculus are widely spreading, the present lecture notes are strongly oriented toward *infinite dimensional harmonic analysis*. The clue to go on is found in the following three topics: (i) infinite dimensional rotation group; (ii) Laplacians; (iii) Fourier transform. Being almost as new as the white noise calculus, they have been so far discussed somehow separately. Since the very beginning of the development Hida has emphasized the importance of the infinite dimensional rotation group $O(E; H)$, that is, the group of automorphisms of the Gelfand triple $E \subset H \subset E^*$. In fact, it played an interesting role in the study of symmetry of Brownian motion and Gaussian random fields. There are various candidates for infinite dimensional Laplacians which possess some typical properties of a finite dimensional Laplacian. So far the Gross Laplacian Δ_G, the number operator N and the Lévy Laplacian Δ_L have been found to be important in white noise calculus, though the Lévy Laplacian is not discussed in these lecture notes. As for Fourier transform, among some candidates that have been discussed Kuo's Fourier transform (simply called the *Fourier transform* hereafter) has been found well suited to white noise calculus.

In these lecture notes the above listed three subjects are treated systematically by means of our operator calculus and are found closely related to each other. For example, the Gross Laplacian Δ_G and the number operator N are characterized by their rotation-invariance. The Fourier transform intertwines the coordinate differential operators and coordinate multiplication oprators just as in the case of finite dimension and, this property actually characterizes the Fourier transform. Moreover, the Fourier transform is imbedded in a one-parameter transformation group of the generalized white noise functionals (called the *Fourier-Mehler transform*) and its infinitesimal generator is expressed with Δ_G and N. These results would suggest a fruitful application of white noise calculus to infinite dimensional harmonic analysis. It is also expected that our operator calculus is useful in some problems in quantum field theory and quantum probability.

As is well known, a lot of efforts to develop distribution theories on an infinite dimensional space equipped with Gaussian measure have been made by many authors. In fact, mathematical study of Brownian motion or equivalently of white noise is now one of the most important and vital fields of mathematics toward infinite dimensional analysis.

Since the main purpose is to develop an operator theory on white noise functionals, the present lecture notes are mostly based on a functional analytic point of view rather than probability theory or stochastic analysis. In Chapter 1 we survey some fundamentals in functional analysis required during the main discussion and propose a notion of a standard countably Hilbert space which makes the discussion clearer. The purpose of Chapter 2 is to establish the well-known Wiener-Itô-Segal isomorphism between $L^2(E^*, \mu)$ and the Fock space. Chapter 3 is devoted to a study of generalized white noise functionals. In Chapter 4 we develop an operator theory on white noise functionals, or equivalently on Fock space, in terms of Hida's differential operators ∂_t and their duals ∂_t^*. By means of the operator theory we discuss in Chapter 5 a few topics toward harmonic analysis including first order differential operators, the number operator, the Gross Laplacians, infinite dimensional rotation group, Fourier transform and certain one-parameter transformation groups. Chapter 6 is added after finishing the first draft of these lecture notes. We discuss integral-sum kernel operators, the finite dimensional calculus derived from our framework and a generalization to cover vector-valued white noise functionals. These topics are expected to open a new area in infinite dimensional analysis.

ACKNOWLEDGEMENTS

First of all I would like to express my sincere gratitude to Professor H. Heyer who invited me to Tübingen to do research within his working group. The stay there was supported by the Alexander von Humboldt Foundation which I appreciate very much. In fact, the present work is based on a series of my lectures in the Arbeitsgemeinschaft "Stochastik und Analysis" during the Summer Semester 1991. The lectures aimed at providing fundamentals of white noise calculus and at introducing some new aspects of harmonic analysis and quantum probability theory. Professor H. Heyer encouraged me to write up the notes and to expand them for publication. My special thanks are extended to Professors B. Kümmerer, E. Siebert and A. Wolff for their kind hospitality in Tübingen.

I owe also special thanks to Professors Ju. G. Kondrat'ev, H.-H. Kuo, L. Streit and J.-A. Yan for interesting discussion. During my writing these lecture notes, I learnt many relevant works made under the name of quantum probability. Let me mention with special gratitude the names of Professors L. Accardi, V. P. Belavkin, R. Hudson, J. M. Lindsay, P. A. Meyer and M. Schürmann.

My basic references have been among others the works of Hida-Potthoff [1], Kuo [7], [9], Lee [3] and Yan [4] which I appreciated highly. The readers are recommended to consult the recently published monograph Hida-Kuo-Potthoff-Streit [1] which contains different topics and various applications. It will complement our discussion certainly.

Finally I am extremely grateful to Professor T. Hida for his constant encouragement. He initiated the white noise calculus around 1975 and remains always a fount of knowledge.

January 1994 Nobuaki Obata
Nagoya, Japan

Chapter 1

Prerequisites

1.1 Locally convex spaces in general

We first agree that all vector spaces under consideration are over the real numbers $\mathbb{R}$ or the complex numbers $\mathbb{C}$. A topological vector space $\mathfrak{X}$ is called *locally convex* if the topology of $\mathfrak{X}$ is Hausdorff and given by a family of seminorms $\{\|\cdot\|_\alpha\}_{\alpha\in A}$. Then the seminorms are called *defining seminorms* for $\mathfrak{X}$. Without changing the topology we may choose a *directed* family of defining seminorms for $\mathfrak{X}$, which means that for any $\alpha, \beta \in A$ there exists $\gamma \in A$ such that $\|\xi\|_\alpha \le \|\xi\|_\gamma$ and $\|\xi\|_\beta \le \|\xi\|_\gamma$ for all $\xi \in \mathfrak{X}$. In that case A becomes a directed set naturally. Unless otherwise stated, $\mathfrak{X} \cong \mathfrak{Y}$ means that two locally convex spaces $\mathfrak{X}$ and $\mathfrak{Y}$ are isomorphic as topological vector spaces.

For a systematic study of locally convex spaces we introduce general notion of projective and inductive systems and their limits. Let $\{\mathfrak{X}_\alpha\}_{\alpha\in A}$ be a family of locally convex spaces. The *direct product*

$$\prod_{\alpha\in A} \mathfrak{X}_\alpha = \{(\xi_\alpha)_{\alpha\in A}\,;\, \xi_\alpha \in \mathfrak{X}_\alpha\}$$

is always equipped with the weakest locally convex topology such that the canonical projection $p_\beta : \prod_{\alpha\in A} \mathfrak{X}_\alpha \to \mathfrak{X}_\beta$ is continuous for all $\beta \in A$. The *direct sum*

$$\bigoplus_{\alpha\in A} \mathfrak{X}_\alpha = \left\{(\xi_\alpha)_{\alpha\in A} \in \prod_{\alpha\in A} \mathfrak{X}_\alpha\,;\, \xi_\alpha = 0 \text{ except finitely many } \alpha \in A\right\}$$

is equipped with the strongest locally convex topology such that the canonical injection $i_\beta : \mathfrak{X}_\beta \to \bigoplus_{\alpha\in A} \mathfrak{X}_\alpha$ is continuous for all $\beta \in A$.

Let $\{\mathfrak{X}_\alpha\}_{\alpha\in A}$ be a family of locally convex spaces, with A being a directed set. Suppose that we are given a continuous linear map $f_{\alpha,\beta} : \mathfrak{X}_\beta \to \mathfrak{X}_\alpha$ for any pair $\alpha, \beta \in A$ with $\alpha \le \beta$. Then $\{\mathfrak{X}_\alpha, f_{\alpha,\beta}\}$ is called a *projective system* of locally convex spaces if (i) $f_{\alpha,\alpha} = \text{id.}$; and (ii) $f_{\alpha,\gamma} = f_{\alpha,\beta} f_{\beta,\gamma}$ whenever $\alpha \le \beta \le \gamma$. Then

$$\operatorname*{proj\,lim}_{\alpha\in A} \mathfrak{X}_\alpha = \left\{(\xi_\alpha)_{\alpha\in A} \in \prod_{\alpha\in A} \mathfrak{X}_\alpha\,;\, f_{\alpha,\beta}(\xi_\beta) = \xi_\alpha \text{ whenever } \alpha \le \beta\right\}$$

with the relative topology induced from $\prod_{\alpha\in A} \mathfrak{X}_\alpha$ is called the *projective limit* of $\{\mathfrak{X}_\alpha, f_{\alpha,\beta}\}$. So far as the projective limit is under consideration, it suffices to consider a *reduced* projective system; namely, every canonical projection $p_\beta : \operatorname{proj\,lim}_{\alpha\in A} \mathfrak{X}_\alpha \to \mathfrak{X}_\beta$ has a dense image.

We now introduce a dual object. Let $\{\mathfrak{X}_\alpha\}_{\alpha\in A}$ be the same as above and suppose that we are given a continuous linear operator $g_{\alpha,\beta} : \mathfrak{X}_\beta \to \mathfrak{X}_\alpha$ for all pair $\alpha, \beta \in A$ with $\alpha \ge \beta$. Then $\{\mathfrak{X}_\alpha, g_{\alpha,\beta}\}$ is called an *inductive system* of locally convex spaces if (i) $g_{\alpha,\alpha} = \text{id.}$; and (ii) $g_{\alpha,\gamma} = g_{\alpha,\beta} g_{\beta,\gamma}$ whenever $\alpha \ge \beta \ge \gamma$. Consider $\sum_{\alpha\ge\beta} \operatorname{Ran}(i_\beta - i_\alpha g_{\alpha,\beta})$ which is a subspace of $\bigoplus_{\alpha\in A} \mathfrak{X}_\alpha$ generated by the ranges of the linear maps $i_\beta - i_\alpha g_{\alpha,\beta}$, where α, β run over all pairs with $\alpha \ge \beta$. If $\sum_{\alpha\ge\beta} \operatorname{Ran}(i_\beta - i_\alpha g_{\alpha,\beta})$ is closed, the quotient space

$$\operatorname*{ind\,lim}_{\alpha\in A} \mathfrak{X}_\alpha = \bigoplus_{\alpha\in A} \mathfrak{X}_\alpha \Big/ \sum_{\alpha\ge\beta} \operatorname{Ran}(i_\beta - i_\alpha g_{\alpha,\beta})$$

equipped with the quotient topology is called the *inductive limit* of $\{\mathfrak{X}_\alpha, g_{\alpha,\beta}\}$.

If $\|\cdot\|$ is a seminorm on a vector space $\mathfrak{X}$, then $\mathfrak{N} = \{\xi \in \mathfrak{X}; \|\xi\| = 0\}$ becomes a subspace of $\mathfrak{X}$ and the quotient space $\mathfrak{X}/\mathfrak{N}$ admits a natural *norm* which is denoted by the same symbol. The completion of $\mathfrak{X}/\mathfrak{N}$ with respect to this norm $\|\cdot\|$ is called the *Banach space associated with the seminorm* $\|\cdot\|$. Now consider two seminorms $\|\cdot\|_\alpha$ and $\|\cdot\|_\beta$ satisfying $\|\xi\|_\alpha \le C\|\xi\|_\beta$, $\xi \in \mathfrak{X}$, for some $C \ge 0$. Note that $\mathfrak{N}_\beta = \{\xi \in \mathfrak{X}; \|\xi\|_\beta = 0\} \subset \mathfrak{N}_\alpha = \{\xi \in \mathfrak{X}; \|\xi\|_\alpha = 0\}$. Let $\mathfrak{X}_\alpha$ and $\mathfrak{X}_\beta$ be the Banach spaces associated with $\|\cdot\|_\alpha$ and $\|\cdot\|_\beta$, respectively. Then, the canonical map from $\mathfrak{X}/\mathfrak{N}_\beta$ onto $\mathfrak{X}/\mathfrak{N}_\alpha$ extends to a continuous linear map $f_{\alpha,\beta} : \mathfrak{X}_\beta \to \mathfrak{X}_\alpha$.

Proposition 1.1.1 *Let $\mathfrak{X}$ be a locally convex space with a directed family of defining seminorms $\{\|\cdot\|_\alpha\}_{\alpha\in A}$. Then, notations being as above, $\{\mathfrak{X}_\alpha, f_{\alpha,\beta}\}$ becomes a reduced projective system of Banach spaces. If in addition $\mathfrak{X}$ is complete, $\mathfrak{X} \cong \operatorname{proj\,lim}_{\alpha\in A} \mathfrak{X}_\alpha$.*

Let $\mathfrak{X}$ be a locally convex space with defining seminorms $\{\|\cdot\|_\alpha\}_{\alpha\in A}$. A subset $S \subset \mathfrak{X}$ is called *bounded* if $\sup_{\xi\in S} \|\xi\|_\alpha < \infty$ for all $\alpha \in A$. Let $\mathfrak{X}^*$ be the *dual space* of $\mathfrak{X}$, i.e., the space of continuous linear functionals on $\mathfrak{X}$ and we denote the canonical bilinear form on $\mathfrak{X}^* \times \mathfrak{X}$ by $\langle\cdot\,,\cdot\rangle$ or similar symbols. Unless otherwise stated, $\mathfrak{X}^*$ always carries the *strong dual topology* or the *topology of bounded convergence*. This topology is defined by the seminorms:

$$\|x\|_S = \sup_{\xi\in S} |\langle x\,,\xi\rangle|, \qquad x \in \mathfrak{X}^*,$$

where S runs over the bounded subsets of $\mathfrak{X}$. In that case $\mathfrak{X}^*$ is called the *strong dual space* as well. For a continuous linear operator T from a locally convex space $\mathfrak{X}$ into another $\mathfrak{Y}$ its adjoint T^* is defined by $\langle T^*y\,,\xi\rangle = \langle y\,,T\xi\rangle$, $y \in \mathfrak{Y}^*$, $\xi \in \mathfrak{X}$. Then T^* becomes a continuous linear operator from $\mathfrak{Y}^*$ into $\mathfrak{X}^*$.

In accord with Proposition 1.1.1 we can discuss the dual space of a locally convex space. We keep the notations there. Since the canonical map $p_\alpha : \mathfrak{X} \to \mathfrak{X}_\alpha$ has a dense image, its adjoint map $p_\alpha^* : \mathfrak{X}_\alpha^* \to \mathfrak{X}^*$ is injective and thereby $\mathfrak{X}_\alpha^*$ is regarded as a subspace of $\mathfrak{X}^*$. In that case $\mathfrak{X}_\alpha^*$ consists of linear functionals on $\mathfrak{X}$ which are continuous with respect to $\|\cdot\|_\alpha$. Therefore,

$$\mathfrak{X}^* = \bigcup_{\alpha\in A} \mathfrak{X}_\alpha^* \qquad \text{as vector spaces.}$$

Note also that $\mathfrak{X}_\alpha^* \subset \mathfrak{X}_\beta^*$ for $\alpha \leq \beta$. Namely, in a purely algebraic sense $\mathfrak{X}^*$ is the inductive limit of $\{\mathfrak{X}_\alpha^*\}$. In general, if $\{\mathfrak{X}_\alpha, f_{\alpha,\beta}\}$ is a projective system of locally convex spaces, $\{\mathfrak{X}_\alpha^*, f_{\alpha,\beta}^*\}$ becomes an inductive system of locally convex spaces in an obvious way. Unfortunately, with respect to the strong dual topology $\mathfrak{X}^* \cong \operatorname{ind\,lim}_{\alpha\in A} \mathfrak{X}_\alpha^*$ does not hold in general. While, it is true whenever $\mathfrak{X}^*$ and $\mathfrak{X}_\alpha^*$ are equipped with the Mackey topologies $\tau(\mathfrak{X}^*, \mathfrak{X})$ and $\tau(\mathfrak{X}_\alpha^*, \mathfrak{X}_\alpha)$, respectively. Instead of going into a detailed topological argument we note a class of locally convex spaces $\mathfrak{X}$ for which the strong dual topology coincides with the Mackey topology $\tau(\mathfrak{X}^*, \mathfrak{X})$.

A locally convex space is called *Fréchet* if it is metrizable and complete. Recall that a locally convex space is metrizable if and only if it admits a countable set of defining seminorms. A locally convex space $\mathfrak{X}$ is called *reflexive* if the canonical injection $\mathfrak{X} \to \mathfrak{X}^{**}$ is a topological isomorphism, where $\mathfrak{X}^{**}$ is the strong bidual of $\mathfrak{X}$. It is known that for a reflexive Fréchet space $\mathfrak{X}$ the strong dual topology on $\mathfrak{X}^*$ coincides with the Mackey topology $\tau(\mathfrak{X}^*, \mathfrak{X})$. Since the projective limit of a sequence of reflexive Fréchet spaces is again a reflexive Fréchet space, we have the following

Proposition 1.1.2 *Let $\{\mathfrak{X}_n\}_{n=1}^\infty$ be a reduced projective sequence of reflexive Fréchet spaces. Then,*

$$\left(\operatorname*{proj\,lim}_{n\to\infty} \mathfrak{X}_n\right)^* \cong \operatorname*{ind\,lim}_{n\to\infty} \mathfrak{X}_n^*,$$

where the strong dual topologies are taken into consideration.

We note another important property of a Fréchet space (in fact, a characteristic property of a barreled topological vector space).

Proposition 1.1.3 *Let $\mathfrak{X}$ be a Fréchet space. Then for a subset $S \subset \mathfrak{X}^*$ the following four properties are equivalent:*

(i) *S is equicontinuous, i.e., if $\{\|\cdot\|_\alpha\}_{\alpha\in A}$ is a directed family of defining seminorms for $\mathfrak{X}$, one may find $C \geq 0$ and $\alpha \in A$ such that $|\langle x, \xi\rangle| \leq C\,\|\xi\|_\alpha$ for all $\xi \in \mathfrak{X}$ and $x \in S$;*

(ii) *S is (strongly) bounded;*

(iii) *S is weakly bounded;*

(iv) *S is relatively weakly compact.*

1.2 Countably Hilbert spaces

A seminorm $\|\cdot\|$ on a vector space $\mathfrak{X}$ over $\mathbf{R}$ (resp. $\mathbf{C}$) is called *Hilbertian* if it is induced by some non-negative, symmetric bilinear (resp. hermitian sesquilinear) form $(\cdot,\cdot)$ on $\mathfrak{X} \times \mathfrak{X}$, namely if $\|\xi\|^2 = (\xi, \xi)$ for all $\xi \in \mathfrak{X}$. Here it is not assumed that $(\xi, \xi) = 0$ implies $\xi = 0$. We further agree that a hermitian sesquilinear form is linear on the right and antilinear on the left. The Banach space associated with a Hilbertian seminorm becomes a Hilbert space in an obvious way.

A complete locally convex space $\mathfrak{X}$ is called a *countably Hilbert space* or a *CH-space* for brevity if it admits a countable set of defining Hilbertian seminorms. We first note the following

Proposition 1.2.1 *Any CH-space is a projective limit of a reduced projective sequence of Hilbert spaces, and therefore, is a reflexive Fréchet space.*

Then, in view of Proposition 1.1.2 we have

Proposition 1.2.2 *Let $\mathfrak{X}$ be a CH-space and let $\{H_n, f_{m,n}\}$ be a reduced projective sequence of Hilbert spaces such that $\mathfrak{X} \cong \operatorname{proj\,lim}_{n\to\infty} H_n$. Then, $\{H_n^*, f_{m,n}^*\}$ becomes an inductive sequence of Hilbert spaces and $\mathfrak{X}^* \cong \operatorname{ind\,lim}_{n\to\infty} H_n^*$. Moreover, $\{H_n^*\}_{n=0}^{\infty}$ is regarded as an increasing family of subspaces of $\mathfrak{X}^*$ and $\mathfrak{X}^* = \bigcup_{n=0}^{\infty} H_n^*$ as vector spaces.*

We shall be mostly concerned with a particular class (or construction) of CH-spaces. The following general result will be useful.

Lemma 1.2.3 *Let A be a positive linear operator in a complex Hilbert space $\mathfrak{H}$. Then A is selfadjoint if and only if $(1 + A)\operatorname{Dom}(A) = \mathfrak{H}$.*

We first consider the complex case. Let $\mathfrak{H}$ be a complex Hilbert space with norm $\|\cdot\|_0$ and let A be a selfadjoint operator in $\mathfrak{H}$ with (dense) domain $\operatorname{Dom}(A) \subset \mathfrak{H}$. Suppose $\inf \operatorname{Spec}(A) > 0$ and put

$$\rho = (\inf \operatorname{Spec}(A))^{-1}. \tag{1.1}$$

According to the spectral theory we may define a (positive) selfadjoint operator A^p for all $p \in \mathbf{R}$ with (maximal) domain $\operatorname{Dom}(A^p) \subset \mathfrak{H}$. For the moment suppose $p \geq 0$. Since $0 \notin \operatorname{Spec}(A^p)$, by definition A^p admits a dense range and bounded inverse. In fact, we see from Lemma 1.2.3 that the range of A^p coincides with the whole $\mathfrak{H}$ because we have $\inf \operatorname{Spec}(A^p) > 0$ by assumption. Therefore $(A^p)^{-1}$ is everywhere defined bounded operator on $\mathfrak{H}$. In that case $(A^p)^{-1} = A^{-p}$, $p \geq 0$, in particular, $\operatorname{Dom}(A^{-p}) = \mathfrak{H}$, and

$$\|A^{-p}\|_{\mathrm{op}} = \rho^p, \qquad p \geq 0. \tag{1.2}$$

Note also that

$$A^{-p}A^{-q} = A^{-(p+q)}, \qquad A^pA^q \subset A^{p+q}, \qquad p, q \geq 0.$$

It is known that the closure of A^pA^q coincides with A^{p+q}, $p, q \geq 0$.

We now introduce a family of Hilbertian norms:

$$\|\xi\|_p = \|A^p\xi\|_0, \qquad \xi \in \operatorname{Dom}(A^p), \quad p \in \mathbf{R}. \tag{1.3}$$

Note that $\operatorname{Dom}(A^q) \subset \operatorname{Dom}(A^p)$ whenever $q \geq p \geq 0$. In fact, by (1.2) we have

$$\|\xi\|_p = \|A^{-(q-p)}A^q\xi\|_0 \leq \rho^{q-p}\|\xi\|_q, \qquad \xi \in \operatorname{Dom}(A^q). \tag{1.4}$$

Equipped with the norm $\|\cdot\|_p$ the vector space $\operatorname{Dom}(A^p)$ becomes a Hilbert space which we shall denote by $\mathfrak{E}_p$. Then, the inclusion $\operatorname{Dom}(A^q) \subset \operatorname{Dom}(A^p)$, $q \geq p \geq 0$, gives rise to a continuous injection $f_{p,q} : \mathfrak{E}_q \to \mathfrak{E}_p$ and $\{\mathfrak{E}_p, f_{p,q}\}$ becomes a projective system of Hilbert spaces. In this case we have also a chain of Hilbert spaces:

$$\cdots \subset \mathfrak{E}_q \subset \cdots \subset \mathfrak{E}_p \subset \cdots \subset \mathfrak{E}_0 = \mathfrak{H}, \qquad q \geq p \geq 0. \tag{1.5}$$

Lemma 1.2.4 *For any $q \ge p \ge 0$ the vector subspece $\mathfrak{E}_q$ is dense in $\mathfrak{E}_p$. In particular, the projective system $\{\mathfrak{E}_p, f_{p,q}\}$ is reduced.*

PROOF. Obviously, $\mathfrak{E}_{q-p} = \mathrm{Dom}(A^{q-p})$ is a dense subspace of $\mathfrak{H}$. Note also from definition (1.3) that A^p is an isometric isomorphism from $\mathfrak{E}_p$ onto $\mathfrak{H}$. Hence the inverse image of $\mathfrak{E}_{q-p}$ is a dense subspace of $\mathfrak{E}_p$. On the other hand,

$$(A^p)^{-1}(\mathfrak{E}_{q-p}) = A^{-p}(A^{-(q-p)}\mathfrak{H}) = A^{-p-(q-p)}\mathfrak{H} = A^{-q}\mathfrak{H} = \mathfrak{E}_q.$$

Consequently, $\mathfrak{E}_q$ is dense in $\mathfrak{E}_p$. qed

By virtue of (1.5) a subspace of $\mathfrak{H}$ defined by

$$\mathfrak{E} = \bigcap_{p \ge 0} \mathfrak{E}_p \tag{1.6}$$

becomes a CH-space equipped with the Hilbertian seminorms $\{\|\cdot\|_p\}_{p\ge 0}$. Obviously $\mathfrak{E}$ is isomorphic to the projective limit:

$$\mathfrak{E} \cong \operatorname*{proj\,lim}_{p\to\infty} \mathfrak{E}_p.$$

To sum up, given a pair $(\mathfrak{H}, A)$ where A is a selfadjoint operator in a complex Hilbert space $\mathfrak{H}$ with $\inf \mathrm{Spec}(A) > 0$, we have constructed a CH-space $\mathfrak{E}$.

Definition 1.2.5 The above $\mathfrak{E}$ is called a *standard CH-space constructed from* $(\mathfrak{H}, A)$.

As may be proved easily from definition, we have

Lemma 1.2.6 *Let A be a positive selfadjoint operator in $\mathfrak{H}$ with $\inf \mathrm{Spec}(A) > 0$. Then, the standard CH-spaces constructed from $(\mathfrak{H}, A)$ and $(\mathfrak{H}, A^s)$ are isomorphic for any $s > 0$.*

As for the dual space of $\mathfrak{E}$, it follows from Proposition 1.2.2 that

$$\mathfrak{E}^* \cong \operatorname*{ind\,lim}_{p\to\infty} \mathfrak{E}_p^* \qquad \text{and} \qquad \mathfrak{E}^* = \bigcup_{p\ge 0} \mathfrak{E}_p^* \qquad \text{as vector spaces.}$$

Recall that $\mathfrak{E}_p^*$ is identified with the space of linear functionals on $\mathfrak{E}$ which are continuous with respect to $\|\cdot\|_p$. With this identification the canonical bilinear forms on $\mathfrak{E}^* \times \mathfrak{E}$ and on $\mathfrak{E}_p^* \times \mathfrak{E}_p$, $p \ge 0$, are denoted by the same symbol $\langle \cdot , \cdot \rangle$.

By virtue of our particular construction $\mathfrak{E}_p^*$ and $\mathfrak{E}^*$ can be described more explicitly. We have already defined in (1.3) a Hilbertian norm $\|\cdot\|_{-p}$ on $\mathfrak{H}$ for $p \ge 0$. Let $\mathfrak{E}_{-p}$ be the completion of $\mathfrak{H}$ with respect to $\|\cdot\|_{-p}$. Then the identity map from $\mathfrak{H}$ onto itself extends to a continuous injection $f_{-q,-p} : \mathfrak{E}_{-p} \to \mathfrak{E}_{-q}$ whenever $q \ge p \ge 0$, and thereby $\{\mathfrak{E}_{-p}, f_{-q,-p}\}$ becomes an inductive system of Hilbert spaces. Furthermore, there is a natural inclusion relation:

$$\mathfrak{H} = \mathfrak{E}_0 \subset \cdots \subset \mathfrak{E}_{-p} \subset \cdots \subset \mathfrak{E}_{-q} \subset \cdots \qquad q \ge p \ge 0. \tag{1.7}$$

Recall that $A^{-p} : \mathfrak{H} \to \mathfrak{E}_p$ is a bounded operator and by definition it satisfies

$$\|A^{-p}\xi\|_0 = \|\xi\|_{-p}, \qquad \xi \in \mathfrak{H}.$$

Therefore A^{-p} extends to an isometric isomorphism $\widetilde{A^{-p}}$ from $\mathfrak{E}_{-p}$ onto $\mathfrak{H}$.

The inner product $(\cdot,\cdot)_p$ of $\mathfrak{E}_p$ is by definition given as

$$(\xi,\eta)_p = (A^p\xi, A^p\eta)_0, \qquad \xi,\eta \in \mathfrak{E}_p. \tag{1.8}$$

It follows from Riesz' theorem that there exists an isometric anti-isomorphism $R_p : \mathfrak{E}_p^* \to \mathfrak{E}_p$ such that

$$\langle x^*, \xi\rangle = (R_p(x^*), \xi)_p, \qquad x^* \in \mathfrak{E}_p^*, \quad \xi \in \mathfrak{E}_p.$$

On the other hand, in view of (1.8) we have

$$(R_p(x^*),\xi)_p = (A^pR_p(x^*), A^p\xi)_0 = \left(\widetilde{A^{-p}}(\widetilde{A^{-p}})^{-1}A^pR_p(x^*), A^p\xi\right)_0.$$

Thus, $h_p = (\widetilde{A^{-p}})^{-1} \circ A^p \circ R_p : \mathfrak{E}_p^* \longrightarrow \mathfrak{E}_{-p}$ becomes an isometric anti-isomorphism such that

$$\langle x^*, \xi\rangle = \left(\widetilde{A^{-p}}h_p(x^*), A^p\xi\right)_0, \qquad x^* \in \mathfrak{E}_p^*, \quad \xi \in \mathfrak{E}_p. \tag{1.9}$$

Moreover, using $A^{-(q-p)}\widetilde{A^{-p}} = \widetilde{A^{-q}}f_{-q,-p}$, one may prove easily that $f_{-q,-p} \circ h_p = h_q \circ f_{p,q}^*$ for any $0 \le p \le q$. Consequently,

Lemma 1.2.7 *Two inductive systems $\{\mathfrak{E}_p^*, f_{p,q}^*\}$ and $\{\mathfrak{E}_{-p}, f_{-q,-p}\}$ of Hilbert spaces are anti-isomorphic under the isometric anti-isomorphisms $\{h_p\}$. Therefore, $\mathfrak{E}^*$ is anti-isomorphic to* $\operatorname{ind\,lim}_{p\to\infty} \mathfrak{E}_{-p}$.

To be sure we shall give the inverse h_p^{-1} more explicitly. Let $x \in \mathfrak{E}_{-p}$, $p \ge 0$. Then $\widetilde{A^{-p}}x \in \mathfrak{H}$ and we obtain a continuous linear function $\xi \mapsto (\widetilde{A^{-p}}x, A^p\xi)_0$, $\xi \in \mathfrak{E}$. In fact,

$$|(\widetilde{A^{-p}}x, A^p\xi)_0| \le \|\widetilde{A^{-p}}x\|_0 \, \|A^p\xi\|_0 = \|x\|_{-p} \, \|\xi\|_p.$$

Therefore there exists $x^* \in \mathfrak{E}_p^*$ such that

$$\langle x^*, \xi\rangle = (\widetilde{A^{-p}}x, A^p\xi)_0, \qquad \xi \in \mathfrak{E}_p.$$

Thus (1.9) is reproduced and, as is easily verified, $x^* = h_p^{-1}(x)$. The correspondence $x \mapsto x^*$ yields an anti-linear isomorphism from $\bigcup_{p\ge0}\mathfrak{E}_{-p}$ onto $\mathfrak{E}^*$. In that case, identifying x with x^*, we come to

$$\mathfrak{E}^* = \bigcup_{p\ge0} \mathfrak{E}_{-p}, \tag{1.10}$$

namely, the union of the increasing chain of Hilbert spaces (1.7). This is a counterpart of (1.5) and (1.6).

Lemma 1.2.8 *Let $\{e_j\}_{j=0}^{\infty}$ be a complete orthonormal basis of $\mathfrak{H}$. Under the identification (1.10) we have*

$$\|x\|_{-p}^2 = \sum_{j=0}^{\infty} \left|\left\langle x, A^{-p}e_j\right\rangle\right|^2, \qquad x \in \mathfrak{E}^*, \quad p \geq 0.$$

PROOF. In fact, identifying x with x^* we see that

$$\begin{aligned} \|x\|_{-p}^2 &= \|\widetilde{A^{-p}}x\|_0^2 = \sum_{j=0}^{\infty} \left|\left(\widetilde{A^{-p}}x, e_j\right)_0\right|^2 \\ &= \sum_{j=0}^{\infty} \left|\left(\widetilde{A^{-p}}x, A^p A^{-p} e_j\right)_0\right|^2 = \sum_{j=0}^{\infty} \left|\left\langle x, A^{-p}e_j\right\rangle\right|^2, \end{aligned}$$

where (1.9) is taken into consideration. qed

From the universal property of an inductive limit we may deduce the following

Proposition 1.2.9 *A linear operator T from $\mathfrak{E}^*$ into a locally convex space $\mathfrak{X}$ is continuous if and only if the restriction of T to $\mathfrak{E}_{-p}$ is continuous for all $p \geq 0$.*

We are now in a position to discuss the real case. Let $\mathfrak{H}$ be a real Hilbert space and its complexification is denoted by $\mathfrak{H}_{\mathbf{C}}$. A densely defined operator A in $\mathfrak{H}$ admits a unique extension to a densely defined operator $A_{\mathbf{C}}$ in $\mathfrak{H}_{\mathbf{C}}$. If $A_{\mathbf{C}}$ is selfadjoint with $\inf \mathrm{Spec}(A_{\mathbf{C}}) > 1$, we say simply that A is a selfadjoint operator in $\mathfrak{H}$ with $\inf \mathrm{Spec}(A) > 1$. Taking the real part of the complex CH-space constructed from $(\mathfrak{H}_{\mathbf{C}}, A_{\mathbf{C}})$, we obtain a real CH-space $\mathfrak{E}$ imbedded in $\mathfrak{H}$. This $\mathfrak{E}$ is called a CH-space constructed from $(\mathfrak{H}, A)$. The above discussion for complex spaces are also valid for real spaces with obvious modification.

1.3 Nuclear spaces and kernel theorem

We begin with the following

Definition 1.3.1 A locally convex space $\mathfrak{X}$ equipped with defining Hilbertian seminorms $\{\|\cdot\|_\alpha\}_{\alpha \in A}$ is called *nuclear* if for any $\alpha \in A$ there is $\beta \in A$ with $\alpha \leq \beta$ such that the canonical map $f_{\alpha,\beta} : \mathfrak{X}_\beta \to \mathfrak{X}_\alpha$ is of Hilbert-Schmidt type.

By definition a nuclear Fréchet space is a CH-space. As for structural characterization of a nuclear Fréchet space we mention the following

Proposition 1.3.2 *A nuclear Fréchet space $\mathfrak{E}$ admits a sequence of defining Hilbertian seminorms $\{|\cdot|_n\}_{n=0}^{\infty}$ such that*

(i) $|\xi|_n \leq C_n |\xi|_{n+1}$, $\xi \in \mathfrak{E}$, *with some $C_n \geq 0$;*

(ii) $f_{n,n+1} : H_{n+1} \to H_n$ *is of Hilbert-Schmidt type, where H_n is the Hilbert space associated with $|\cdot|_n$;*

(iii) $\{H_n, f_{m,n}\}$ *is a reduced projective sequence of Hilbert spaces;*

(iv) $\mathfrak{E} \cong \operatorname{proj\,lim}_{n\to\infty} H_n$.
Conversely, if $\{H_n, f_{m,n}\}$ *is a (reduced) projective sequence of Hilbert spaces with* $f_{n,n+1}$ *being of Hilbert-Schmidt type, then* $\operatorname{proj\,lim}_{n\to\infty} H_n$ *becomes a nuclear Fréchet space.*

Proposition 1.3.3 *A Fréchet space* $\mathfrak{X}$ *is nuclear if and only if so is* $\mathfrak{X}^*$.

Proposition 1.3.4 *A standard CH-space* $\mathfrak{E}$ *constructed from* $(\mathfrak{H}, A)$ *is nuclear if and only if* A^{-r} *is of Hilbert-Schmidt type for some* $r > 0$.

PROOF. Let $\|\cdot\|_p$ be the defining seminorms of $\mathfrak{E}$ given as $\|\xi\|_p = \|A^p\xi\|_0$ and denote by $\mathfrak{E}_p$ the associated Hilbert space.

Suppose first that A^{-r} is of Hilbert-Schmidt type with $r > 0$. Then, there exists a complete orthonormal basis $\{e_j\}_{j=0}^{\infty}$ for $\mathfrak{H}$ contained in $\mathrm{Dom}(A)$ such that $Ae_j = \lambda_j e_j$ with $\lambda_j > 0$ satisfying $\sum_{j=0}^{\infty} \lambda_j^{-2r} < \infty$. Note that $\{\lambda_j^{-(p+r)} e_j\}_{j=0}^{\infty}$ is a complete orthonormal basis for $\mathfrak{E}_{p+r}$ and

$$\sum_{j=0}^{\infty} \left\| \lambda_j^{-(p+r)} e_j \right\|_p^2 = \sum_{j=0}^{\infty} \lambda_j^{-2r} < \infty.$$

Hence the canonical map $f_{p,p+r} : \mathfrak{E}_{p+r} \to \mathfrak{E}_p$ is of Hilbert-Schmidt type for all $p \ge 0$. Therefore, $\mathfrak{E}$ is nuclear.

Conversely, suppose that $\mathfrak{E}$ is nuclear. Let $\{|\cdot|_n\}_{n=0}^{\infty}$ be a sequence of Hilbertian seminorms described as in Proposition 1.3.2. Since $\mathfrak{E} \cong \operatorname{proj\,lim}_{p\to\infty} \mathfrak{E}_p$ as well, we may find $n \ge 0$ and $r \ge 0$ such that

$$\|\xi\|_0 \le C\,|\xi|_n, \qquad |\xi|_{n+1} \le C'\,\|\xi\|_r, \qquad \xi \in \mathfrak{E},$$

with some $C, C' \ge 0$. Then we have a chain of canonical maps:

$$\mathfrak{E}_0 = \mathfrak{H} \longleftarrow H_n \stackrel{f_{n,n+1}}{\longleftarrow} H_{n+1} \longleftarrow \mathfrak{E}_r.$$

Since $f_{n,n+1}$ is of Hilbert-Schmidt type, so is the composion of the three which is nothing but the canonical map $\mathfrak{E}_r \to \mathfrak{E}_0 = \mathfrak{H}$. Let $\{e_j\}_{j=0}^{\infty}$ be a complete orthonormal basis of $\mathfrak{H}$. The obvious relation

$$(A^{-r}e_i, A^{-r}e_j)_r = (A^r A^{-r} e_i, A^r A^{-r} e_j)_0 = (e_i, e_j)_0 = \delta_{ij}$$

means that $\{A^{-r}e_j\}_{j=0}^{\infty}$ is an orthonormal sequence in $\mathfrak{E}_r$. Since the canonical map $\mathfrak{E}_r \to \mathfrak{E}_0 = \mathfrak{H}$ is of Hilbert-Schmidt type, $\sum_{j=0}^{\infty} \|A^{-r}e_j\|_0^2 < \infty$, that is, A^{-r} is of Hilbert-Schmidt type. qed

In particular,

Corollary 1.3.5 *A standard CH-space constructed from* $(\mathfrak{H}, A)$ *is nuclear if* A *is a positive selfadjoint operator with Hilbert-Schmidt inverse.*

For two vector spaces $\mathfrak{X}$ and $\mathfrak{Y}$ we denote by $\mathfrak{X}\otimes_{\mathrm{alg}}\mathfrak{Y}$ their algebraic tensor product. Various locally convex topologies can be introduced into $\mathfrak{X}\otimes_{\mathrm{alg}}\mathfrak{Y}$ if $\mathfrak{X}$ and $\mathfrak{Y}$ are locally convex spaces. Among others we shall be mostly concerned with π-topology. The *π-topology* on $\mathfrak{X}\otimes_{\mathrm{alg}}\mathfrak{Y}$ is the strongest locally convex topology such that the canonical map $\mathfrak{X}\times\mathfrak{Y}\to\mathfrak{X}\otimes_{\mathrm{alg}}\mathfrak{Y}$ is continuous. Let $\{\|\cdot\|_\alpha\}$ and $\{\|\cdot\|'_\beta\}$ be defining seminorms of $\mathfrak{X}$ and $\mathfrak{Y}$, respectively. Then the π-topology is defined by the seminorms

$$\|\omega\|_{\alpha,\beta}=\inf\sum_j\|\xi_j\|_\alpha\,\|\eta_j\|'_\beta\,,\qquad \omega\in\mathfrak{X}\otimes_{\mathrm{alg}}\mathfrak{Y},$$

where the infimum is taken over all finite pairs (ξ_j,η_j) satisfying $\omega=\sum_j\xi_j\otimes\eta_j$, $\xi_j\in\mathfrak{X}$, $\eta_j\in\mathfrak{Y}$. The completion of $\mathfrak{X}\otimes_{\mathrm{alg}}\mathfrak{Y}$ with respect to the π-topology is called the *π-tensor product* and is denoted by $\mathfrak{X}\otimes_\pi\mathfrak{Y}$. The following facts are useful.

Proposition 1.3.6 *Let $\mathfrak{X}$ and $\mathfrak{Y}$ be Fréchet spaces. Then every element $\omega\in\mathfrak{X}\otimes_\pi\mathfrak{Y}$ is the sum of an absolutely convergent series $\omega=\sum_{j=1}^\infty\xi_j\otimes\eta_j$, where $(\xi_j)_{j=1}^\infty$ and $(\eta_j)_{j=1}^\infty$ are respectively sequences in $\mathfrak{X}$ and $\mathfrak{Y}$ converging to zero.*

Proposition 1.3.7 *If $\mathfrak{X}$ and $\mathfrak{Y}$ are nuclear spaces, so is $\mathfrak{X}\otimes_\pi\mathfrak{Y}$.*

Let H and K be Hilbert spaces with inner products $(\cdot,\cdot)_H$ and $(\cdot,\cdot)_K$, respectively. Then one may define an inner product in $H\otimes_{\mathrm{alg}}K$ by

$$\left(\sum_i\xi_i\otimes\eta_i\,,\sum_j\xi'_j\otimes\eta'_j\right)=\sum_{i,j}\left(\xi_i\,,\xi'_j\right)_H\left(\eta_i\,,\eta'_j\right)_K.$$

The completion of $H\otimes_{\mathrm{alg}}K$ with respect to this inner product is denoted simply by $H\otimes K$. Obviously, $H\otimes K$ becomes a Hilbert space again and is called the *Hilbert space tensor product.* It is noted that the Hilbert space tensor product does not coincide with the π-tensor product (as locally convex spaces) whenever both H and K are infinite dimensional.

Proposition 1.3.8 *Let $\mathfrak{X}$ and $\mathfrak{Y}$ be locally convex spaces with defining Hilbertian seminorms $\{\|\cdot\|_\alpha\}_{\alpha\in A}$ and $\{\|\cdot\|'_\beta\}_{\beta\in B}$, respectively. Let $\mathfrak{X}_\alpha$ and $\mathfrak{Y}_\beta$ be the Hilbert spaces associated with $\|\cdot\|_\alpha$ and $\|\cdot\|'_\beta$, respectively. Then, $\{\mathfrak{X}_\alpha\otimes\mathfrak{Y}_\beta\}_{\alpha\in A,\beta\in B}$ becomes a projective system of Hilbert spaces. If in addition $\mathfrak{X}$ or $\mathfrak{Y}$ is nuclear, we have*

$$\mathfrak{X}\otimes_\pi\mathfrak{Y}\cong\operatorname*{proj\,lim}_{\alpha,\beta}\,(\mathfrak{X}_\alpha\otimes\mathfrak{Y}_\beta)\,.$$

Let $\mathfrak{X}$ and $\mathfrak{Y}$ be locally convex spaces with defining seminorms $\{\|\cdot\|_\alpha\}_{\alpha\in A}$ and $\{\|\cdot\|'_\beta\}_{\beta\in B}$, respectively. We denote by $\mathcal{L}(\mathfrak{X},\mathfrak{Y})$ the space of continuous linear operators from $\mathfrak{X}$ into $\mathfrak{Y}$ equipped with the *topology of bounded convergence,* namely, the locally convex topology defined by the seminorms:

$$\|T\|_{S,\beta}=\sup_{\xi\in S}\|T\xi\|'_\beta\,,\qquad T\in\mathcal{L}(\mathfrak{X},\mathfrak{Y}),$$

where S runs over the bounded subsets of $\mathfrak{X}$ and $\beta\in B$. By definition $\mathfrak{X}^*=\mathcal{L}(\mathfrak{X},\mathbf{R})$ or $=\mathcal{L}(\mathfrak{X},\mathbf{C})$ according as $\mathfrak{X}$ is a real or complex space.

Let $\mathfrak{Z}$ be another locally convex space with defining seminorms $\{\|\cdot\|''_\gamma\}_{\gamma\in\Gamma}$. We denote by $\mathcal{B}(\mathfrak{X},\mathfrak{Y};\mathfrak{Z})$ be the space of jointly continuous bilinear maps from $\mathfrak{X}\times\mathfrak{Y}$ into $\mathfrak{Z}$. It is equipped with the *topology of bi-bounded convergence*, namely, the locally convex topology defined by the seminorms:

$$\|\phi\|_{S_1,S_2,\gamma} = \sup_{\xi\in S_1,\eta\in S_2} \|\phi(\xi,\eta)\|''_\gamma, \qquad \phi\in\mathcal{B}(\mathfrak{X},\mathfrak{Y};\mathfrak{Z}),$$

where S_1 and S_2 run over the bounded subsets of $\mathfrak{X}$ and $\mathfrak{Y}$, respectively, and $\gamma\in\Gamma$. When $\mathfrak{Z}=\mathbf{C}$ or $=\mathbf{R}$, we put $\mathcal{B}(\mathfrak{X},\mathfrak{Y})=\mathcal{B}(\mathfrak{X},\mathfrak{Y};\mathfrak{Z})$ for simplicity.

There is a canonical correspondence between $T\in\mathcal{L}(\mathfrak{X}\otimes_\pi\mathfrak{Y},\mathfrak{Z})$ and $\phi\in\mathcal{B}(\mathfrak{X},\mathfrak{Y};\mathfrak{Z})$ given by

$$\phi(\xi,\eta)=T(\xi\otimes\eta), \qquad \xi\in\mathfrak{X}, \quad \eta\in\mathfrak{Y}.$$

This gives rise to a universal property of the π-tensor product.

Proposition 1.3.9 *For any locally convex spaces $\mathfrak{X},\mathfrak{Y},\mathfrak{Z}$ it holds that*

$$\mathcal{B}(\mathfrak{X},\mathfrak{Y};\mathfrak{Z})\cong\mathcal{L}(\mathfrak{X}\otimes_\pi\mathfrak{Y},\mathfrak{Z}) \qquad \textit{as vector spaces.}$$

We then recall the canonical correspondence among tensor products, spaces of continuous linear operators and spaces of continuous bilinear forms. Let $\mathfrak{X},\mathfrak{Y}$ be two locally convex spaces.

1. $\mathfrak{X}\otimes_{\mathrm{alg}}\mathfrak{Y}\to\mathcal{L}(\mathfrak{Y}^*,\mathfrak{X})$. With each $\zeta=\sum\xi_i\otimes\eta_i\in\mathfrak{X}\otimes_{\mathrm{alg}}\mathfrak{Y}$ we associate a continuous linear operator $T_\zeta\in\mathcal{L}(\mathfrak{Y}^*,\mathfrak{X})$ defined by

$$T_\zeta y=\sum\langle y\,,\eta_i\rangle\,\xi_i, \qquad y\in\mathfrak{Y}^*.$$

2. $\mathfrak{X}^*\otimes_{\mathrm{alg}}\mathfrak{Y}\to\mathcal{L}(\mathfrak{X},\mathfrak{Y})$. With each $\omega=\sum x_i\otimes\eta_i\in\mathfrak{X}^*\otimes_{\mathrm{alg}}\mathfrak{Y}$ we associate a continuous linear operator $T_\omega\in\mathcal{L}(\mathfrak{X},\mathfrak{Y})$ defined by

$$T_\omega\xi=\sum\langle x_i\,,\xi\rangle\,\eta_i, \qquad \xi\in\mathfrak{X}.$$

3. $\mathfrak{X}^*\otimes_{\mathrm{alg}}\mathfrak{Y}^*\to\mathcal{B}(\mathfrak{X},\mathfrak{Y})$. With each $z=\sum x_i\otimes y_i\in\mathfrak{X}^*\otimes_{\mathrm{alg}}\mathfrak{Y}^*$ we associate a continuous bilinear form $\phi_z\in\mathcal{B}(\mathfrak{X},\mathfrak{Y})$ defined by

$$\phi_z(\xi,\eta)=\sum\langle x_i\,,\xi\rangle\,\langle y_i\,,\eta\rangle\,, \qquad \xi\in\mathfrak{X}, \quad \eta\in\mathfrak{Y}.$$

4. $\mathcal{B}(\mathfrak{X},\mathfrak{Y})\to\mathcal{L}(\mathfrak{X},\mathfrak{Y}^*)$. With each $\phi\in\mathcal{B}(\mathfrak{X},\mathfrak{Y})$ we associate $T_\phi\in\mathcal{L}(\mathfrak{X},\mathfrak{Y}^*)$ by

$$\phi(\xi,\eta)=\langle T_\phi\xi\,,\eta\rangle\,, \qquad \xi\in\mathfrak{X}, \quad \eta\in\mathfrak{Y}.$$

5. $(\mathfrak{X}\otimes_\pi\mathfrak{Y})^*\to\mathcal{B}(\mathfrak{X},\mathfrak{Y})$. With each $\omega\in(\mathfrak{X}\otimes_\pi\mathfrak{Y})^*$ we associate a continuous bilinear form $\phi_\omega\in\mathcal{B}(\mathfrak{X},\mathfrak{Y})$ defined by

$$\phi_\omega(\xi,\eta)=\langle\omega\,,\xi\otimes\eta\rangle\,, \qquad \xi\in\mathfrak{X}, \quad \eta\in\mathfrak{Y}.$$

   This is a particular case of Proposition 1.3.9 and, in fact, $\mathcal{B}(\mathfrak{X},\mathfrak{Y})\cong(\mathfrak{X}\otimes_\pi\mathfrak{Y})^*$ as vector spaces.

The above listed maps are all injective. From that viewpoint a nuclear space enjoys a significant property stated in the following

Theorem 1.3.10 (KERNEL THEOREM) *Let $\mathfrak{X}$ be a nuclear Fréchet space and let $\mathfrak{Y}$ be a Fréchet space. Then the above listed five linear injections yields (topological) isomorphisms:*

$$\mathfrak{X} \otimes_\pi \mathfrak{Y} \cong \mathcal{L}(\mathfrak{Y}^*, \mathfrak{X}),$$
$$\mathfrak{X}^* \otimes_\pi \mathfrak{Y} \cong \mathcal{L}(\mathfrak{X}, \mathfrak{Y}),$$
$$\mathfrak{X}^* \otimes_\pi \mathfrak{Y}^* \cong (\mathfrak{X} \otimes_\pi \mathfrak{Y})^* \cong \mathcal{B}(\mathfrak{X}, \mathfrak{Y}) \cong \mathcal{L}(\mathfrak{X}, \mathfrak{Y}^*).$$

Here is a quick remark on separately continuous bilinear maps. Let $\mathcal{B}_{\text{sep}}(\mathfrak{X}, \mathfrak{Y}; \mathfrak{Z})$ be the space of separately continuous bilinear maps from $\mathfrak{X} \times \mathfrak{Y}$ into $\mathfrak{Z}$. If $\mathfrak{Z} = \mathbb{C}$ or $= \mathbb{R}$, we write simply $\mathcal{B}_{\text{sep}}(\mathfrak{X}, \mathfrak{Y})$ for $\mathcal{B}_{\text{sep}}(\mathfrak{X}, \mathfrak{Y}; \mathfrak{Z})$. In general, there is a crucial difference between $\mathcal{B}_{\text{sep}}(\mathfrak{X}, \mathfrak{Y}; \mathfrak{Z})$ and $\mathcal{B}(\mathfrak{X}, \mathfrak{Y}; \mathfrak{Z})$. In this connection the following result is useful.

Proposition 1.3.11 *Let $\mathfrak{X}, \mathfrak{Y}, \mathfrak{Z}$ be locally convex spaces. Then, $\mathcal{B}_{\text{sep}}(\mathfrak{X}, \mathfrak{Y}; \mathfrak{Z}) = \mathcal{B}(\mathfrak{X}, \mathfrak{Y}; \mathfrak{Z})$ holds if both $\mathfrak{X}$ and $\mathfrak{Y}$ are Fréchet spaces; or if $\mathfrak{X}, \mathfrak{Y}$ and $\mathfrak{Z}$ are all strong dual spaces of reflexive Fréchet spaces.*

In general, for any locally convex spaces $\mathfrak{X}$ and $\mathfrak{Y}$ there is a canonical isomorphism:

$$\mathcal{B}_{\text{sep}}(\mathfrak{X}, \mathfrak{Y}) \cong \mathcal{L}(\mathfrak{X}, \mathfrak{Y}^*_\sigma) \qquad \text{as vector spaces,}$$

where $\mathfrak{Y}^*_\sigma$ is the dual space of $\mathfrak{Y}$ equipped with the weak topology, i.e., the topology of pointwise convergence. Therefore,

Proposition 1.3.12 *If both $\mathfrak{X}$ and $\mathfrak{Y}$ are Fréchet spaces, then*

$$(\mathfrak{X} \otimes_\pi \mathfrak{Y})^* \cong \mathcal{B}(\mathfrak{X}, \mathfrak{Y}) \cong \mathcal{B}_{\text{sep}}(\mathfrak{X}, \mathfrak{Y}) \cong \mathcal{L}(\mathfrak{X}, \mathfrak{Y}^*) \cong \mathcal{L}(\mathfrak{X}, \mathfrak{Y}^*_\sigma) \qquad \textit{as vector spaces.}$$

1.4 Standard CH-spaces of functions

In §1.2 we discussed construction of a standard CH-space from a pair $(\mathfrak{H}, A)$, where A is a selfadjoint operator in a Hilbert space $\mathfrak{H}$ with $\inf \operatorname{Spec}(A) > 0$. We discuss in this section the case where $\mathfrak{H}$ is an L^2-space.

Let Ω be a topological space equipped with a σ-finite Borel measure ν. (All the measures under consideration are assumed to be σ-finite in these lecture notes.) Suppose that we are given a selfadjoint operator A in the (real or complex) Hilbert space $\mathfrak{H} = L^2(\Omega, \nu)$ with $\inf \operatorname{Spec}(A) > 0$. Then the standard CH-space constructed from $(\mathfrak{H}, A)$ is written explicitly as $\mathcal{S}_A(\Omega)$. As usual, we understand that $\mathcal{S}_A(\Omega)$ and $\mathcal{S}^*_A(\Omega)$ are spaces of test functions and generalized functions (or distributions) on Ω, respectively.

That the delta function (evaluation map) is a member of $\mathcal{S}^*_A(\Omega)$ is not clear at all. On the contrary, continuity of a test function does not follow automatically. By construction each element of $\mathcal{S}_A(\Omega)$ is merely a function on Ω which is determined up to ν-null functions. We thus need the first hypothesis:

(H1) For each function $\xi \in S_A(\Omega)$ there exists a unique continuous function $\tilde{\xi}$ on Ω such that $\xi(\omega) = \tilde{\xi}(\omega)$ for ν-a.e. $\omega \in \Omega$.

Once this condition is satisfied, we always regard $S_A(\Omega)$ as a space of continuous functions on Ω and we do not use the exclusive symbol $\tilde{\xi}$. The uniqueness in (H1) is equivalent to that any continuous function on Ω which is zero ν-a.e. is identically zero. Under (H1) we put two more hypotheses to keep a delta function in $S_A^*(\Omega)$:

(H2) For each $\omega \in \Omega$ the evaluation map $\delta_\omega : \xi \mapsto \xi(\omega)$, $\xi \in S_A(\Omega)$, is a continuous linear functional, i.e., $\delta_\omega \in S_A^*(\Omega)$.

(H3) The map $\omega \mapsto \delta_\omega \in S_A^*(\Omega)$, $\omega \in \Omega$, is continuous with respect to the strong dual topology of $S_A^*(\Omega)$.

As will be clear in Proposition 1.4.3, hypothesis (H3) relates to a certain property of tensor products.

Proposition 1.4.1 *Let $S_A(\Omega)$ be a standard CH-space constructed as above and let $\xi_n \in S_A(\Omega)$, $n = 1, 2, \cdots$, be a sequence converging to 0 in $S_A(\Omega)$. If (H1) and (H2) are satisfied, then the sequence converges pointwisely, i.e., $\lim_{n\to\infty} \xi_n(\omega) = \xi(\omega)$ for any $\omega \in \Omega$. Moreover, if (H3) is satisfied in addition, the pointwise convergence is uniform on every compact subset of Ω.*

PROOF. By (H2) the delta function $\delta_\omega \in S_A^*(\Omega)$, and hence $\|\delta_\omega\|_{-p} < \infty$ for some $p \ge 0$ depending on $\omega \in \Omega$. Then,

$$|\xi_n(\omega) - \xi(\omega)| = |\langle \delta_\omega, \xi_n - \xi \rangle| \le \|\delta_\omega\|_{-p} \|\xi_n - \xi\|_p .$$

Hence $\lim_{n\to\infty} \xi_n(\omega) = \xi(\omega)$.

Suppose (H3) is satisfied and let $\Omega_0 \subset \Omega$ be a compact subset. Then, by assumption $\{\delta_\omega; \omega \in \Omega_0\}$ is a compact subset of $S_A^*(\Omega)$, and therefore it is bounded. It follows from Proposition 1.1.3 that there exists $p \ge 0$ and $C \ge 0$ such that

$$|\langle \delta_\omega, \xi \rangle| \le C \|\xi\|_p , \qquad \xi \in S_A(\Omega), \quad \omega \in \Omega_0.$$

Hence we have

$$\sup_{\omega \in \Omega_0} |\xi_n(\omega) - \xi(\omega)| \le C \|\xi_n - \xi\|_p ,$$

and therefore $\lim_{n\to\infty} \xi_n(\omega) = \xi(\omega)$ uniformly on Ω_0. qed

Lemma 1.4.2 *Let A be a selfadjoint operator in $L^2(\Omega, \nu)$ with A^{-r} being of Hilbert-Schmidt type for some $r > 0$. Assume that $S_A(\Omega)$ satisfies (H1) and (H2). Then,*

$$\|A^{-r}\|_{\mathrm{HS}}^2 = \int_\Omega \|\delta_\omega\|_{-r}^2 \nu(d\omega) = \sum_{j=0}^{\infty} \lambda_j^{-2r} < \infty,$$

where $0 < \lambda_1 \le \lambda_2 \le \cdots$ are the eigenvalues of A.

PROOF. By assumption there exists a complete orthonormal basis $\{e_j\}_{j=0}^{\infty} \subset L^2(\Omega, \nu)$ such that $Ae_j = \lambda_j e_j$. Then, by Lemma 1.2.8,

$$\|\delta_\omega\|_{-r}^2 = \sum_{j=0}^{\infty} |\langle \delta_\omega , A^{-r} e_j \rangle|^2 = \sum_{j=0}^{\infty} \lambda_j^{-2r} |\langle \delta_\omega , e_j \rangle|^2 = \sum_{j=0}^{\infty} \lambda_j^{-2r} |e_j(\omega)|^2, \qquad \omega \in \Omega.$$

We then integrate both sides to get the result. qed

It is noteworthy that properties (H1)–(H3) are preserved under forming a tensor product. Recall that for two selfadjoint operators A_i in $\mathfrak{H}_i$, $i = 1, 2$, their tensor product $A_1 \otimes A_2$ becomes a selfadjoint operator in $\mathfrak{H}_1 \otimes \mathfrak{H}_2$ in a canonical way. Moreover, note that if $\inf \operatorname{Spec}(A_i) > 0$, $i = 1, 2$, then $\inf \operatorname{Spec}(A_1 \otimes A_2) > 0$ as well.

Proposition 1.4.3 *For $i = 1, 2$ let Ω_i be a topological space with a Borel measure ν_i and let A_i be a positive selfadjoint operator in $L^2(\Omega_i, \nu_i)$ with Hilbert-Schmidt inverse. Then*

$$\mathcal{S}_{A_1 \otimes A_2}(\Omega_1 \times \Omega_2) \cong \mathcal{S}_{A_1}(\Omega_1) \otimes_\pi \mathcal{S}_{A_2}(\Omega_2) \tag{1.11}$$

under the identification: $L^2(\Omega_1 \times \Omega_2, \nu_1 \times \nu_2; \mathbf{R}) \cong L^2(\Omega_1, \nu_1; \mathbf{R}) \otimes L^2(\Omega_2, \nu_2; \mathbf{R})$. Moreover, if both $\mathcal{S}_{A_1}(\Omega_1)$ and $\mathcal{S}_{A_2}(\Omega_2)$ satisfy hypotheses (H1)–(H3), so does $\mathcal{S}_{A_1 \otimes A_2}(\Omega_1 \times \Omega_2)$.

PROOF. We see from Corollary 1.3.5 that $\mathcal{S}_{A_i}(\Omega_i)$ is nuclear. It then follows from Proposition 1.3.8 that the standard CH-space constructed from $(L^2(\Omega_1, \nu_1) \otimes L^2(\Omega_2, \nu_2), A_1 \otimes A_2)$ is isomorphic to the π-tensor product of the standard CH-spaces constructed from $(L^2(\Omega_1, \nu_1), A_1)$ and $(L^2(\Omega_2, \nu_2), A_2)$. This proves (1.11).

We now suppose that both $\mathcal{S}_{A_1}(\Omega_1)$ and $\mathcal{S}_{A_2}(\Omega_2)$ satisfy hypotheses (H1)–(H3). For $\zeta \in \mathcal{S}_{A_1 \otimes A_2}(\Omega_1 \times \Omega_2) \cong \mathcal{S}_{A_1}(\Omega_1) \otimes_\pi \mathcal{S}_{A_2}(\Omega_2)$ we put

$$\tilde{\zeta}(\omega_1, \omega_2) = \langle \delta_{\omega_1} \otimes \delta_{\omega_2} , \zeta \rangle, \qquad \omega_1 \in \Omega_1, \quad \omega_2 \in \Omega_2.$$

We note that $\tilde{\zeta}$ is a continuous function on $\Omega_1 \times \Omega_2$. Indeed, $\tilde{\zeta}$ is a composition of continuous maps:

$$\omega_i \mapsto \delta_{\omega_i} \in \mathcal{S}_{A_i}^*(\Omega_i), \qquad \omega_i \in \Omega_i, \quad i = 1, 2,$$

is continuous by (H3);

$$x, y \mapsto x \otimes y \in \mathcal{S}_{A_1}^*(\Omega_1) \otimes_\pi \mathcal{S}_{A_2}^*(\Omega_2), \qquad x \in \mathcal{S}_{A_1}^*(\Omega_1), \quad y \in \mathcal{S}_{A_2}^*(\Omega_2),$$

is continuous by the definition of π-tensor product;

$$\mathcal{S}_{A_1}^*(\Omega_1) \otimes_\pi \mathcal{S}_{A_2}^*(\Omega_2) \cong \mathcal{S}_{A_1 \otimes A_2}^*(\Omega_1 \times \Omega_2)$$

by the kernel theorem (Theorem 1.3.10).

We then prove that $\tilde{\zeta}$ coincides with ζ for $\nu_1 \times \nu_2$-a. e. Take an approximating sequence $\zeta_n \in \mathcal{S}_{A_1}(\Omega_1) \otimes_{\mathrm{alg}} \mathcal{S}_{A_2}(\Omega_2)$ converging to ζ. We see from Lemma 1.4.2 that

$$\begin{aligned}
\|\zeta_n - \tilde{\zeta}\|_0^2 &= \int_{\Omega_1 \times \Omega_2} |\zeta_n(\omega_1, \omega_2) - \tilde{\zeta}(\omega_1, \omega_2)|^2 \nu_1(d\omega_1) \nu_2(d\omega_2) \\
&= \int_{\Omega_1 \times \Omega_2} |\langle \delta_{\omega_1} \otimes \delta_{\omega_2} , \zeta_n - \zeta \rangle|^2 \nu_1(d\omega_1) \nu_2(d\omega_2) \\
&\le \|\zeta_n - \zeta\|_1^2 \int_{\Omega_1} \|\delta_{\omega_1}\|_{-1}^2 \, \nu_1(d\omega_1) \int_{\Omega_2} \|\delta_{\omega_2}\|_{-1}^2 \, \nu_2(d\omega_2) \\
&\le \|A_1^{-1}\|_{\mathrm{HS}}^2 \|A_2^{-1}\|_{\mathrm{HS}}^2 \, \|\zeta_n - \zeta\|_1^2 .
\end{aligned}$$

Hence,

$$\begin{aligned}\|\zeta-\tilde{\zeta}\|_0 &\le \|\zeta-\zeta_n\|_0+\|\zeta_n-\tilde{\zeta}\|_0\\ &\le \|\zeta-\zeta_n\|_0+\|A_1^{-1}\|_{\mathrm{HS}}\|A_2^{-1}\|_{\mathrm{HS}}\,\|\zeta_n-\zeta\|_1\to 0, \qquad \text{as } n\to\infty.\end{aligned}$$

Therefore, $\|\zeta-\tilde{\zeta}\|_0=0$ which implies that ζ and $\tilde{\zeta}$ coincide almost everywhere.

In order to complete the proof of $\mathcal{S}_{A_1\otimes A_2}(\Omega_1\times\Omega_2)$ satisfying (H1) we must show the uniqueness. For that purpose it is sufficient to prove that any continuous function ζ on $\Omega_1\times\Omega_2$ which is zero $\nu_1\times\nu_2$-a.e. is identically zero. Without loss of generality we may assume that ζ is bounded. For a measurable subset $U\subset\Omega_2$ with $\nu_2(U)<\infty$ we consider

$$Z_U(\omega_1)=\int_U \zeta(\omega_1,\omega_2)\nu_2(d\omega_2). \tag{1.12}$$

Then Z_U is a continuous function on Ω_1 by Lebesgue's theorem. On the other hand, it follows from Fubini's theorem that $Z_U(\omega_1)=0$ for ν_1-a.e. $\omega_1\in\Omega_1$. Since (H1) is satisfied for $\mathcal{S}_{A_1}(\Omega_1)$ by assumption, Z_U is identically zero. Now fix $\omega_1\in\Omega_1$. Then in view of (1.12) we see that $\zeta(\omega_1,\omega_2)=0$ for ν_2-a.e. $\omega_2\in\Omega_2$. Since $\omega_2\mapsto\zeta(\omega_1,\omega_2)$ is continuous by assumption, it follows from (H1) for $\mathcal{S}_{A_2}(\Omega_2)$ that $\zeta(\omega_1,\omega_2)=0$ for all $\omega_2\in\Omega_2$. Consequently, ζ is identically zero as desired.

Since $\delta_{(\omega_1,\omega_2)}=\delta_{\omega_1}\otimes\delta_{\omega_2}$, (H2) and (H3) follow immediately. qed

In the rest of this section we discuss a sufficient condition for (H1)–(H3).

Lemma 1.4.4 *Let A be a positive selfadjoint operator in $L^2(\Omega,\nu)$ with Hilbert-Schmidt inverse. Let $\{\lambda_j\}_{j=0}^{\infty}$ and $\{e_j\}_{j=0}^{\infty}$ be its eigenvalues and eigenfunctions, respectively. Assume that every e_j is continuous on a Borel subset $\Omega_0\subset\Omega$ with respect to the relative topology. If*

$$M\equiv\sup_{j\ge0}\lambda_j^{-\alpha}\sup_{\omega\in\Omega_0}|e_j(\omega)|<\infty$$

for some $\alpha\ge0$, then for any $\phi\in\mathcal{S}_A(\Omega)$ the series

$$\tilde{\phi}(\omega)\equiv\sum_{j=0}^{\infty}\langle\phi,e_j\rangle\,e_j(\omega),\qquad \omega\in\Omega_0,$$

is absolutely convergent and becomes a continuous function on Ω_0 such that $\phi(\omega)=\tilde{\phi}(\omega)$ for ν-a.e. $\omega\in\Omega_0$.

Proof. For $\omega\in\Omega_0$ we have

$$\begin{aligned}\sum_{j=0}^{\infty}|\langle\phi,e_j\rangle\,e_j(\omega)| &= \sum_{j=0}^{\infty}|\langle\phi,e_j\rangle|\lambda_j^{\alpha}\lambda_j^{-\alpha}|e_j(\omega)|\\ &\le M\sum_{j=0}^{\infty}\lambda_j^{\alpha}|\langle\phi,e_j\rangle|\\ &\le M\left(\sum_{j=0}^{\infty}\lambda_j^{2(\alpha+1)}\langle\phi,e_j\rangle^2\right)^{1/2}\left(\sum_{j=0}^{\infty}\lambda_j^{-2}\right)^{1/2}\\ &= M\,\|\phi\|_{\alpha+1}\,\|A^{-1}\|_{\mathrm{HS}}<\infty. \end{aligned}\tag{1.13}$$

Therefore $\tilde{\phi}(\omega)$ is absolutely convergent. We next prove that $\tilde{\phi}$ is continuous on Ω_0. For $\omega_1, \omega_2 \in \Omega_0$ we have

$$\begin{aligned}
&\left|\tilde{\phi}(\omega_1) - \tilde{\phi}(\omega_2)\right| \\
&\leq \left|\sum_{j=0}^{n} \langle \phi, e_j \rangle (e_j(\omega_1) - e_j(\omega_2))\right| + \left|\sum_{j>n} \langle \phi, e_j \rangle (e_j(\omega_1) - e_j(\omega_2))\right| \\
&\leq \sum_{j=0}^{n} |\langle \phi, e_j \rangle| |e_j(\omega_1) - e_j(\omega_2)| + \sum_{j>n} 2M\lambda_j^{\alpha} |\langle \phi, e_j \rangle| \\
&\leq \sum_{j=0}^{n} |\langle \phi, e_j \rangle| |e_j(\omega_1) - e_j(\omega_2)| + 2M \|\phi\|_{\alpha+1} \left(\sum_{j>n} \lambda_j^{-2}\right)^{1/2}. \qquad (1.14)
\end{aligned}$$

Then it is easy to see that $\tilde{\phi}$ is continuous on Ω_0. Since the Fourier expansion $\phi = \sum_{j=0}^{\infty} \langle \phi, e_j \rangle e_j$ converges in L^2-sense, $\phi(\omega) = \tilde{\phi}(\omega)$ for ν-a.e. $\omega \in \Omega_0$ qed

Theorem 1.4.5 *Let A be a positive selfadjoint operator in $L^2(\Omega, \nu)$ with Hilbert-Schmidt inverse. Assume the following three conditions are satisfied.*

(i) $\nu(U) > 0$ *for any non-empty open subset $U \subset \Omega$;*

(ii) *every e_j is a continuous function on Ω;*

(iii) *there exists an open covering $\Omega = \bigcup_\gamma \Omega_\gamma$ and $\alpha(\gamma) \geq 0$ such that*

$$\sup_{j \geq 0} \lambda_j^{-\alpha(\gamma)} \sup_{\omega \in \Omega_\gamma} |e_j(\omega)| < \infty.$$

Then (H1)-(H3) are satisfied. Moreover, for $\phi \in S_A(\Omega)$ the unique continuous function $\tilde{\phi}$ on Ω such that $\phi(\omega) = \tilde{\phi}(\omega)$ for ν-a.e. $\omega \in \Omega$ is given by the absolutely convergent series:

$$\tilde{\phi}(\omega) = \sum_{j=0}^{\infty} \langle \phi, e_j \rangle e_j(\omega).$$

PROOF. It follows from Lemma 1.4.4 that $\tilde{\phi}$ is a continuous function on Ω. Since the Fourier expansion converges in L^2-sense, $\phi(\omega) = \tilde{\phi}(\omega)$ for ν-a.e. $\omega \in \Omega$. The uniqueness following from the assumption (i), we have proved (H1). In what follows, according to our convention, we do not use the symbol $\tilde{\phi}$.

We next prove (H2). Since $\Omega = \bigcup_\gamma \Omega_\gamma$ is an open covering, for a given $\omega \in \Omega$ we take an open subset containing it, say, $\omega \in \Omega_0$. It then follows from (1.13) that $\delta_\omega \in S_A^*(\Omega)$.

We finally prove (H3), i.e., the continuity of $\omega \mapsto \delta_\omega$. For $\omega_1, \omega_2 \in \Omega_0$ and $\phi \in S_A(\Omega)$ we have

$$\begin{aligned}
|\langle \delta_{\omega_1} - \delta_{\omega_2}, \phi \rangle| &= |\phi(\omega_1) - \phi(\omega_2)| \\
&\leq \sum_{j=0}^{n} |\langle \phi, e_j \rangle| |e_j(\omega_1) - e_j(\omega_2)| + 2M \|\phi\|_{\alpha(0)+1} \left(\sum_{j>n} \lambda_j^{-2}\right)^{1/2},
\end{aligned}$$

which follows from (1.14). Then for any bounded subset $S \subset E$, we have

$$\sup_{\phi \in S} |\langle \delta_{\omega_1} - \delta_{\omega_2}, \phi \rangle| \leq M_0 \sum_{j=0}^{n} |e_j(\omega_1) - e_j(\omega_2)| + 2MM_1 \left(\sum_{j>n} \lambda_j^{-2} \right)^{1/2},$$

where $M_0 = \sup_{\phi \in S} \|\phi\|_0$ and $M_1 = \sup_{\phi \in S} \|\phi\|_{\alpha(0)+1}$. This shows that the map $\omega \mapsto \delta_\omega$ is continuous on Ω_0, and therefore on Ω. qed

A prototype of our consideration is the case where $\Omega = \mathbb{R}$ with the Lebesgue measure $\nu(dt) = dt$ and

$$A = 1 + t^2 - \frac{d^2}{dt^2} = \left(t + \frac{d}{dt} \right)^* \left(t + \frac{d}{dt} \right) + 2. \tag{1.15}$$

Then, equipped with the maximal L^2-domain, A becomes a selfadjoint operator in $L^2(\mathbb{R})$. In fact, H_j being the Hermite polynomial of degree j (see Appendix A.2),

$$e_j(t) = \left(\sqrt{\pi} 2^j j! \right)^{-1/2} H_j(t) e^{-t^2/2}, \qquad t \in \mathbb{R}, \quad j = 0, 1, 2, \cdots,$$

form a complete orthonormal basis for $L^2(\mathbb{R})$ with $Ae_j = (2j + 2)e_j$. In particular, A admits a Hilbert-Schmidt inverse. The standard CH-space constructed from $(L^2(\mathbb{R}), A)$ coincides with $\mathcal{S}(\mathbb{R})$ of rapidly decreasing functions of Schwartz. In that case the conditions (i)-(iii) in Theorem 1.4.5 are satisfied with trivial open covering of $\mathbb{R}$. (Only for construction of $\mathcal{S}(\mathbb{R})$ one may replace $+2$ in (1.15) with an arbitrary positive number. When we discuss white noise functionals, $\inf \mathrm{Spec}(A) = 2 > 1$ is required, see §3.1.)

1.5 Bochner-Minlos theorem

Let E be a real nuclear space, E^* its dual space and $\langle \cdot, \cdot \rangle$ the canonical bilinear form on $E^* \times E$. Let $\mathfrak{B}$ be the so-called cylindrical σ-field on E^*, i.e., the smallest σ-field such that the function

$$x \mapsto (\langle x, \xi_1 \rangle, \cdots, \langle x, \xi_n \rangle) \in \mathbb{R}^n, \qquad x \in E^*,$$

is measurable for any choice of $\xi_1, \cdots, \xi_n \in E$ and $n = 1, 2, \cdots$, where $\mathbb{R}^n$ is equipped with the topological (Borel) σ-field. We recall a one-to-one correspondence between characteristic functions and probability measures on $(E^*, \mathfrak{B})$.

Definition 1.5.1 A $\mathbb{C}$-valued function C on E is called a *characteristic function* if
(i) C is continuous;
(ii) C is positive definite, i.e.,

$$\sum_{j,k=1}^{n} \alpha_j \overline{\alpha_k} C(\xi_j - \xi_k) \geq 0$$

for any choice of $\alpha_1, \cdots, \alpha_n \in \mathbb{C}$, $\xi_1, \cdots, \xi_n \in E$ and $n = 1, 2, \cdots$;

(iii) (normalization) $C(0) = 1$.

Theorem 1.5.2 (BOCHNER-MINLOS) *If ν is a probability measure on E^*, its Fourier transform*

$$\widehat{\nu}(\xi) = \int_{E^*} e^{i\langle x, \xi \rangle} \nu(dx), \qquad \xi \in E, \tag{1.16}$$

is a characteristic function. Conversely, for a characteristic function C on a nuclear space E there exists a unique probability measure ν on E^* such that $C = \widehat{\nu}$.

As for the support of ν we only mention a simple result. Recall that $E_\alpha^* \subset E_\beta^* \subset E^*$ whenever $\|\xi\|_\alpha \le \|\xi\|_\beta$ for all $\xi \in E$.

Theorem 1.5.3 *Let E be a nuclear space equipped with defining Hilbertian semi-norms $\{\|\cdot\|_\alpha\}_{\alpha \in A}$ and let ν be a probability measure on E^*. If $\widehat{\nu}$ admits a continuous extension to E_α for some $\alpha \in A$, then $\nu(E_\beta^*) = 1$ whenever the canonical map $E_\beta \to E_\alpha$, $\alpha \le \beta$, is of Hilbert-Schmidt type.*

As we have agreed in §1.1, E^* carries always the strong dual topology. If E is a nuclear Fréchet space, the topological σ-field coincides with the cylindrical σ-field $\mathfrak{B}$.

1.6 Further notational remarks

Here are some notational remarks used throughout.

When there is no danger of confusion, the π-tensor product $\mathfrak{X} \otimes_\pi \mathfrak{Y}$ is denoted by $\mathfrak{X} \otimes \mathfrak{Y}$ for simplicity. Let $\mathfrak{X}$ be a Hilbert space or a nuclear Fréchet space. For $\xi_1, \cdots, \xi_n \in \mathfrak{X}$ the symmetrization of $\xi_1 \otimes \cdots \otimes \xi_n$ is defined by

$$\xi_1 \widehat{\otimes} \cdots \widehat{\otimes} \xi_n = \frac{1}{n!} \sum_{\sigma \in \mathfrak{S}_n} \xi_{\sigma(1)} \otimes \cdots \otimes \xi_{\sigma(n)}, \tag{1.17}$$

where $\mathfrak{S}_n$ stands for the group of permutations of $\{1, 2, \cdots, n\}$. Let $\mathfrak{X}^{\widehat{\otimes} n}$ denote the closed subspace of $\mathfrak{X}^{\otimes n} = \mathfrak{X} \otimes \cdots \otimes \mathfrak{X}$ (n times) spanned by $\xi_1 \widehat{\otimes} \cdots \widehat{\otimes} \xi_n$, where $\xi_1, \cdots, \xi_n$ run over $\mathfrak{X}$. By virtue of the polarization formula (A.2), $\mathfrak{X}^{\widehat{\otimes} n}$ coincides with the closed subspace of $\mathfrak{X}^{\otimes n}$ spanned by $\xi^{\otimes n}$ with ξ running over $\mathfrak{X}$.

For $F \in (\mathfrak{X}^{\otimes n})^*$ and $\sigma \in \mathfrak{S}_n$ let F^σ be an element in $(\mathfrak{X}^{\otimes n})^*$ uniquely determined by

$$\langle F^\sigma, \xi_1 \otimes \cdots \otimes \xi_n \rangle = \left\langle F, \xi_{\sigma^{-1}(1)} \otimes \cdots \otimes \xi_{\sigma^{-1}(n)} \right\rangle, \qquad \xi_1, \cdots, \xi_n \in \mathfrak{X}. \tag{1.18}$$

For $F \in (\mathfrak{X}^{\otimes n})^*$ the *symmetrization* $\widehat{F}$ is defined by

$$\widehat{F} = \frac{1}{n!} \sum_{\sigma \in \mathfrak{S}_n} F^\sigma. \tag{1.19}$$

If $\widehat{F} = F$, it is called *symmetric*. Let $(\mathfrak{X}^{\otimes n})^*_{\mathrm{sym}}$ denote the subspace of $(\mathfrak{X}^{\otimes n})^*$ consisting of symmetric elements. Since

$$(f_1 \otimes \cdots \otimes f_n)^\wedge = f_1 \widehat{\otimes} \cdots \widehat{\otimes} f_n, \qquad f_1, \cdots, f_n \in \mathfrak{X}^*,$$

two definitions (1.17) and (1.19) are consistent. For $F \in (\mathfrak{X}^{\otimes m})^*_{\mathrm{sym}}$ and $G \in (\mathfrak{X}^{\otimes n})^*_{\mathrm{sym}}$, we write $F \widehat{\otimes} G$ for the symmetrization of $F \otimes G$. Note that $(\mathfrak{X}^{\otimes n})^* \cong (\mathfrak{X}^*)^{\otimes n}$. This is apparent when $\mathfrak{X}$ is a Hilbert space and is also valid for a nuclear Fréchet space by the kernel theorem (Theorem 1.3.10). Therefore, in both cases we have $(\mathfrak{X}^{\otimes n})^*_{\mathrm{sym}} \cong (\mathfrak{X}^*)^{\widehat{\otimes} n}$. However, we prefer to the symbol $(\mathfrak{X}^{\otimes n})^*_{\mathrm{sym}}$ in order to avoid some notational ambiguity appearing in the later discussion.

If $\mathfrak{X}$ is a locally convex space over $\mathbf{R}$, its complexification is denoted by $\mathfrak{X}_{\mathbf{C}}$. The canonical bilinear form on $\mathfrak{X}^* \times \mathfrak{X}$, which is an $\mathbf{R}$-bilinear form, is naturally extended to a $\mathbf{C}$-bilinear form on $\mathfrak{X}^*_{\mathbf{C}} \times \mathfrak{X}_{\mathbf{C}}$ which is denoted by the same symbol. Namely,

$$\langle x + iy\,, \xi + i\eta \rangle = \langle x\,, \xi \rangle - \langle y\,, \eta \rangle + i(\langle x\,, \eta \rangle + \langle y\,, \xi \rangle), \qquad x, y \in \mathfrak{X}^*, \quad \xi, \eta \in \mathfrak{X}.$$

Then special attention is needed when a Hilbert space is discussed. Let $\mathfrak{X} = H$ be a real Hilbert space with norm $\|\cdot\|$ and inner product $(\cdot,\cdot)$. In that case the canonical bilinear form denoted by $\langle \cdot\,, \cdot \rangle$ on $H^* \times H$ is identified with the inner product by Riesz' theorem since $H \rightarrow H^*$ is linear isomorphism. On the other hand, both $(\cdot,\cdot)$ and $\|\cdot\|$ are naturally extended to the inner product and the norm of $H_{\mathbf{C}}$ which are denoted by the same symbol usually. Then, due to the above mentioned convention, we see that

$$\|\zeta\|^2 = (\zeta, \zeta) = \langle \overline{\zeta}\,, \zeta \rangle, \qquad \zeta \in H_{\mathbf{C}},$$

where $\overline{\zeta}$ stands for the conjugation induced from the complex structure of $H_{\mathbf{C}}$. From the next chapter on we seldom use the hermitian inner product of $H_{\mathbf{C}}$ but use the canonical bilinear form.

Bibliographical Notes

For general theory of topological vector spaces, see Schaefer [1] and Treves [1] where a nuclear space is defined not using Hilbertian seminorms but using nuclear operators between Banach spaces. Their definition is seemingly more general, however, coincides with ours, see Yamasaki [2]. A concise account of countably Hilbert spaces is also found in Gelfand-Vilenkin [1].

The essential idea of a standard CH-space was given by Gelfand-Vilenkin [1] and Berezansky-Kondrat'ev [1], and its systematic application to white noise calculus started in Hida-Obata-Saitô [1] and Obata [8], [9]. As for basic properties of selfadjoint operators, see e.g., Strătilă-Zsidó [1].

Hypotheses (H1)–(H3) in §1.4 were first introduced by Kubo-Takenaka [2] in a slightly different way. More generally, the evaluation map (delta function) δ_ω is discussed in Gelfand-Vilenkin [1] without topological structure of Ω.

For the proof of Bochner-Minlos theorem, see e.g., Gelfand-Vilenkin [1], Hida [2] and Yamasaki [2]. In particular, Yamasaki [2] discussed measures on an infinite dimensional vector space in full detail. Closely related matters are discussed concisely also in Glimm-Jaffe [1] and Reed-Simon [1].

Chapter 2

White Noise Space

2.1 Gaussian measure

Let H be a real separable (infinite dimensional) Hilbert space and let E be a real nuclear space which is densely and continuously imbedded in H. Then, with the identification $H \cong H^*$ by Riesz' theorem, we come to a Gelfand triple:

$$E \subset H \subset E^*.$$

Let $\langle \cdot , \cdot \rangle$ denote the canonical bilinear form on $E^* \times E$ and let $|\cdot|$ be the norm of H. Then $|\xi|^2 = \langle \xi , \xi \rangle$ for $\xi \in E$. We keep these notation throughout the chapter.

Lemma 2.1.1 *For $\sigma \geq 0$,*

$$C_\sigma(\xi) = \exp\left(-\frac{\sigma^2}{2}|\xi|^2\right), \qquad \xi \in E,$$

is a characteristic function.

PROOF. Obviously, C_σ is continuous on E and $C_\sigma(0) = 1$. We shall prove that C_σ is positive definite. Given $\alpha_1, \cdots, \alpha_n \in \mathbb{C}$ and $\xi_1, \cdots, \xi_n \in E$, we see from definition that

$$\sum_{j,k=1}^{n} \alpha_j \overline{\alpha_k} C_\sigma(\xi_j - \xi_k) = \sum_{j,k=1}^{n} \beta_j \overline{\beta_k} \exp\left(\sigma^2 \langle \xi_j , \xi_k \rangle\right), \tag{2.1}$$

where $\beta_j = \alpha_j \exp(-\sigma^2 |\xi_j|^2 /2)$. In general, if (λ_{jk}) is a positive definite matrix, so is $(e^{\lambda_{jk}})$. Since the matrix $(\sigma^2 \langle \xi_j , \xi_k \rangle)_{1 \leq j,k \leq n}$ is obviously positive definite, the right hand side of (2.1) is non-negative, i.e., C_σ is positive definite. qed

An application of the Bochner-Minlos theorem (Theorem 1.5.2) leads us to the following

Definition 2.1.2 The probability measure of which characteristic function is C_σ, $\sigma \geq 0$, is called the *Gaussian measure with variance* σ^2 and denoted by μ_σ. If $\sigma = 1$, we denote it by μ and call it the *(standard) Gaussian measure.* The probability space (E^*, μ) is called the *(standard) Gaussian space.*

By definition the Gaussian measure μ satisfies

$$e^{-|\xi|^2/2} = \int_{E^*} e^{i\langle x,\xi\rangle}\mu(dx), \qquad \xi \in E. \tag{2.2}$$

Most practical computations owe to the following two facts.

Lemma 2.1.3 *Let $\xi_1, \cdots, \xi_n \in E$ be an orthonormal system for H. Then the image of the Gaussian measure μ under the map*

$$x \mapsto (\langle x,\xi_1\rangle, \cdots, \langle x,\xi_n\rangle) \in \mathbb{R}^n, \qquad x \in E^*,$$

is a product of the standard Gaussian measure on $\mathbb{R}$, namely,

$$\left(\frac{1}{\sqrt{2\pi}}\right)^n e^{-(t_1^2+\cdots+t_n^2)/2}\, dt_1 \cdots dt_n. \tag{2.3}$$

PROOF. Let us denote by ν the image of μ under consideration. Since ν is a probability measure on $\mathbb{R}^n$, it is uniquely determined by its Fourier transform $\hat{\nu}$. By definition, for $s_1, \cdots, s_n \in \mathbb{R}$ we have

$$\begin{aligned} \hat{\nu}(s_1, \cdots, s_n) &= \int_{-\infty}^{+\infty} \cdots \int_{-\infty}^{+\infty} \exp\left(i\sum_{k=1}^n s_k t_k\right) \nu(dt_1 \cdots dt_n) \\ &= \int_{E^*} \exp\left(i\sum_{k=1}^n s_k \langle x,\xi_k\rangle\right) \mu(dx). \end{aligned}$$

Hence, in view of (2.2) we obtain

$$\hat{\nu}(s_1, \cdots, s_n) = \exp\left(-\frac{1}{2}\left|\sum_{k=1}^n s_k\xi_k\right|^2\right) = \exp\left(-\frac{1}{2}\sum_{k=1}^n s_k^2\right).$$

Therefore $\nu(dt_1 \cdots dt_n)$ is given by (2.3) as desired. qed

Lemma 2.1.4 *Let $\xi_1, \cdots, \xi_n \in E$ be an orthogonal system for H and let $f_1, \cdots, f_n$ be integrable functions on $\mathbb{R}$ with respect to the 1-dimensional standard Gaussian measure. Then,*

$$\int_{E^*} f_1(\langle x,\xi_1\rangle) \cdots f_n(\langle x,\xi_n\rangle)\mu(dx) = \prod_{k=1}^n \int_{E^*} f_k(\langle x,\xi_k\rangle)\, \mu(dx).$$

The proof is immediate by a simple application of Lemma 2.1.3. We next show that the canonical bilinear form $\langle\cdot,\cdot\rangle$ on $E^* \times E$ is extended to $E^* \times H_{\mathbb{C}}$ as an L^2-function on E^*. To this end we need the following

Lemma 2.1.5 *For any $\xi \in E_{\mathbb{C}}$ it holds that*

$$\int_{E^*} |\langle x,\xi\rangle|^2 \mu(dx) = |\xi|^2. \tag{2.4}$$

PROOF. Since (2.4) is obvious for $\xi = 0$, we assume first that $\xi \in E$, $\xi \neq 0$. By Lemma 2.1.3 we obtain

$$\int_{E^*} |\langle x, \xi\rangle|^2 \mu(dx) = |\xi|^2 \int_{E^*} \left\langle x, \frac{\xi}{|\xi|}\right\rangle^2 \mu(dx) = \frac{|\xi|^2}{\sqrt{2\pi}} \int_{-\infty}^{+\infty} t^2 e^{-t^2/2} dt = |\xi|^2 .$$

This proves that (2.4) is true for $\xi \in E$. Using this fact, we may show the identity (2.4) for an arbitrary $\xi = \xi_1 + i\xi_2 \in E_{\mathbf{C}}$ with $\xi_1, \xi_2 \in E$. In fact,

$$\int_{E^*} |\langle x, \xi\rangle|^2 \mu(dx) = \int_{E^*} \left(\langle x, \xi_1\rangle^2 + \langle x, \xi_2\rangle^2\right) \mu(dx) = |\xi_1|^2 + |\xi_2|^2 = |\xi|^2 .$$

This completes the proof. qed

The norm of $\phi \in L^2(E^*, \mu; \mathbf{C})$ is denoted by

$$\|\phi\| = \left(\int_{E^*} |\phi(x)|^2 \mu(dx)\right)^{1/2} .$$

Then Lemma 2.1.5 says that a linear function $x \mapsto \langle x, \xi\rangle$, $\xi \in E_{\mathbf{C}}$, belongs to $L^2(E^*, \mu; \mathbf{C})$. We now define $\langle x, \xi\rangle$ for any $\xi \in H_{\mathbf{C}}$ as an L^2-function of $x \in E^*$. Taking an approximating sequence $\{\xi_n\}_{n=1}^{\infty} \subset E_{\mathbf{C}}$ such that $\lim_{n\to\infty} |\xi_n - \xi| = 0$, we put $\phi_n(x) = \langle x, \xi_n\rangle$. It then follows from Lemma 2.1.5 that

$$\lim_{m,n\to\infty} \|\phi_m - \phi_n\| = \lim_{m,n\to\infty} |\xi_m - \xi_n| = 0.$$

Hence there exists $\phi \in L^2(E^*, \mu; \mathbf{C})$ such that $\lim_{n\to\infty} \|\phi_n - \phi\| = 0$. This ϕ is independent of the choice of an approximate sequence and is denoted by $\phi(x) = \langle x, \xi\rangle$. If $\xi \in H$, then $\phi(x) = \langle x, \xi\rangle$ belongs to $L^2(E^*, \mu; \mathbf{R})$. The assertions in Lemmas 2.1.3 and 2.1.4 are also true for $\xi_1, \cdots, \xi_n, \xi \in H$.

Here are some useful formulae, where $\xi, \xi_1, \xi_2, \cdots \in H_{\mathbf{C}}$ and $n = 0, 1, 2, \cdots$.

$$\int_{E^*} |\langle x, \xi\rangle|^2 \mu(dx) = |\xi|^2 . \tag{2.5}$$

$$\int_{E^*} e^{\langle x, \xi\rangle} \mu(dx) = e^{\langle \xi, \xi\rangle/2}. \tag{2.6}$$

$$\int_{E^*} \langle x, \xi\rangle^{2n} \mu(dx) = \frac{(2n)!}{2^n n!} \langle \xi, \xi\rangle^n . \tag{2.7}$$

$$\int_{E^*} \langle x, \xi\rangle^{2n+1} \mu(dx) = 0. \tag{2.8}$$

$$\int_{E^*} \prod_{j=1}^{2n} \langle x, \xi_j\rangle \mu(dx) = \frac{1}{2^n n!} \sum_{\sigma \in \mathfrak{S}_{2n}} \left\langle \xi_{\sigma(1)}, \xi_{\sigma(2)}\right\rangle \cdots \left\langle \xi_{\sigma(2n-1)}, \xi_{\sigma(2n)}\right\rangle . \tag{2.9}$$

$$\int_{E^*} \langle x, \xi_1\rangle \langle x, \xi_2\rangle \mu(dx) = \langle \xi_1, \xi_2\rangle . \tag{2.10}$$

PROOF OF (2.5) has been already established during the above discussion. qed

PROOF OF (2.6). Using a property of 1-dimensional Gaussian measure, we see easily that

$$\int_{E^*} e^{\alpha\langle x, \xi\rangle} \mu(dx) = e^{\alpha^2|\xi|^2/2}, \qquad \xi \in H, \quad \alpha \in \mathbf{C}.$$

Then, with the help of Lemma 2.1.3 we obtain

$$\int_{E^*} e^{\langle x, \xi + i\eta \rangle} \mu(dx) = e^{\langle \xi + i\eta, \xi + i\eta \rangle / 2}, \qquad \xi, \eta \in H,$$

where an orthogonal decomposition: $\eta = \eta' + \langle \xi, \eta \rangle |\xi|^{-2} \xi$ is useful. qed

PROOF OF (2.7) AND (2.8). It follows from (2.6) that

$$\int_{E^*} e^{z\langle x, \xi \rangle} \mu(dx) = e^{z^2 \langle \xi, \xi \rangle / 2}, \qquad z \in \mathbf{C}, \quad \xi \in H_{\mathbf{C}}.$$

The Taylor expansion of the left hand side is computed by Cauchy integral formula and Lebesgue's convergence theorem to get

$$\sum_{n=0}^{\infty} \frac{z^n}{n!} \int_{E^*} \langle x, \xi \rangle^n \mu(dx) = \sum_{m=0}^{\infty} \frac{z^{2m}}{2^m m!} \langle \xi, \xi \rangle^m, \qquad z \in \mathbf{C}.$$

From this the desired relations follow immediately. qed

PROOF OF (2.9) AND (2.10). By the polarization formula (A.2) we have

$$\prod_{j=1}^{2n} \langle x, \xi_j \rangle = \frac{1}{(2n)! 2^{2n}} \sum_{\epsilon} \epsilon_1 \cdots \epsilon_{2n} \left\langle x, \sum_{j=1}^{2n} \epsilon_j \xi_j \right\rangle^{2n},$$

where the sum is taken over $\epsilon_1 = \pm 1, \cdots, \epsilon_{2n} = \pm 1$. It then follows from (2.7) that

$$\begin{aligned} \int_{E^*} \prod_{j=1}^{2n} \langle x, \xi_j \rangle \mu(dx) &= \frac{1}{(2n)! 2^{2n}} \sum_{\epsilon} \epsilon_1 \cdots \epsilon_{2n} \frac{(2n)!}{2^n n!} \left\langle \sum_{j=1}^{2n} \epsilon_j \xi_j, \sum_{j=1}^{2n} \epsilon_j \xi_j \right\rangle^n \\ &= \frac{1}{2^{2n} 2^n n!} \sum_{\epsilon} \epsilon_1 \cdots \epsilon_{2n} \left\langle \sum_{j=1}^{2n} \epsilon_j \xi_j, \sum_{j=1}^{2n} \epsilon_j \xi_j \right\rangle^n. \end{aligned} \tag{2.11}$$

Define a symmetric $2n$-form on $H_{\mathbf{C}}$ by

$$F(\xi_1, \cdots, \xi_{2n}) = \frac{1}{(2n)!} \sum_{\sigma \in \mathfrak{S}_{2n}} \left\langle \xi_{\sigma(1)}, \xi_{\sigma(2)} \right\rangle \cdots \left\langle \xi_{\sigma(2n-1)}, \xi_{\sigma(2n)} \right\rangle$$

and put

$$A(\eta) = F(\eta, \cdots, \eta) = \langle \eta, \eta \rangle^n.$$

Then, applying the polarization formula again we have

$$\begin{aligned} F(\xi_1, \cdots, \xi_{2n}) &= \frac{1}{2^{2n} (2n)!} \sum_{\epsilon} \epsilon_1 \cdots \epsilon_{2n} A\left(\sum_{j=1}^{2n} \epsilon_j \xi_j \right) \\ &= \frac{1}{2^{2n} (2n)!} \sum_{\epsilon} \epsilon_1 \cdots \epsilon_{2n} \left\langle \sum_{j=1}^{2n} \epsilon_j \xi_j, \sum_{j=1}^{2n} \epsilon_j \xi_j \right\rangle^n \end{aligned} \tag{2.12}$$

Consequently, comparing (2.11) and (2.12) we see that

$$\begin{aligned}\int_{E^*} \prod_{j=1}^{2n} \langle x, \xi_j \rangle \, \mu(dx) &= \frac{(2n)!}{2^n n!} F(\xi_1, \cdots, \xi_{2n}) \\ &= \frac{1}{2^n n!} \sum_{\sigma \in \mathfrak{S}_{2n}} \left\langle \xi_{\sigma(1)}, \xi_{\sigma(2)} \right\rangle \cdots \left\langle \xi_{\sigma(2n-1)}, \xi_{\sigma(2n)} \right\rangle.\end{aligned}$$

This proves (2.9) and (2.10) is its immediate consequence. qed

Finally we note the following

Proposition 2.1.6 *The Gaussian measure μ is quasi-invariant under the translation by any $\xi \in H$ and the Radon-Nikodym derivative is given by*

$$\frac{\mu(dx - \xi)}{\mu(dx)} = e^{\langle x, \xi \rangle - \langle \xi, \xi \rangle / 2}, \qquad x \in E^*. \tag{2.13}$$

PROOF. Suppose $\xi \in H$ and denote by $\phi(x)$ the right hand side of (2.13). In view of (2.6) one can easily verify that $\phi(x)\mu(dx)$ is a probability measure on E^*. We compute its Fourier transform. By definition, for $\eta \in E$

$$\widehat{\phi\mu}(\eta) = \int_{E^*} e^{i\langle x, \eta \rangle} \phi(x) \mu(dx) = \int_{E^*} e^{i\langle x, \eta \rangle + \langle x, \xi \rangle - \langle \xi, \xi \rangle / 2} \mu(dx).$$

Then, using (2.6) we obtain

$$\widehat{\phi\mu}(\eta) = e^{\langle \xi + i\eta, \xi + i\eta \rangle / 2 - \langle \xi, \xi \rangle / 2} = e^{i\langle \xi, \eta \rangle - \langle \eta, \eta \rangle / 2}, \qquad \eta \in E.$$

On the other hand, since

$$\int_{E^*} e^{i\langle x, \eta \rangle} \mu(dx - \xi) = \int_{E^*} e^{i\langle x + \xi, \eta \rangle} \mu(dx) = e^{i\langle \xi, \eta \rangle - \langle \eta, \eta \rangle / 2},$$

we come to

$$\widehat{\phi\mu}(\eta) = \int_{E^*} e^{i\langle x, \eta \rangle} \mu(dx - \xi).$$

Since the characteristic functional of a probability measure on E^* is unique as is stated in Theorem 1.5.2, we conclude that

$$\mu(dx - \xi) = \phi(x) \mu(dx).$$

Hence, μ is quasi-invariant under translation by $\xi \in H$ and (2.13) holds. qed

2.2 Wick-ordered polynomials

We first introduce polynomial functions on the Gaussian space E^*. Let $\mathcal{P}_n(\mathbb{R})$ (resp. $\mathcal{P}_n(\mathbb{C})$) be the space of finite linear combinations of functions of the form

$$x \mapsto \langle x, \xi_1 \rangle \cdots \langle x, \xi_n \rangle = \left\langle x^{\otimes n}, \xi_1 \otimes \cdots \otimes \xi_n \right\rangle, \qquad x \in E^*, \tag{2.14}$$

where $\xi_1, \cdots, \xi_n \in E$ (resp. $\xi_1, \cdots, \xi_n \in E_{\mathbf{C}}$). Since

$$\left\langle x^{\otimes n}, \xi_1 \otimes \cdots \otimes \xi_n \right\rangle = \left\langle x^{\otimes n}, \xi_1 \hat{\otimes} \cdots \hat{\otimes} \xi_n \right\rangle,$$

it follows from the polarization formula (A.2) that $\mathcal{P}_n(\mathbf{R})$ (resp. $\mathcal{P}_n(\mathbf{C})$) coincides with the space of finite linear combinations of the functions of the form

$$x \mapsto \langle x, \xi \rangle^n = \left\langle x^{\otimes n}, \xi^{\otimes n} \right\rangle, \tag{2.15}$$

where ξ runs over E (resp. $E_{\mathbf{C}}$). Note that $\mathcal{P}_n(\mathbf{C}) = \mathcal{P}_n(\mathbf{R}) + i\mathcal{P}_n(\mathbf{R})$.

An element of the algebraic sums

$$\mathcal{P}(\mathbf{R}) = \sum_{n=0}^{\infty} \mathcal{P}_n(\mathbf{R}), \qquad \mathcal{P}(\mathbf{C}) = \sum_{n=0}^{\infty} \mathcal{P}_n(\mathbf{C})$$

is called a *polynomial* on the Gaussian space E^*. Although polynomials are elementary and important functions on the Gaussian space, expression (2.14) or (2.15) is not useful mainly because they do not satisfy a simple orthogonality relation. Thus we need another expression called *Wick-ordered polynomials.*

Taking the canonical correspondence between $\mathcal{B}(E, E)$ and $(E \otimes E)^*$ into account (Proposition 1.3.9), let $\tau \in (E \otimes E)^*_{\mathrm{sym}}$ be determined uniquely by the formula:

$$\langle \tau, \xi \otimes \eta \rangle = \langle \xi, \eta \rangle, \qquad \xi, \eta \in E.$$

This τ is called the *trace* and the notation is fixed throughout.

Proposition 2.2.1 *For any complete orthonormal basis $\{e_j\}_{j=0}^{\infty}$ of H it holds that*

$$\langle \tau, \omega \rangle = \sum_{j=0}^{\infty} \langle e_j \otimes e_j, \omega \rangle, \qquad \omega \in E \otimes E.$$

Moreover, $\tau = \sum_{j=0}^{\infty} e_j \otimes e_j$ with respect to the strong dual topology of $(E \otimes E)^*$.

PROOF. Since E is a nuclear space imbedded in H, there exists a Hilbertian seminorm (in fact, a norm) $|\cdot|_1$ on E such that $|\cdot| \le |\cdot|_1$ and the canonical map $H_1 \to H$ is of Hilbert-Schmidt type, where H_1 is the associated Hilbert space with $|\cdot|_1$. By a similar discussion as in §1.2 we see that the canonical map $H \to H_{-1} \cong H_1^*$ is of Hilbert-Schmidt type and, therefore, $\sum_{j=0}^{\infty} |e_j|_{-1}^2 < \infty$. We now put

$$\tau_N = \sum_{j=0}^{N} e_j \otimes e_j \in H \otimes H.$$

As is easily seen,

$$\langle \tau - \tau_N, \omega \rangle = \sum_{j>N} \langle e_j \otimes e_j, \omega \rangle, \qquad \omega \in E \otimes E.$$

Hence for a bounded subset $S \subset E \otimes E$ we have

$$\sup_{\omega \in S} |\langle \tau - \tau_N, \omega \rangle| \le \sup_{\omega \in S} |\omega|_1 \sum_{j>N} |e_j \otimes e_j|_{-1} = \sup_{\omega \in S} |\omega|_1 \sum_{j>N} |e_j|_{-1}^2 \longrightarrow 0,$$

as $N \to \infty$. It follows that $\tau_N \to \tau$ with respect to the strong dual topology of $(E \otimes E)^*$. qed

Definition 2.2.2 For $x \in E^*$ and $n = 0, 1, 2, \cdots$ we define $:x^{\otimes n}: \in (E^{\otimes n})^*_{\mathrm{sym}}$ inductively as follows:

$$\begin{cases} :x^{\otimes 0}: & = 1; \\ :x^{\otimes 1}: & = x; \\ :x^{\otimes n}: & = x\widehat{\otimes} :x^{\otimes (n-1)}: -(n-1)\tau\widehat{\otimes} :x^{\otimes (n-2)}:, \qquad n \ge 2. \end{cases}$$

Proposition 2.2.3 *Let $x \in E^*$ and $\xi \in E_{\mathbb{C}}$. Then*

$$\left\langle :x^{\otimes n}:, \xi^{\otimes n} \right\rangle = \sum_{k=0}^{[n/2]} \frac{n!}{k!(n-2k)!} \left(-\frac{1}{2} \langle \xi, \xi \rangle \right)^k \langle x, \xi \rangle^{n-2k}, \tag{2.16}$$

$$\langle x, \xi \rangle^n = \sum_{k=0}^{[n/2]} \frac{n!}{k!(n-2k)!} \left(\frac{1}{2} \langle \xi, \xi \rangle \right)^k \left\langle :x^{\otimes (n-2k)}:, \xi^{\otimes (n-2k)} \right\rangle. \tag{2.17}$$

These are easily verified by induction and Definition 2.2.2. Then, in view of the definition of the trace τ one can easily deduce the following

Corollary 2.2.4 *For $x \in E^*$ we have*

$$:x^{\otimes n}: = \sum_{k=0}^{[n/2]} \frac{(-1)^k n!}{(n-2k)!k!2^k} \tau^{\widehat{\otimes} k} \widehat{\otimes} x^{\otimes (n-2k)}, \tag{2.18}$$

$$x^{\otimes n} = \sum_{k=0}^{[n/2]} \frac{n!}{(n-2k)!k!2^k} \tau^{\widehat{\otimes} k} \widehat{\otimes} :x^{\otimes (n-2k)}:. \tag{2.19}$$

Let $\mathcal{Q}_n(\mathbb{R})$ (resp. $\mathcal{Q}_n(\mathbb{C})$) be the space of finite linear combinations of functions

$$x \mapsto \left\langle :x^{\otimes n}:, \xi_1 \otimes \cdots \otimes \xi_n \right\rangle, \qquad x \in E^*,$$

where $\xi_1, \cdots, \xi_n \in E$ (resp. $\xi_1, \cdots, \xi_n \in E_{\mathbb{C}}$) or equivalently,

$$x \mapsto \left\langle :x^{\otimes n}:, \xi^{\otimes n} \right\rangle, \qquad x \in E^*,$$

where $\xi \in E$ (resp. $\xi \in E_{\mathbb{C}}$). Then, by Proposition 2.2.3 we have

Corollary 2.2.5 *For $n = 0, 1, 2, \cdots$ it holds that*

$$\sum_{k=0}^{n} \mathcal{P}_k(\mathbb{R}) = \sum_{k=0}^{n} \mathcal{Q}_k(\mathbb{R}), \qquad \textit{and} \qquad \sum_{k=0}^{n} \mathcal{P}_k(\mathbb{C}) = \sum_{k=0}^{n} \mathcal{Q}_k(\mathbb{C}).$$

Namely, every polynomial $\phi \in \mathcal{P}(\mathbb{R})$ (resp. $\mathcal{P}(\mathbb{C})$) is expressed as

$$\phi(x) = \sum_{n=0}^{\infty} \left\langle :x^{\otimes n}:, f_n \right\rangle, \tag{2.20}$$

where f_n belongs to the symmetric n-fold algebraic tensor product of E (resp. $E_{\mathbb{C}}$) and is non-zero except finitely many n.

Definition 2.2.6 A polynomial expressed as in (2.20) is called a *Wick-ordered polynomial.*

As will be seen in Corollary 2.2.11, for a given polynomial ϕ expression (2.20) is unique, namely, if $\phi(x)=0$ for all $x\in E^*$, then $f_n=0$ for all $n=0,1,2,\cdots$.

Wick-ordered polynomials are closely related to the Hermite polynomials, for their definition and formulae see Appendix B. In fact, particular Wick-ordered polynomials are expressed directly in terms of Hermite polynomials. The next result follows from formula (B.2) for Hermite polynomials.

Lemma 2.2.7 *For $\xi\in E$, $\xi\neq 0$, it holds that*

$$\left\langle :x^{\otimes n}:,\xi^{\otimes n}\right\rangle=\frac{|\xi|^n}{2^{n/2}}H_n\left(\frac{\langle x,\xi\rangle}{\sqrt{2}\,|\xi|}\right).$$

Lemma 2.2.8 *Let $\xi\in E$ with $|\xi|=1$. Then,*

$$\int_{E^*}H_m\left(\frac{\langle x,\xi\rangle}{\sqrt{2}}\right)H_n\left(\frac{\langle x,\xi\rangle}{\sqrt{2}}\right)\mu(dx)=2^n n!\delta_{mn}.$$

PROOF. It follows from Lemma 2.1.3 that

$$\int_{E^*}H_m\left(\frac{\langle x,\xi\rangle}{\sqrt{2}}\right)H_n\left(\frac{\langle x,\xi\rangle}{\sqrt{2}}\right)\mu(dx)=\frac{1}{\sqrt{2\pi}}\int_{-\infty}^{+\infty}H_m\left(\frac{t}{\sqrt{2}}\right)H_n\left(\frac{t}{\sqrt{2}}\right)e^{-t^2/2}dt.$$

We then need only to apply (B.8). qed

Lemma 2.2.9 *Let $\xi,\eta\in E_{\mathbf{C}}$. Then,*

$$\int_{E^*}\left\langle :x^{\otimes m}:,\xi^{\otimes m}\right\rangle\left\langle :x^{\otimes n}:,\eta^{\otimes n}\right\rangle\mu(dx)=n!\,\langle\xi,\eta\rangle^n\,\delta_{mn}. \tag{2.21}$$

PROOF. We first consider the case where $\xi,\eta\in E$. It is sufficient to prove the identity under the assumption $|\xi|=|\eta|=1$. Taking a unit vector $\zeta\in E$ such that $\langle\xi,\zeta\rangle=0$, we may write

$$\eta=\alpha\xi+\beta\zeta,\qquad \alpha^2+\beta^2=1.$$

It follows from (B.7) that

$$\begin{aligned}H_n\left(\frac{\langle x,\eta\rangle}{\sqrt{2}}\right)&=H_n\left(\alpha\frac{\langle x,\xi\rangle}{\sqrt{2}}+\beta\frac{\langle x,\zeta\rangle}{\sqrt{2}}\right)\\&=\sum_{j=0}^{n}\binom{n}{j}\alpha^{n-j}\beta^jH_{n-j}\left(\frac{\langle x,\xi\rangle}{\sqrt{2}}\right)H_j\left(\frac{\langle x,\zeta\rangle}{\sqrt{2}}\right).\end{aligned} \tag{2.22}$$

In view of Lemma 2.2.7 and (2.22) we obtain

$$\begin{aligned}&\left\langle :x^{\otimes m}:,\xi^{\otimes m}\right\rangle\left\langle :x^{\otimes n}:,\eta^{\otimes n}\right\rangle=\frac{1}{2^{(m+n)/2}}H_m\left(\frac{\langle x,\xi\rangle}{\sqrt{2}}\right)H_n\left(\frac{\langle x,\eta\rangle}{\sqrt{2}}\right)\\&=\frac{1}{2^{(m+n)/2}}\sum_{j=0}^{n}\binom{n}{j}\alpha^{n-j}\beta^jH_m\left(\frac{\langle x,\xi\rangle}{\sqrt{2}}\right)H_{n-j}\left(\frac{\langle x,\xi\rangle}{\sqrt{2}}\right)H_j\left(\frac{\langle x,\zeta\rangle}{\sqrt{2}}\right).\end{aligned}$$

On the other hand, since $\xi \perp \zeta$, applying Lemma 2.1.4 we obtain

$$\int_{E^*} H_m\left(\frac{\langle x,\xi\rangle}{\sqrt{2}}\right) H_{n-j}\left(\frac{\langle x,\xi\rangle}{\sqrt{2}}\right) H_j\left(\frac{\langle x,\zeta\rangle}{\sqrt{2}}\right) \mu(dx)$$
$$= \int_{E^*} H_m\left(\frac{\langle x,\xi\rangle}{\sqrt{2}}\right) H_{n-j}\left(\frac{\langle x,\xi\rangle}{\sqrt{2}}\right) \mu(dx) \int_{E^*} H_j\left(\frac{\langle x,\zeta\rangle}{\sqrt{2}}\right) \mu(dx).$$

By Lemma 2.2.8 the last integral is equal to 0 unless $j = 0$ and is equal to 1 if $j = 0$. Hence,

$$\int_{E^*} \left\langle :x^{\otimes m}:, \xi^{\otimes m}\right\rangle \left\langle :x^{\otimes n}:, \eta^{\otimes n}\right\rangle \mu(dx)$$
$$= \frac{\alpha^n}{2^{(m+n)/2}} \int_{E^*} H_m\left(\frac{\langle x,\xi\rangle}{\sqrt{2}}\right) H_n\left(\frac{\langle x,\xi\rangle}{\sqrt{2}}\right) \mu(dx).$$

Applying Lemma 2.2.8 again, we conclude that

$$\int_{E^*} \left\langle :x^{\otimes m}:, \xi^{\otimes m}\right\rangle \left\langle :x^{\otimes n}:, \eta^{\otimes n}\right\rangle \mu(dx) = \frac{2^n\alpha^n n!}{2^{(m+n)/2}}\delta_{mn} = n!\alpha^n\delta_{mn}.$$

Since $\alpha = \langle \xi, \eta\rangle$, we have completed the proof of (2.21) for $\xi, \eta \in E$. As for general case, writing as $\xi = \xi_1 + i\xi_2$, $\eta = \eta_1 + i\eta_2$ with $\xi_1, \xi_2, \eta_1, \eta_2 \in E$, we can derive (2.21) from the real case by means of polarization formula. qed

Proposition 2.2.10 *For two polynomials $\phi, \psi \in \mathcal{P}(\mathbb{C})$ given respectively by*

$$\phi(x) = \sum_{n=0}^{\infty} \left\langle :x^{\otimes n}:, f_n\right\rangle, \qquad \psi(x) = \sum_{n=0}^{\infty} \left\langle :x^{\otimes n}:, g_n\right\rangle,$$

it holds that

$$\int_{E^*} \phi(x)\psi(x)\mu(dx) = \sum_{n=0}^{\infty} n!\,\langle f_n, g_n\rangle. \tag{2.23}$$

In particular,

$$\|\phi\|^2 = \sum_{n=0}^{\infty} n!\,|f_n|^2. \tag{2.24}$$

PROOF. First note that $f_n = g_n = 0$ except finitely many n by definition. Hence the sums in the statement are all finite. Since f_n and g_n are linear combination of elements of the form $\xi^{\otimes n}$, $\xi \in E_{\mathbb{C}}$, (2.23) is an immediate consequence of Lemma 2.2.9. Since

$$\overline{\phi(x)} = \sum_{n=0}^{\infty} \left\langle :x^{\otimes n}:, \overline{f_n}\right\rangle,$$

(2.24) follows immediately from (2.23). qed

Taking (2.24) into consideration, we immediately obtain

Corollary 2.2.11 *Each $\phi \in \mathcal{P}(\mathbb{C})$ is uniquely expressed as a Wick-ordered polynomial.*

2.3 Wiener-Itô-Segal isomorphism and Fock space

We first consider a particular Wick-ordered polynomial

$$\phi(x) = \left\langle :x^{\otimes n}:, f \right\rangle, \qquad x \in E^*, \tag{2.25}$$

where $f \in E_{\mathbb{C}}^{\widehat{\otimes} n}$ is a finite linear combination of elements of the form $\eta^{\otimes n}$, $\eta \in E_{\mathbb{C}}$. Note that such polynomials belong to $(L^2) = L^2(E^*, \mu; \mathbb{C})$. In fact, it follows from Proposition 2.2.10 that

$$\|\phi\|^2 = n! \, |f|^2 \tag{2.26}$$

We now define (2.25) for an arbitrary $f \in H_{\mathbb{C}}^{\widehat{\otimes} n}$ as a function in (L^2). The idea is similar to that employed when we extended the canonical bilinear form on $E^* \times E$ to $E^* \times H_{\mathbb{C}}$ as L^2-function, see §2.1. For $f \in H_{\mathbb{C}}^{\widehat{\otimes} n}$ we take an approximate sequence $f^{(j)}$ in symmetric n-fold algebraic tensor product of $E_{\mathbb{C}}$ such that $\lim_{j\to\infty} |f^{(j)} - f| = 0$. Define a sequence of Wick-ordered polynomials $\phi_j(x) = \left\langle :x^{\otimes n}:, f^{(j)} \right\rangle$. It then follows from Proposition 2.2.10 that

$$\lim_{j,k\to\infty} \|\phi_j - \phi_k\|^2 = n! \lim_{j,k\to\infty} |f^{(j)} - f^{(k)}|^2 = 0.$$

Hence there exists $\phi \in L^2(E^*, \mu; \mathbb{C})$ such that $\lim_{j\to\infty} \|\phi_j - \phi\| = 0$. It is easily verified that ϕ is independent of the choice of an approximate sequence $f^{(j)}$. We then denote it by (2.25). Obviously, (2.26) holds again.

Let $\mathcal{H}_n(\mathbb{R}) \subset L^2(E^*, \mu; \mathbb{R})$ and $\mathcal{H}_n(\mathbb{C}) \subset L^2(E^*, \mu; \mathbb{C})$ be the closed subspaces spanned by $\mathcal{Q}_n(\mathbb{R})$ and $\mathcal{Q}_n(\mathbb{C})$, respectively. During the above argument we have already established the following

Proposition 2.3.1 *$\mathcal{H}_n(\mathbb{R})$ (resp. $\mathcal{H}_n(\mathbb{C})$) coincides with the space of functions of the form $x \mapsto \langle :x^{\otimes n}:, f \rangle$ with $f \in H^{\widehat{\otimes} n}$ (resp. $f \in H_{\mathbb{C}}^{\widehat{\otimes} n}$).*

We next prove the following fundamental fact.

Proposition 2.3.2 *The polynomials $\mathcal{P}(\mathbb{R})$ and $\mathcal{P}(\mathbb{C})$ are dense subspaces of $L^2(E^*, \mu; \mathbb{R})$ and $L^2(E^*, \mu; \mathbb{C})$, respectively.*

PROOF. It is sufficient to prove that $\mathcal{P}(\mathbb{R})$ is a dense subspace of $L^2(E^*, \mu; \mathbb{R})$. Suppose that $\phi \in L^2(E^*, \mu; \mathbb{R})$ is orthogonal to $\mathcal{P}(\mathbb{R})$. We take and fix a complete orthonormal basis $\{e_j\}_{j=0}^{\infty} \subset E$ for H. Then, in particular, ϕ satisfies

$$\int_{E^*} \phi(x) \langle x, e_1 \rangle^{p_1} \cdots \langle x, e_n \rangle^{p_n} \mu(dx) = 0 \tag{2.27}$$

for any choice of $p_1, \cdots, p_n = 0, 1, 2, \cdots$ and $n = 0, 1, 2, \cdots$. Let $\mathfrak{B}_n$ be the smallest σ-field on E^* such that the map

$$x \mapsto (\langle x, e_1 \rangle, \cdots, \langle x, e_n \rangle) \in \mathbb{R}^n$$

is measurable. Then, the cylindrical σ-field $\mathfrak{B}$ on E^* (see §1.5) is generated by $\mathfrak{B}_n$, i.e.,

$$\mathfrak{B}_1 \subset \mathfrak{B}_2 \subset \cdots \qquad \text{and} \qquad \mathfrak{B} = \bigvee_{n=0}^{\infty} \mathfrak{B}_n.$$

Let P_n be the conditional expectation subject to $\mathfrak{B}_n$, i.e., the orthogonal projection from $L^2(E^*, \mathfrak{B}, \mu; \mathbb{R})$ onto $L^2(E^*, \mathfrak{B}_n, \mu; \mathbb{R})$. It is easily verified that $P_n \uparrow I$, i.e.,

$$\lim_{n\to\infty} \|P_n\phi - \phi\| = 0, \qquad \phi \in L^2(E^*, \mathfrak{B}, \mu; \mathbb{R}). \tag{2.28}$$

(Among probabilists $P_n\phi$ is called the *conditional expectation relative to* $\mathfrak{B}_n$ and is often denoted by $E(\phi|\mathfrak{B}_n)$.) On the other hand, by (2.27) we have

$$\int_{E^*} P_n\phi(x) \langle x, e_1\rangle^{p_1} \cdots \langle x, e_n\rangle^{p_n} \mu(dx) = 0. \tag{2.29}$$

Since $P_n\phi$ is $\mathfrak{B}_n$-measurable, there exists a measurable function f_n on $\mathbb{R}^n$ such that

$$P_n\phi(x) = f_n(\langle x, e_1\rangle, \cdots, \langle x, e_n\rangle).$$

Moreover, f_n is square-integrable with respect to the Gaussian measure on $\mathbb{R}^n$, see (2.3). In view of (2.29) we obtain

$$\begin{aligned} 0 &= \int_{E^*} f_n(\langle x, e_1\rangle \cdots \langle x, e_n\rangle) \langle x, e_1\rangle^{p_1} \cdots \langle x, e_n\rangle^{p_n} \mu(dx) \\ &= \left(\frac{1}{\sqrt{2\pi}}\right)^n \int_{-\infty}^{+\infty} \cdots \int_{-\infty}^{+\infty} f_n(t_1, \cdots, t_n) t_1^{p_1} \cdots t_n^{p_n} e^{-(t_1^2+\cdots+t_n^2)/2} dt_1 \cdots dt_n \end{aligned}$$

for all $p_1, \cdots, p_n = 0, 1, 2, \cdots$. Hence $f_n = 0$ a.e. on $\mathbb{R}^n$, namely, $P_n\phi(x) = 0$ for μ-a.e. $x \in E^*$. Consequently, $\phi = 0$ by (2.28). qed

Theorem 2.3.3 (Wiener-Itô Decomposition) *It holds that*

$$L^2(E^*, \mu; \mathbb{R}) = \sum_{n=0}^{\infty} \oplus \mathcal{H}_n(\mathbb{R}) \qquad \text{and} \qquad L^2(E^*, \mu; \mathbb{C}) = \sum_{n=0}^{\infty} \oplus \mathcal{H}_n(\mathbb{C}),$$

where the right hand sides mean the orthogonal direct sum.

Proof. Suppose that $m \neq n$. It follows from Proposition 2.2.10 that $\mathcal{Q}_m(\mathbb{R}) \perp \mathcal{Q}_n(\mathbb{R})$ and therefore $\mathcal{H}_m(\mathbb{R}) \perp \mathcal{H}_n(\mathbb{R})$. On the other hand, by Corollary 2.2.5,

$$\mathcal{P}(\mathbb{R}) = \sum_{n=0}^{\infty} \mathcal{P}_n(\mathbb{R}) = \sum_{n=0}^{\infty} \mathcal{Q}_n(\mathbb{R}) \subset \sum_{n=0}^{\infty} \oplus \mathcal{H}_n(\mathbb{R}).$$

Since $\mathcal{P}(\mathbb{R}) \subset L^2(E^*, \mu; \mathbb{R})$ is a dense subspace by Proposition 2.3.2, we obtain the orthogonal direct sum decomposition of $L^2(E^*, \mu; \mathbb{R})$ as desired. The complex case is discussed similarly. qed

Definition 2.3.4 Let $\mathfrak{H}$ be a real or complex Hilbert space with the norm $|\cdot|$. Let $\Gamma(\mathfrak{H})$ be the space of all sequences $\mathbf{f} = (f_n)_{n=0}^{\infty}$, $f_n \in \mathfrak{H}^{\widehat{\otimes} n}$, such that $\sum_{n=0}^{\infty} n!\, |f_n|^2 < \infty$. Equipped with the norm

$$\|\mathbf{f}\|_{\Gamma(\mathfrak{H})}^2 = \sum_{n=0}^{\infty} n!\, |f_n|^2 ,$$

the Hilbert space $\Gamma(\mathfrak{H})$ is called the *(Boson) Fock space* or the *symmetric Hilbert space* over $\mathfrak{H}$.

As is easily seen, if $\mathfrak{H}$ is a real Hilbert space it holds that $\Gamma(\mathfrak{H}_{\mathbf{C}}) = \Gamma(\mathfrak{H})_{\mathbf{C}}$. With this notation together with Proposition 2.3.1 and Theorem 2.3.3 we can claim the following

Theorem 2.3.5 (Wiener-Itô-Segal) *For each $\phi \in L^2(E^*, \mu; \mathbf{C})$ there exists a unique $\mathbf{f} = (f_n)_{n=0}^{\infty} \in \Gamma(H_{\mathbf{C}})$ such that*

$$\phi(x) = \sum_{n=0}^{\infty} \left\langle :x^{\otimes n}: , f_n \right\rangle , \tag{2.30}$$

in the L^2-sense. Conversely, for any $\mathbf{f} = (f_n)_{n=0}^{\infty} \in \Gamma(H_{\mathbf{C}})$, (2.30) defines a function in $L^2(E^*, \mu; \mathbf{C})$. In that case,

$$\|\phi\|^2 = \sum_{n=0}^{\infty} n!\, |f_n|^2 = \|\mathbf{f}\|_{\Gamma(H_{\mathbf{C}})}^2 . \tag{2.31}$$

A similar assertion is also true for the real case. In short, we have canonical isomorphisms:

$$L^2(E^*, \mu; \mathbf{R}) \cong \Gamma(H) \qquad \textit{and} \qquad L^2(E^*, \mu; \mathbf{C}) \cong \Gamma(H_{\mathbf{C}}).$$

Definition 2.3.6 The canonical isomorphisms established in the above theorem is called the *Wiener-Itô-Segal isomorphism.* The expression as in (2.30) is called the *Wiener-Itô expansion* of $\phi \in L^2(E^*, \mu; \mathbf{C})$.

Here are some general properties of Fock space $\Gamma(\mathfrak{H})$. A complete orthonormal basis of $\Gamma(\mathfrak{H})$ is easily obtained.

Proposition 2.3.7 *Let $\{e_j\}_{j=0}^{\infty}$ be a complete orthonormal basis for $\mathfrak{H}$. Then,*

$$\sqrt{\frac{n!}{n_0! n_1! \cdots}}\; e_0^{\otimes n_0} \widehat{\otimes} e_1^{\otimes n_1} \widehat{\otimes} \cdots, \qquad n_0 + n_1 + \cdots = n,$$

form a complete orthonormal basis for $\mathfrak{H}^{\widehat{\otimes} n}$. Hence,

$$\left(0, \cdots, 0, \frac{e_0^{\otimes n_0} \widehat{\otimes} e_1^{\otimes n_1} \widehat{\otimes} \cdots}{\sqrt{n_0! n_1! \cdots}}, 0, \cdots \right), \qquad n_0 + n_1 + \cdots = n, \quad n = 0, 1, 2, \cdots,$$

where the non-zero element occurs in the n-th place, form a complete orthonormal basis for $\Gamma(\mathfrak{H})$.

PROOF. The first assertion can be checked by a direct calculation. The second is obvious by the definition of $\Gamma(\mathfrak{H})$. qed

For $\xi \in \mathfrak{H}$ we put

$$\phi_\xi = \left(1, \xi, \frac{\xi^{\otimes 2}}{2!}, \cdots, \frac{\xi^{\otimes n}}{n!}, \cdots\right). \tag{2.32}$$

Then,

$$\|\phi_\xi\|^2_{\Gamma(\mathfrak{H})} = \sum_{n=0}^{\infty} n! \left|\frac{\xi^{\otimes n}}{n!}\right|^2 = \sum_{n=0}^{\infty} \frac{|\xi|^{2n}}{n!} = e^{|\xi|^2}, \tag{2.33}$$

and, in particular, $\phi_\xi \in \Gamma(\mathfrak{H})$.

Definition 2.3.8 The above ϕ_ξ is called an *exponential vector* or a *coherent state.*

Proposition 2.3.9 *The exponential vectors $\{\phi_\xi\,;\,\xi \in \mathfrak{H}\}$ are linearly independent.*

PROOF. We prove the assertion for a real $\mathfrak{H}$. The complex case is discussed almost similarly. Choose distinct $\xi_1, \cdots, \xi_N \in \mathfrak{H}$. We must prove that

$$\sum_{j=1}^{N} \alpha_j \phi_{\xi_j} = 0 \tag{2.34}$$

with $\alpha_j \in \mathbb{R}$ implies $\alpha_j = 0$. Since $\xi_1, \cdots, \xi_N \in \mathfrak{H}$ are mutually distinct, there is $\eta \in \mathfrak{H}$ such that

$$\langle \xi_j - \xi_k\,, \eta\rangle \neq 0, \qquad j \neq k. \tag{2.35}$$

By assumption (2.34)

$$0 = \sum_{j=1}^{N} \alpha_j \left\langle \phi_{\xi_j}\,, \phi_{t\eta}\right\rangle = \sum_{j=1}^{N} \alpha_j \sum_{n=0}^{\infty} \frac{\langle \xi_j\,, t\eta\rangle^n}{n!} = \sum_{j=1}^{N} \alpha_j e^{t\langle \xi_j\,,\eta\rangle}, \qquad t \in \mathbb{R},$$

and differentiating by t we obtain

$$\sum_{j=1}^{N} \alpha_j \langle \xi_j\,, \eta\rangle^m = 0, \qquad m = 0, 1, 2, \cdots.$$

In particular,

$$\begin{pmatrix} 1 & \cdots & 1 \\ \langle \xi_1\,, \eta\rangle & \cdots & \langle \xi_N\,, \eta\rangle \\ \vdots & \ddots & \vdots \\ \langle \xi_1\,, \eta\rangle^{N-1} & \cdots & \langle \xi_N\,, \eta\rangle^{N-1} \end{pmatrix} \begin{pmatrix} \alpha_1 \\ \alpha_2 \\ \vdots \\ \alpha_N \end{pmatrix} = 0.$$

The determinant of the above matrix (known as Vandermonde's determinant) is

$$\prod_{1\le k<j\le N} \langle \xi_j - \xi_k\,, \eta\rangle,$$

which is non-zero by (2.35). Therefore $\alpha_1 = \cdots = \alpha_N = 0$. qed

Combining Theorem 2.3.5 and Proposition 2.3.7 we obtain

Proposition 2.3.10 *Let $\{e_j\}_{j=0}^{\infty}$ be a complete orthonormal basis for H. For $\mathbf{n} = (n_0, n_1, \cdots)$ with $|\mathbf{n}| = n_0 + n_1 + \cdots = n < \infty$ we put*

$$\phi_{\mathbf{n}}(x) = (n_0! n_1! \cdots)^{-1/2} \left\langle :x^{\otimes n}:, e_0^{\otimes n_0} \widehat{\otimes} e_1^{\otimes n_1} \widehat{\otimes} \cdots \right\rangle.$$

Then, $\{\phi_{\mathbf{n}}; |\mathbf{n}| = n\}$ becomes a complete orthonormal basis for $\mathcal{H}_n(\mathbf{R})$ and $\mathcal{H}_n(\mathbf{C})$. Therefore, $\{\phi_{\mathbf{n}}; |\mathbf{n}| < \infty\}$ forms a complete orthonormal basis for $L^2(E^*, \mu; \mathbf{R})$ and $L^2(E^*, \mu; \mathbf{C})$.

It will be proved in Proposition 3.5.11 that

$$\phi_{\mathbf{n}}(x) = \prod_{j=0}^{\infty} (n_j! 2^{n_j})^{-1/2} H_{n_j}\left(\frac{\langle x, e_j\rangle}{\sqrt{2}}\right),$$

where $\mathbf{n} = (n_0, n_1, \cdots)$ with $|\mathbf{n}| = n_0 + n_1 + \cdots < \infty$.

Taking the Wiener-Itô-Segal isomorphism into account, we define an exponential vector in $L^2(E^*, \mu; \mathbf{C})$ by

$$\phi_\xi(x) = \sum_{n=0}^{\infty} \left\langle :x^{\otimes n}:, \frac{\xi^{\otimes n}}{n!} \right\rangle, \qquad x \in E^*, \quad \xi \in E_{\mathbf{C}}.$$

It will be proved in Proposition 3.3.10 that

$$\phi_\xi(x) = e^{\langle x, \xi\rangle - \langle \xi, \xi\rangle/2}, \qquad x \in E^*, \quad \xi \in H_{\mathbf{C}}.$$

In particular, it follows from Proposition 2.1.6 that

$$\phi_\xi(x) = \frac{\mu(dx - \xi)}{\mu(dx)}, \qquad x \in E^*, \quad \xi \in H.$$

Bibliographical Notes

The present chapter is mostly devoted to assembling basic results which are more or less common apart from notations. We mention only some classical references.

The theory of Wiener-Itô-Segal isomorphism is originally due to Wiener [2]. Itô [1] developed the theory of multiple Wiener integrals and established Theorem 1.3.3 in essence. Segal [1] investigated the canonical isomorphism between Fock space and $L^2(E^*, \mu; \mathbf{C})$ using a weak distribution on H instead of the Gaussian measure on E^*. For further relevant discussion see also Gross [1], [2] and Segal [2], [3].

The name of Fock space originated in Fock [1] and it has become a most basic concept of the operator formalism of quantum physics. Historical literatures include also Dirac [1] and Jordan-Wigner [1]. A mathematical treatment was also discussed by Guichardet [1]. The notion of Wick-ordering traces back to Wick [1] and has been widely utilized, see e.g., Simon [1] where a slightly different introduction of the Wick products is found as well as a number of relevant references.

Chapter 3

White Noise Functionals

3.1 Standard Construction

We first fix the notation used thoughout the rest of the lecture notes. Let T be a topological space equipped with a Borel measure $\nu(dt) = dt$ and consider the real Hilbert space $H = L^2(T, \nu; \mathbb{R})$. The norm of H is denoted by $|\cdot|_0$ instead of $|\cdot|$. Let A be a positive selfadjoint operator in H with Hilbert-Schmidt inverse. Its eigenvalues and normalized eigenvectors are denoted by $\{\lambda_j\}_{j=0}^{\infty}$ and $\{e_j\}_{j=0}^{\infty}$, respectively. Then, $Ae_j = \lambda_j e_j$ and $\{e_j\}_{j=0}^{\infty}$ becomes a complete orthonormal basis of H. Hereafter we further assume that

$$1 < \lambda_0 \le \lambda_1 \le \cdots \longrightarrow \infty.$$

(Note that $\lambda_0 = \inf \operatorname{Spec}(A) > 1$ is newly assumed.) We define two constant numbers:

$$\delta = \|A^{-1}\|_{\rm HS} = \left(\sum_{j=0}^{\infty} \lambda_j^{-2}\right)^{1/2}, \qquad \rho = \|A^{-1}\|_{\rm op} = \lambda_0^{-1},$$

which will be often used together with the following obvious inequalities:

$$0 < \rho < 1, \qquad \rho < \delta.$$

Let $E = \mathcal{S}_A(T)$ be the standard CH-space constructed from (H, A), see §1.4. Since A admits a Hilbert-Schmidt inverse, E becomes a nuclear space and we come to a Gelfand triple:

$$E = \mathcal{S}_A(T) \subset H = L^2(T, \nu; \mathbb{R}) \subset E^* = \mathcal{S}_A^*(T). \tag{3.1}$$

By construction,

$$E = \bigcap_{p \ge 0} E_p \cong \operatorname*{proj\,lim}_{p\to\infty} E_p, \qquad E^* = \bigcup_{p\ge 0} E_{-p} \cong \operatorname*{ind\,lim}_{p\to\infty} E_{-p},$$

where E_p is the Hilbert space equipped with the norm $|\xi|_p = |A^p \xi|_0$, see §1.2. We assume in addition that $E = \mathcal{S}_A(T)$ satisfies hypotheses (H1)–(H3) introduced in §1.4. Let μ be the Gaussian measure on E^* defined by

$$e^{-|\xi|_0^2/2} = \int_{E^*} e^{i\langle x, \xi\rangle} \mu(dx), \qquad \xi \in E, \tag{3.2}$$

see §2.1. The framework of our discussion will be built on the particular Gaussian space (E^*, μ). We fix the above setup hereafter throughout.

Theorem 3.1.1 $\mu(E_{-1}) = 1$.

PROOF. By definition (3.2) the characteristic functional of μ is continuous on H. On the other hand, since A^{-1} is of Hilbert-Schmidt type by assumption, so is the injection $E_1 \to H$, see the proof of Proposition 1.3.4. It then follows from Theorem 1.5.3 that $\mu(E_{-1}) = 1$. qed

Again by $|\cdot|_p$ we denote the norm of the Hilbert space $E_p^{\otimes n}$, $p \in \mathbb{R}$. Then by virtue of Propositions 1.3.8 and 1.2.2 we have

$$E^{\otimes n} = \bigcap_{p\geq 0} E_p^{\otimes n} \cong \operatorname{proj\,lim}_{p\to\infty} E_p^{\otimes n}, \qquad (E^{\otimes n})^* = \bigcup_{p\geq 0} E_{-p}^{\otimes n} \cong \operatorname{ind\,lim}_{p\to\infty} E_{-p}^{\otimes n}.$$

It follows from Proposition 1.4.3 that every element in $E^{\otimes n}$ is a continuous function on $T \times \cdots \times T$ (n times). Since $|\omega|_p = |(A^{\otimes n})^p \omega|_0$ for $\omega \in E^{\otimes n}$ and $p \in \mathbb{R}$,

$$|\omega|_p \leq \rho^n \, |\omega|_{p+1}, \qquad \omega \in E^{\otimes n}, \quad p \in \mathbb{R}.$$

In particular,

$$\lim_{p\to\infty} |F|_{-p} = 0, \qquad F \in (E^{\otimes n})^*. \tag{3.3}$$

Now consider $L^2(E^*, \mu; \mathbb{R})$ of which norm is denoted by $\|\cdot\|_0$. If $\phi \in L^2(E^*, \mu; \mathbb{R})$ is given with Wiener-Itô expansion:

$$\phi(x) = \sum_{n=0}^{\infty} \left\langle :x^{\otimes n}:, f_n \right\rangle, \qquad x \in E^*, \quad (f_n)_{n=0}^{\infty} \in \Gamma(H), \tag{3.4}$$

it holds that

$$\|\phi\|_0^2 = \sum_{n=0}^{\infty} n! \, |f_n|_0^2 .$$

We now define an operator $\Gamma(A)$ densely defined in $L^2(E^*, \mu; \mathbb{R})$. The domain, denoted by $\mathrm{Dom}(\Gamma(A))$, is the space of functions ϕ of the form (3.4) such that $f_n \in \mathrm{Dom}(A^{\otimes n})$ and $\sum_{n=0}^{\infty} n! \, |A^{\otimes n} f_n|_0^2 < \infty$. Then, the operator $\Gamma(A)$ is defined by

$$\Gamma(A)\phi(x) = \sum_{n=0}^{\infty} \left\langle :x^{\otimes n}:, A^{\otimes n} f_n \right\rangle, \qquad \phi \in \mathrm{Dom}(\Gamma(A)), \tag{3.5}$$

which is apparently densely defined operator in $L^2(E^*, \mu; \mathbb{R})$. Furthermore,

Lemma 3.1.2 *$\Gamma(A)$ is a positive selfadjoint operator with Hilbert-Schmidt inverse.*

PROOF. It is sufficient to prove that $\Gamma(A)$ admits a Hilbert-Schmidt inverse. It follows from Proposition 2.3.10 that

$$\phi_{\mathbf{n}}(x) = (n_0! n_1! \cdots)^{-1/2} \left\langle :x^{\otimes n}:, e_0^{\otimes n_0} \hat{\otimes} e_1^{\otimes n_1} \hat{\otimes} \cdots \right\rangle,$$

where $\mathbf{n} = (n_0, n_1, \cdots)$, $|\mathbf{n}| = n_0 + n_1 + \cdots = n < \infty$, form a complete orthonormal basis for $L^2(E^*, \mu; \mathbb{R})$. Obviously, these are eigenfunctions of $\Gamma(A)$ with eigenvalues $\lambda_0^{n_0}\lambda_1^{n_1}\cdots$. Since $\sum_{j=0}^{\infty} \lambda_j^{-2} < \infty$ and $1 < \lambda_0 \le \lambda_1 \le \cdots$ by assumption, we have

$$\sum_{n=0}^{\infty} \sum_{n_0+n_1+\cdots=n} (\lambda_0^{n_0}\lambda_1^{n_1}\cdots)^{-2} = \prod_{j=0}^{\infty} \sum_{n_j=0}^{\infty} \lambda_j^{-2n_j} = \prod_{j=0}^{\infty} (1-\lambda_j^{-2})^{-1} < \infty.$$

This means that $\Gamma(A)^{-1}$ is of Hilbert-Schmidt type. qed

We then obtain again a standard CH-space constructed from $(L^2(E^*, \mu; \mathbb{R}), \Gamma(A))$, which is nuclear by Lemma 3.1.2. In other words, we have obtained another Gelfand triple:

$$\mathcal{S}_{\Gamma(A)}(E^*) \subset L^2(E^*, \mu; \mathbb{R}) \subset \mathcal{S}^*_{\Gamma(A)}(E^*).$$

Following tradition the complexification is denoted by

$$(E) \subset (L^2) \equiv L^2(E^*, \mu; \mathbb{C}) \subset (E)^*.$$

Although immediate from Corollary 1.3.5, we note the following

Theorem 3.1.3 *(E) is a nuclear Fréchet space.*

Definition 3.1.4 Elements $\phi \in (E)$ and $\Phi \in (E)^*$ are called a *test (white noise) functional* and a *generalized (white noise) functional*, respectively.

By construction (E) is equipped with the Hilbertian norms

$$\|\phi\|_p = \|\Gamma(A)^p \phi\|_0, \qquad \phi \in (E), \quad p \in \mathbb{R}. \tag{3.6}$$

These norms are linearly ordered: $\|\phi\|_p \le \|\phi\|_q$ whenever $p \le q$. The canonical bilinear form on $(E)^* \times (E)$ is denoted by $\langle\langle \Phi, \phi \rangle\rangle$. Then,

$$\|\phi\|_0^2 = \langle\langle \overline{\phi}, \phi \rangle\rangle \qquad \phi \in (E),$$

due to our convention, see the last paragraph in §1.6.

The space $(E) \subset (L^2)$ is characterized in terms of Wiener-Itô-Segal isomorphism.

Theorem 3.1.5 *Let $\phi \in (L^2)$ be given with Wiener-Itô expansion:*

$$\phi(x) = \sum_{n=0}^{\infty} \left\langle :x^{\otimes n}:, f_n \right\rangle, \qquad x \in E^*, \quad (f_n)_{n=0}^{\infty} \in \Gamma(H_{\mathbb{C}}).$$

Then, $\phi \in (E)$ if and only if $f_n \in E_{\mathbb{C}}^{\widehat{\otimes} n}$ for all $n = 0, 1, 2, \cdots$ and $\sum_{n=0}^{\infty} n! \, |f_n|_p^2 < \infty$ for all $p \ge 0$. In that case, it holds that

$$\|\phi\|_p^2 = \sum_{n=0}^{\infty} n! \, |f_n|_p^2, \qquad p \in \mathbb{R}. \tag{3.7}$$

Furthermore, letting $\phi_n(x) = \langle :x^{\otimes n}:, f_n \rangle$, the infinite series $\sum_{n=0}^{\infty} \phi_n$ converges to ϕ in (E).

PROOF. It follows from the definitions (3.6) and (3.5) that

$$\|\phi\|_p^2 = \|\Gamma(A)^p\phi\|_0^2 = \sum_{n=0}^{\infty} n! \left|(A^{\otimes n})^p f_n\right|_0^2 = \sum_{n=0}^{\infty} n! \, |f_n|_p^2 .$$

This proves (3.7) and the rest is immediate. qed

Theorem 3.1.6 *For each $\Phi \in (E)^*$ there exists a unique sequence $(F_n)_{n=0}^{\infty}$, $F_n \in (E_{\mathbb{C}}^{\otimes n})^*_{\mathrm{sym}}$ such that*

$$\langle\!\langle \Phi , \phi \rangle\!\rangle = \sum_{n=0}^{\infty} n! \, \langle F_n , f_n \rangle , \qquad \phi \in (E), \tag{3.8}$$

where ϕ and $(f_n)_{n=0}^{\infty}$ are related as in Theorem 3.1.5. Conversely, given a sequence $(F_n)_{n=0}^{\infty}$ such that $F_n \in (E_{\mathbb{C}}^{\otimes n})^*_{\mathrm{sym}}$ and $\sum_{n=0}^{\infty} n! \, |F_n|_{-p}^2 < \infty$ for some $p \geq 0$, a generalized functional $\Phi \in (E)^*$ is defined by (3.8). In that case, it holds that

$$\|\Phi\|_{-p}^2 = \sum_{n=0}^{\infty} n! \, |F_n|_{-p}^2 . \tag{3.9}$$

PROOF. Let $\Phi \in (E)^*$ be given. By construction there exists some $p \geq 0$ such that $\|\Phi\|_{-p} < \infty$. For $f \in E_{\mathbb{C}}^{\otimes n}$, $n = 0, 1, 2, \cdots$, we put

$$\phi_f(x) = \left\langle :x^{\otimes n}: , \widehat{f} \right\rangle ,$$

where $\widehat{f}$ is the symmetrization (see §1.6). Then $\phi_f \in (E)$ by Theorem 3.1.5 and a linear functional

$$f \mapsto \langle\!\langle \Phi , \phi_f \rangle\!\rangle , \qquad f \in E_{\mathbb{C}}^{\otimes n},$$

is continuous. In fact,

$$| \langle\!\langle \Phi , \phi_f \rangle\!\rangle | \leq \|\Phi\|_{-p} \|\phi_f\|_p \leq \sqrt{n!} \, \|\Phi\|_{-p} |\widehat{f}|_p \leq \sqrt{n!} \, \|\Phi\|_{-p} |f|_p .$$

Therefore there exists $F_n \in (E_{\mathbb{C}}^{\otimes n})^*$ such that

$$\langle\!\langle \Phi , \phi_f \rangle\!\rangle = n! \, \langle F_n , f \rangle , \qquad f \in E_{\mathbb{C}}^{\otimes n}.$$

Since $\langle\!\langle \Phi , \phi_f \rangle\!\rangle = \langle\!\langle \Phi , \phi_{\widehat{f}} \rangle\!\rangle$, we see that

$$\langle F_n , f \rangle = \langle F_n , \widehat{f} \rangle, \qquad f \in E_{\mathbb{C}}^{\otimes n}.$$

Hence $\widehat{F_n} = F_n$. Thus, given Φ we have found $F_n \in (E_{\mathbb{C}}^{\otimes n})^*_{\mathrm{sym}}$, $n = 0, 1, 2, \cdots$. We next prove (3.8). Let $\phi \in (E)$ be given as in the statement, namely,

$$\phi = \sum_{n=0}^{\infty} \phi_{f_n} .$$

Since the series converges in (E) by Theorem 3.1.5, we have

$$\langle\langle \Phi , \phi \rangle\rangle = \lim_{N\to\infty} \sum_{n=0}^{N} \langle\langle \Phi , \phi_{f_n} \rangle\rangle = \sum_{n=0}^{\infty} n! \left\langle F_n , f_n \right\rangle ,$$

as desired. The converse assertion is straightforward.

Finally we prove identity (3.9) by Fourier series expansion. Let $\mathbf{n} = (n_0, n_1, \cdots)$ with $|\mathbf{n}| = n_0 + n_1 + \cdots = n$ and consider

$$\phi_{\mathbf{n}}(x) = (n_0! n_1! \cdots)^{-1/2} \left\langle :x^{\otimes n}: , e_0^{\otimes n_0} \widehat{\otimes} e_1^{\otimes n_1} \widehat{\otimes} \cdots \right\rangle .$$

These functions form a complete orthonormal basis of (L^2), see Proposition 2.3.10. Noting that $\phi_{\mathbf{n}}$ is a real function, we compute

$$\begin{aligned} \|\Phi\|_{-p}^2 &= \left\| \Gamma(A)^{-p} \Phi \right\|_0^2 \\ &= \sum_{n=0}^{\infty} \sum_{|\mathbf{n}|=n} \left\langle\!\left\langle \Gamma(A)^{-p}\Phi , \phi_{\mathbf{n}} \right\rangle\!\right\rangle^2 \\ &= \sum_{n=0}^{\infty} \sum_{|\mathbf{n}|=n} \left\langle\!\left\langle \Phi , \Gamma(A)^{-p}\phi_{\mathbf{n}} \right\rangle\!\right\rangle^2 . \end{aligned}$$

Note that

$$\Gamma(A)^{-p}\phi_{\mathbf{n}}(x) = (n_0! n_1! \cdots)^{-1/2} \left\langle :x^{\otimes n}: , (A^{\otimes n})^{-p} (e_0^{\otimes n_0} \widehat{\otimes} e_1^{\otimes n_1} \widehat{\otimes} \cdots) \right\rangle .$$

Hence by (3.8) we have

$$\left\langle\!\left\langle \Phi , \Gamma(A)^{-p}\phi_{\mathbf{n}} \right\rangle\!\right\rangle = \frac{n!}{\sqrt{n_0! n_1! \cdots}} \left\langle F_n , (A^{\otimes n})^{-p} (e_0^{\otimes n_0} \widehat{\otimes} e_1^{\otimes n_1} \widehat{\otimes} \cdots) \right\rangle .$$

In view of Proposition 2.3.7 we obtain

$$\begin{aligned} \|\Phi\|_{-p}^2 &= \sum_{n=0}^{\infty} n! \sum_{|\mathbf{n}|=n} \frac{n!}{n_0! n_1! \cdots} \left\langle F_n , (A^{\otimes n})^{-p} (e_0^{\otimes n_0} \widehat{\otimes} e_1^{\otimes n_1} \widehat{\otimes} \cdots) \right\rangle^2 \\ &= \sum_{n=0}^{\infty} n! \sum_{|\mathbf{n}|=n} \left\langle (A^{\otimes n})^{-p} F_n , \sqrt{\frac{n!}{n_0! n_1! \cdots}} e_0^{\otimes n_0} \widehat{\otimes} e_1^{\otimes n_1} \widehat{\otimes} \cdots \right\rangle^2 \\ &= \sum_{n=0}^{\infty} n! \left| (A^{\otimes n})^{-p} F_n \right|_0^2 \\ &= \sum_{n=0}^{\infty} n! \left| F_n \right|_{-p}^2 , \end{aligned}$$

which proves (3.9). qed

According to Theorem 3.1.6, we adopt a (formal) expression for $\Phi \in (E)^*$:

$$\Phi(x) = \sum_{n=0}^{\infty} \left\langle :x^{\otimes n}: , F_n \right\rangle , \tag{3.10}$$

where $F_n \in (E_{\mathbb{C}}^{\otimes n})^*_{\mathrm{sym}}$ and $\sum_{n=0}^{\infty} n! \, |F_n|_{-p}^2 < \infty$ for some $p \geq 0$. Note that $\langle :x^{\otimes n}: , F_n \rangle$ is not a function of $x \in E^*$ but a generalized function. Hence (3.10) is only understood through the canonical pairing with test functionals in (E). Expression (3.10) is also called the *Wiener-Itô expansion* of Φ.

Here is a simple example of generalized functionals. Let $t_1, \cdots, t_n \in T$ be fixed. Since $\delta_{t_1} \otimes \cdots \otimes \delta_{t_n} \in (E^{\otimes n})^*$ by hypothesis (H2),

$$\Phi_{t_1,\cdots,t_n}(x) = \left\langle :x^{\otimes n}: , \delta_{t_1} \otimes \cdots \otimes \delta_{t_n} \right\rangle$$

is a generalized functional, namely, $\Phi_{t_1,\cdots,t_n} \in (E)^*$. These are called *white noise polynomials*. In particular, *white noise* is a family of generalized functionals

$$\Phi_t(x) = \langle : x : , \delta_t \rangle = \langle x , \delta_t \rangle \equiv x(t), \qquad t \in T. \tag{3.11}$$

When T is a time interval, $x(t)$ may be considered as time derivative of Brownian motion and intuitively understood as coordinate of white noise space E^*

3.2 Continuous version theorem

By construction each $\phi \in (E)$ is determined only up to μ-null functions. In this section we shall prove that (E) enjoys properties (H1)–(H3) introduced in §1.4.

Theorem 3.2.1 (CONTINUOUS VERSION THEOREM) *For each $\phi \in (E)$ there exists a unique continuous function $\tilde{\phi}$ on E^* such that $\phi(x) = \tilde{\phi}(x)$ for μ-a.e. $x \in E^*$. Moreover, $\tilde{\phi}(x)$ is given by the absolutely convergent series:*

$$\tilde{\phi}(x) = \sum_{n=0}^{\infty} \left\langle :x^{\otimes n}: , f_n \right\rangle, \qquad x \in E^*,$$

where $(f_n)_{n=0}^{\infty} \in \Gamma(H_{\mathbb{C}})$ corresponds to the given ϕ under the Wiener-Itô-Segal isomorphism between (L^2) and $\Gamma(H_{\mathbb{C}})$.

Recall that E^* always carries the strong dual topology. The proof is devided into three steps (Propositions 3.2.2, 3.2.3 and 3.2.11) and we start with the proof of uniqueness.

Proposition 3.2.2 *Let $\tilde{\phi}$ be a continuous function on E^*. If $\tilde{\phi}(x) = 0$ for μ-a.e. $x \in E^*$, then $\tilde{\phi}(x) = 0$ for all $x \in E^*$.*

PROOF. Suppose otherwise, say, $\tilde{\phi}(x_0) > 0$ for some point $x_0 \in E^*$. Take $p \geq 1$ such that $x_0 \in E_{-p}$ and consider the restriction of $\tilde{\phi}$ to E_{-p} which is denoted by the same symbol. Since the canonical injection $E_{-p} \to E^*$ is continuous, $\tilde{\phi}$ is a continuous function on E_{-p} with $\tilde{\phi}(x_0) > 0$. Hence there exists an open neighborhood $U \subset E_{-p}$ of x_0 such that $\tilde{\phi}(x) > 0$ for all $x \in U$. On the other hand, since $\tilde{\phi}(x) = 0$ for μ-a.e. $x \in E^*$ by assumption, we see that $\mu(U) = 0$. But this will yield contradiction.

In fact, we shall prove that any non-empty open subset $U \subset E_{-p}$ is of positive measure. Since $H \subset E_{-p}$ is a dense subspace, we may choose a countable subset

$\{\xi_k\}_{k=0}^{\infty} \subset H$ which is dense in E_{-p}. Let $\epsilon > 0$ be an arbitrary positive number and put

$$B_k(\epsilon) = \left\{x \in E_{-p}; |x - \xi_k|_{-p} < \epsilon\right\}.$$

Obviously,

$$E_{-p} = \bigcup_{k=0}^{\infty} B_k(\epsilon). \tag{3.12}$$

Since the Gaussian measure μ is quasi-invariant under translations by H (Proposition 2.1.6), if $\mu(B_{k_1}(\epsilon)) = 0$ for some k_1 then $\mu(B_k(\epsilon)) = 0$ for all k. But this is impossible because of (3.12) and the fact $\mu(E_{-p}) = 1$, which follows from Theorem 3.1.1. Hence $\mu(B_k(\epsilon)) > 0$ for all k and $\epsilon > 0$. Since $U \subset E_{-p}$ is a non-empty open subset, it contains an open ball $B_k(\epsilon)$ for some k and ϵ. Hence $\mu(U) > 0$. qed

We next prove

Proposition 3.2.3 *Assume that a sequence $f_n \in E_{\mathbb{C}}^{\widehat{\otimes} n}$, $n = 0, 1, 2, \cdots$, satisfies $\sum_{n=0}^{\infty} n! \, |f_n|_p^2 < \infty$ for all $p \geq 0$. Then the series*

$$\sum_{n=0}^{\infty} \left\langle :x^{\otimes n}:, f_n \right\rangle$$

converges absolutely at each $x \in E^*$.

For the proof we need a simple lemma.

Lemma 3.2.4 *If $x \in E^*$ satisfies $|x|_{-p} < \infty$ with $p > 1/2$, then*

$$|:x^{\otimes n}:|_{-p} \leq \sqrt{n!}\left(|\tau|_{-p}^{1/2} + |x|_{-p}\right)^n.$$

PROOF. First note that $|\tau|_{-p} < \infty$ for $p > 1/2$. In fact, since $\tau = \sum_{j=0}^{\infty} e_j \otimes e_j$ by Proposition 2.2.1,

$$|\tau|_{-p}^2 = \sum_{j=0}^{\infty} |e_j \otimes e_j|_{-p}^2 = \sum_{j=0}^{\infty} \lambda_j^{-4p} < \lambda_0^{-4p+2} \sum_{j=0}^{\infty} \lambda_j^{-2} = \rho^{4p-2}\delta^2 < \infty,$$

whenever $p > 1/2$. Since

$$:x^{\otimes n}: := \sum_{k=0}^{[n/2]} \frac{(-1)^k n!}{(n-2k)!k!2^k} \tau^{\widehat{\otimes} k} \widehat{\otimes} x^{\otimes(n-2k)}, \tag{3.13}$$

which is shown in Corollary 2.2.4, we have

$$|:x^{\otimes n}:|_{-p} \leq \sum_{k=0}^{[n/2]} \frac{n!}{(n-2k)!k!2^k} |\tau|_{-p}^k |x|_{-p}^{n-2k}.$$

Using an obvious inequality:

$$\frac{1}{k!2^k} \leq \frac{\sqrt{n!}}{(2k)!}, \qquad 0 \leq k \leq \left[\frac{n}{2}\right], \tag{3.14}$$

we have

$$
\begin{aligned}
|:x^{\otimes n}:|_{-p} &\leq \sum_{k=0}^{[n/2]} \frac{n!}{(n-2k)!} \frac{\sqrt{n!}}{(2k)!} |\tau|_{-p}^{k} |x|_{-p}^{n-2k} \\
&\leq \sqrt{n!} \sum_{k=0}^{n} \frac{n!}{(n-k)!k!} \left(|\tau|_{-p}^{1/2}\right)^{k} |x|_{-p}^{n-k} \\
&\leq \sqrt{n!} \left(|\tau|_{-p}^{1/2} + |x|_{-p}\right)^{n}.
\end{aligned}
$$

This completes the proof. qed

PROOF OF PROPOSITION 3.2.3. Let $x \in E^*$ be fixed. Since

$$
\lim_{p \to \infty} (|\tau|_{-p}^{1/2} + |x|_{-p}) = 0
$$

by (3.3), we may choose $p > 1/2$ such that $|\tau|_{-p}^{1/2} + |x|_{-p} < 1$. It then follows from Lemma 3.2.4 that

$$
\begin{aligned}
\sum_{n=0}^{\infty} \left|\left\langle :x^{\otimes n}:, f_n \right\rangle\right| &\leq \sum_{n=0}^{\infty} |:x^{\otimes n}:|_{-p} |f_n|_p \\
&\leq \sum_{n=0}^{\infty} \sqrt{n!} \left(|\tau|_{-p}^{1/2} + |x|_{-p}\right)^{n} |f_n|_p \\
&\leq \left(\sum_{n=0}^{\infty} n! \, |f_n|_p^2\right)^{1/2} \left(\sum_{n=0}^{\infty} \left(|\tau|_{-p}^{1/2} + |x|_{-p}\right)^{2n}\right)^{1/2}.
\end{aligned}
$$

Consequently,

$$
\sum_{n=0}^{\infty} \left|\left\langle :x^{\otimes n}:, f_n \right\rangle\right| \leq \left(\sum_{n=0}^{\infty} n! \, |f_n|_p^2\right)^{1/2} \left\{1 - \left(|\tau|_{-p}^{1/2} + |x|_{-p}\right)^{2}\right\}^{-1/2} < \infty, \tag{3.15}
$$

as desired. qed

Suppose we are given $\phi \in (E)$ with Wiener-Itô expansion:

$$
\phi(x) = \sum_{n=0}^{\infty} \left\langle :x^{\otimes n}:, f_n \right\rangle.
$$

Then it follows from Theorem 3.1.5 that $f_n \in E_{\mathbb{C}}^{\widehat{\otimes} n}$ and $\sum_{n=0}^{\infty} n! \, |f_n|_p^2 < \infty$ for all $p \geq 0$. On the other hand, it follows from Proposition 3.2.3 that

$$
\widetilde{\phi}(x) \equiv \sum_{n=0}^{\infty} \left\langle :x^{\otimes n}:, f_n \right\rangle
$$

converges at every $x \in E^*$. Therefore, $\phi(x) = \widetilde{\phi}(x)$ for μ-a.e. $x \in E^*$. Thus, for the proof of Theorem 3.2.1 it is sufficient to prove that $\widetilde{\phi}$ is a continuous function on E^* whenever $f_n \in E_{\mathbb{C}}^{\widehat{\otimes} n}$ and $\sum_{n=0}^{\infty} n! \, |f_n|_p^2 < \infty$ for all $p \geq 0$. The statement will be formulated in Proposition 3.2.11.

It is much simpler to show that the restriction of $\widetilde{\phi}$ to E_{-p} is continuous with respect to the norm $|\cdot|_{-p}$. However, this is not enough to assert the continuity of $\widetilde{\phi}$ on E^* (equipped with the strong dual topology), because the inductive system $\{E_{-p}\}_{p=0}^{\infty}$ is not strict.

We start with construction of defining Hilbertian seminorms of E^*. Let $\mathcal{C}$ be the set of sequences $C = (C_p)_{p=0}^{\infty}$ such that $C_0 \geq C_1 \geq \cdots > 0$. For $C \in \mathcal{C}$ we put

$$\| \xi \|_C^2 = \sum_{p=0}^{\infty} C_p^2 \, |\xi|_p^2, \qquad \xi \in E, \tag{3.16}$$

though possibly $\|\xi\|_C = \infty$. Then define

$$|\!|x|\!|_C = \sup\{|\langle x, \xi\rangle| \,;\, \|\xi\|_C \leq 1, \, \xi \in E\}, \qquad x \in E^*, \tag{3.17}$$

which is always finite. Obviously, for any $C \in \mathcal{C}$

$$|\langle x, \xi\rangle| \leq |\!|x|\!|_C \|\xi\|_C, \qquad x \in E^*, \quad \xi \in E, \tag{3.18}$$

though $\|\xi\|_C = \infty$ can happen.

Lemma 3.2.5 *$\{|\!|\cdot|\!|_C\}_{C \in \mathcal{C}}$ is a set of defining Hilbertian seminorms of E^*.*

PROOF. For $C = (C_p)_{p=0}^{\infty} \in \mathcal{C}$ let $E(C)$ be the subspace of all $\xi \in E$ with $\|\xi\|_C < \infty$. Then, $E(C)$ becomes a real Hilbert space with the norm $\|\cdot\|_C$. Let $f : E(C) \to E$ be the natural injection which is apparently continuous (but does not have dense image in general). Then, $f^* : E^* \to E(C)^*$ becomes a continuous linear operator. We denote by $\langle \cdot, \cdot \rangle_{E(C)}$ and $|\cdot|_{E(C)}$ the inner product and the norm of $E(C)$, respectively. Note that $E(C) \cong E(C)^*$ through the inner product $\langle \cdot, \cdot \rangle_{E(C)}$.

Then, by definition, for $x \in E^*$,

$$\begin{aligned} |\!|x|\!|_C &= \sup\{|\langle x, f(\xi)\rangle| \,;\, \|\xi\|_C \leq 1, \, \xi \in E(C)\} \\ &= \sup\{|\langle f^*(x), \xi\rangle_{E(C)}| \,;\, \|\xi\|_C \leq 1, \, \xi \in E(C)\} \\ &= |f^*(x)|_{E(C)} \,. \end{aligned}$$

Therefore $|\!|\cdot|\!|_C$ is a Hilbertian seminorm on E^*.

The strong dual topology of E^* is defined by the seminorms

$$x \mapsto \sup\{|\langle x, \xi\rangle| \,;\, \xi \in S\}, \qquad x \in E^*,$$

where S runs over all bounded subset of E. Note first that $\{\xi \in E; \|\xi\|_C \leq 1\}$ is a bounded subset of E for any $C \in \mathcal{C}$. Hence, in order to prove that $\{|\!|\cdot|\!|_C\}_{C \in \mathcal{C}}$ is a set of defining seminorms of E^* it is sufficient to show that for any bounded subset $S \subset E$ with $S \neq \{0\}$ there is $C \in \mathcal{C}$ such that $S \subset \{\xi \in E; \|\xi\|_C \leq 1\}$. In fact, given a bounded subset $S \subset E$, we put

$$\sigma_p = \sup_{\xi \in S} |\xi|_p \,, \qquad p = 0, 1, 2, \cdots,$$

and define inductively a sequence $C = (C_p)_{p=0}^{\infty}$ by

$$C_0 = \frac{1}{\sqrt{2}\,\sigma_0}, \qquad C_p = \min\left\{\frac{1}{\sqrt{2^{p+1}}\,\sigma_p}, C_{p-1}\right\}, \quad p = 1, 2, \cdots.$$

As is easily verified, this satisfies the desired property. qed

Using the Fourier expansion, we may give $|x|_C$ explicitly.

Lemma 3.2.6 *For $C = (C_p)_{p=0}^{\infty} \in \mathcal{C}$ it holds that*

$$|x|_C^2 = \sum_{j=0}^{\infty} \langle x, e_j\rangle^2 \|e_j\|_C^{-2} = \sum_{j=0}^{\infty} \langle x, e_j\rangle^2 \left(\sum_{p=0}^{\infty} C_p^2 \lambda_j^{2p}\right)^{-1}, \qquad x \in E^*.$$

PROOF. Since $\{e_j\}_{j=0}^{\infty}$ is an orthogonal set for every $|\cdot|_p$, it follows from (3.16) that $\langle e_j, e_k\rangle_{E(C)} = 0$ whenever $e_j, e_k \in E(C)$. It is then easy to see that $\{\|e_j\|_C^{-1} e_j\}_{j=0}^{\infty}$ is an orthonormal basis for $E(C) \cong E(C)^*$, where we understand $\|e_j\|_C^{-1} e_j = 0$ if $\|e_j\|_C = \infty$. On the other hand, each $x \in E^*$ admits a Fourier expansion:

$$x = \sum_{j=0}^{\infty} \langle x, e_j\rangle\, e_j,$$

which converges in E^*. Thus the result follows immediately. qed

The topology of $(E^{\otimes n})^*$ is defined in a similar way. Namely, for $C \in \mathcal{C}$ put

$$\|\omega\|_C^2 = \sum_{p_1,\cdots,p_n=0}^{\infty} C_{p_1}^2 \cdots C_{p_n}^2 \, |\omega|_{p_1,\cdots,p_n}^2, \qquad \omega \in E^{\otimes n}, \tag{3.19}$$

where

$$\begin{aligned} |\omega|_{p_1,\cdots,p_n}^2 &= |(A^{p_1} \otimes \cdots \otimes A^{p_n})\omega|_0^2 \\ &= \sum_{j_1,\cdots,j_n=0}^{\infty} |\langle \omega, e_{j_1} \otimes \cdots \otimes e_{j_n}\rangle|^2 \, |e_{j_1}|_{p_1}^2 \cdots |e_{j_n}|_{p_n}^2 . \end{aligned} \tag{3.20}$$

Then for $F \in (E^{\otimes n})^*$ we put

$$|F|_C = \sup\{|\langle F, \omega\rangle| \,;\, \|\omega\|_C \le 1, \omega \in E^{\otimes n}\}.$$

Then, in a similar way to Lemma 3.2.5 we see that $\{|\cdot|_C\}_{C\in\mathcal{C}}$ is a set of defining Hilbertian seminorms of $(E^{\otimes n})^*$. We next prove the following

Lemma 3.2.7 *Let $C = (C_p)_{p=0}^{\infty} \in \mathcal{C}$. Then*

$$|F|_C \le C_p^{-n} |F|_{-p}, \qquad F \in (E^{\otimes n})^*,$$

though $|F|_{-p} = \infty$ may happen. Moreover, $|\cdot|_C$ is a cross norm, i.e.,

$$|x_1 \otimes \cdots \otimes x_n|_C = |x_1|_C \cdots |x_n|_C, \qquad x_1, \cdots, x_n \in E^*.$$

PROOF. A similar argument as in Lemma 3.2.6 yields

$$|F|^2_C = \sum_{j_1,\cdots,j_n=0}^{\infty} \langle F, e_{j_1} \otimes \cdots \otimes e_{j_n} \rangle^2 \, [\![e_{j_1}]\!]_C^{-2} \cdots [\![e_{j_n}]\!]_C^{-2}.$$

It is then obvious that $|\cdot|_C$ is a cross norm. Since

$$[\![e_j]\!]_C^{-2} = \left(\sum_{p=0}^{\infty} C_p^2 \lambda_j^{2p} \right)^{-1} \le C_p^{-2} \lambda_j^{-2p}$$

for any $p \ge 0$, we obtain

$$|F|^2_C \le C_p^{-2n} \sum_{j_1,\cdots,j_n=0}^{\infty} \langle F, e_{j_1} \otimes \cdots \otimes e_{j_n} \rangle^2 \, \lambda_{j_1}^{-2p} \cdots \lambda_{j_n}^{-2p} = C_p^{-2n} |F|^2_{-p}$$

as desired.   qed

Lemma 3.2.8 *Let $\phi \in (E)$ be given with Wiener-Itô expansion:*

$$\phi(x) = \sum_{n=0}^{\infty} \langle :x^{\otimes n}:, f_n \rangle.$$

Then for any $C = (C_p)_{p=0}^{\infty} \in \mathcal{C}$ we have

$$n! [\![f_n]\!]_C^2 \le C_0^{-2} \left(\frac{C_0^2}{1-\rho^2} \right)^n [\![\phi]\!]_C^2, \qquad [\![\phi]\!]_C^2 = \sum_{p=0}^{\infty} C_p^2 \|\phi\|_p^2.$$

PROOF. Since $\|\phi\|_p^2 = \sum_{n=0}^{\infty} n! |f_n|_p^2$,

$$|f_n|_p^2 \le \frac{1}{n!} \|\phi\|_p^2. \tag{3.21}$$

On the other hand, by definition (3.19) we obtain

$$\begin{aligned}
[\![f_n]\!]_C^2 &= \sum_{p_1,\cdots,p_n=0}^{\infty} C_{p_1}^2 \cdots C_{p_n}^2 |f_n|_{p_1,\cdots,p_n}^2 \\
&= \sum_{p=0}^{\infty} \sum_{\max\{p_1,\cdots,p_n\}=p} C_{p_1}^2 \cdots C_{p_n}^2 |f_n|_{p_1,\cdots,p_n}^2.
\end{aligned} \tag{3.22}$$

By virtue of (3.20), if $\max\{p_1,\cdots,p_n\} = p$,

$$C_{p_1}^2 \cdots C_{p_n}^2 |f_n|_{p_1,\cdots,p_n}^2 \le C_0^{2(n-1)} C_p^2 \rho^{2(p-p_1)+\cdots+2(p-p_n)} |f_n|_p^2.$$

Therefore,

$$\begin{aligned}
\sum_{\max\{p_1,\cdots,p_n\}=p} & C_{p_1}^2 \cdots C_{p_n}^2 |f_n|_{p_1,\cdots,p_n}^2 \\
&\le \sum_{\max\{p_1,\cdots,p_n\}=p} C_0^{2(n-1)} C_p^2 \rho^{2(p-p_1)+\cdots+2(p-p_n)} |f_n|_p^2 \\
&\le C_0^{2(n-1)} C_p^2 |f_n|_p^2 \left(\sum_{q=0}^{p} \rho^{2(p-q)} \right)^n \\
&\le C_0^{2(n-1)} C_p^2 |f_n|_p^2 (1-\rho^2)^{-n}.
\end{aligned}$$

Then, in view of (3.21) and (3.22),

$$\|f_n\|_C^2 \le C_0^{-2}\left(\frac{C_0^2}{1-\rho^2}\right)^n \sum_{p=0}^{\infty} C_p^2 |f_n|_p^2 \le \frac{C_0^{-2}}{n!}\left(\frac{C_0^2}{1-\rho^2}\right)^n \sum_{p=0}^{\infty} C_p^2 \|\phi\|_p^2 .$$

This completes the proof. qed

We now prepare two elementary lemmas.

Lemma 3.2.9 *For $k = 0, 1, 2, \cdots$*

$$\sum_{n=0}^{\infty} \frac{(n+k)!}{n!n!} t^n \le (t+k)^k e^t, \qquad t \ge 0.$$

PROOF. We put

$$P_k(t) = e^{-t} \sum_{n=0}^{\infty} \frac{(n+k)!}{n!n!} t^n.$$

It is then easy to see that $P_k(t)$ is a polynomial of degree k. (In fact, P_k is a small modification of the Laguerre polynomial.) Accordingly we may put $P_k(t) = \sum_{l=0}^{k} a_{kl} t^l$. Then, it is straightforward to see that

$$0 \le a_{kl} \le \binom{k}{l} k^{k-l}, \qquad 0 \le l \le k,$$

from which the assertion follows. qed

Lemma 3.2.10 *Put*

$$\lambda(z, w) = \sum_{k=0}^{\infty} \frac{(z+k)^{k+1} w^k}{k!}.$$

Then the series converges in $\mathbf{C} \times \{|w| < e^{-1}\}$ and $\lambda(z, w)$ becomes a holomorphic function in two variables.

For the proof we only apply the Cauchy-Hadamard formula. Assuming that $C = (C_p)_{p=0}^{\infty} \in \mathcal{C}$ satisfies the conditions $C_0^2 < 1 - \rho^2$, we put

$$\Lambda_C(z, w) = \frac{2}{\sqrt{1-\rho^2-C_0^2}} \exp\left(\frac{z^2}{2}\right) \lambda\left(\frac{z^2}{2}+1, \frac{C_0^2 w}{1-\rho^2}\right).$$

Obviously, $\Lambda_C(z, w)$ is holomorphic in $\mathbf{C} \times \{|w| < e^{-1} C_0^{-2}(1-\rho^2)\}$. We are then able to claim

Proposition 3.2.11 *For each $n = 0, 1, 2, \cdots$ let $f_n \in E_{\mathbf{C}}^{\widehat{\otimes} n}$ be given and assume that $\sum_{n=0}^{\infty} n! \, |f_n|_p^2 < \infty$ for all $p \ge 0$. Put*

$$\tilde{\phi}(x) = \sum_{n=0}^{\infty} \left\langle :x^{\otimes n}:, f_n \right\rangle, \qquad x \in E^*.$$

If $C = (C_p)_{p=0}^{\infty} \in \mathcal{C}$ *satisfies*

$$C_0^2 < 1 - \rho^2 \qquad and \qquad C_0^2 |\tau|_C < \frac{1-\rho^2}{e}, \tag{3.23}$$

then

$$|\tilde{\phi}(x) - \tilde{\phi}(y)| \le |x - y|_C \|\phi\|_C A_C(|x|_C + |y|_C, |\tau|_C), \qquad x, y \in E^*. \tag{3.24}$$

In particular, $\tilde{\phi}$ *is a continuous function on* E^*.

PROOF. Let $x, y \in E^*$ and suppose $C = (C_p)_{p=0}^{\infty} \in \mathcal{C}$ satisfies (3.23). Then, in view of (3.13) and Lemma 3.2.7, we observe

$$\begin{aligned}
&\left|\left\langle :x^{\otimes n}: - :y^{\otimes n}:, f_n \right\rangle\right| \\
&\le \sum_{k=0}^{[n/2]} \frac{n!}{(n-2k)!k!2^k} \left|\left\langle \tau^{\hat{\otimes} k} \hat{\otimes} (x^{\otimes(n-2k)} - y^{\otimes(n-2k)}), f_n \right\rangle\right| \\
&\le \sum_{k=0}^{[n/2]} \frac{n!}{(n-2k)!k!2^k} |\tau|_C^k |x^{\otimes(n-2k)} - y^{\otimes(n-2k)}|_C \|f_n\|_C.
\end{aligned} \tag{3.25}$$

For simplicity we put

$$L = |x|_C + |y|_C.$$

Using an obvious inequality:

$$|x^{\otimes m} - y^{\otimes m}|_C \le |x - y|_C (|x|_C + |y|_C)^{m-1} = |x - y|_C L^{m-1}, \qquad m \ge 1,$$

and summing up both sides of (3.25), we obtain

$$|\tilde{\phi}(x) - \tilde{\phi}(y)| \le |x - y|_C J_C(x, y), \tag{3.26}$$

where

$$J_C(x, y) = \sum_{k=0}^{\infty} \frac{|\tau|_C^k}{k!2^k} \sum_{n=0}^{\infty} \frac{(n+2k+1)!}{(n+1)!} L^n \|f_{n+2k+1}\|_C.$$

We must estimate $J_C(x, y)$. By the Schwartz inequality,

$$\begin{aligned}
&\sum_{n=0}^{\infty} \frac{(n+2k+1)!}{(n+1)!} L^n \|f_{n+2k+1}\|_C \\
&\le \left(\sum_{n=0}^{\infty} (n+2k+1)! \|f_{n+2k+1}\|_C^2\right)^{1/2} \left(\sum_{n=0}^{\infty} \frac{(n+2k+1)!}{(n+1)!(n+1)!} L^{2n}\right)^{1/2} \\
&\le \left(\sum_{n=0}^{\infty} (n+2k+1)! \|f_{n+2k+1}\|_C^2\right)^{1/2} \left(\sum_{n=0}^{\infty} \frac{(n+2k+2)!}{n!n!} L^{2n}\right)^{1/2}.
\end{aligned}$$

Applying Lemmas 3.2.8 and 3.2.9, we have

$$\begin{aligned}
\sum_{n=0}^{\infty} (n+2k+1)! \|f_{n+2k+1}\|_C^2 &\le \sum_{n=0}^{\infty} C_0^{-2} \left(\frac{C_0^2}{1-\rho^2}\right)^{n+2k+1} \|\phi\|_C^2 \\
&= \left(\frac{C_0^2}{1-\rho^2}\right)^{2k} \frac{\|\phi\|_C^2}{1-\rho^2-C_0^2}, \\
\sum_{n=0}^{\infty} \frac{(n+2k+2)!}{n!n!} L^{2n} &\le \left(L^2 + 2k + 2\right)^{2k+2} e^{L^2},
\end{aligned}$$

where we used the assumption $C_0^2 < 1 - \rho^2$. Then

$$\begin{aligned} J_C(x,y) &\le \sum_{k=0}^{\infty} \frac{|\tau|_C^k}{k!2^k} \frac{2\|\phi\|_C}{\sqrt{1-\rho^2-C_0^2}} \left(\frac{2C_0^2}{1-\rho^2}\right)^k \left\{\frac{L^2}{2}+k+1\right\}^{k+1} e^{L^2/2} \\ &= \frac{2\|\tilde{\phi}\|_C}{\sqrt{1-\rho^2-C_0^2}} e^{L^2/2} \sum_{k=0}^{\infty} \frac{1}{k!} \left(\frac{C_0^2|\tau|_C}{1-\rho^2}\right)^k \left\{\frac{L^2}{2}+k+1\right\}^{k+1} \\ &= \frac{2\|\phi\|_C}{\sqrt{1-\rho^2-C_0^2}} \exp\left(\frac{L^2}{2}\right) \lambda\left(\frac{L^2}{2}+1, \frac{C_0^2|\tau|_C}{1-\rho^2}\right) \\ &= \|\phi\|_C \Lambda_C(L, |\tau|_C). \end{aligned}$$

This proves (3.24) and the continuity of $\tilde{\phi}$ on E^*. In fact, taking $C = (C_p)_{p=0}^{\infty} \in \mathcal{C}$ satisfying properties (3.23) (such a C exists certainly), we need only to note the continuity of $z \mapsto \Lambda_C(z, |\tau|_C)$. qed

We have thus completed the proof of Theorem 3.2.1. By this theorem we always assume that $\phi \in (E)$ admits an expression by an absolutely convergent series:

$$\phi(x) = \sum_{n=0}^{\infty} \left\langle :x^{\otimes n}:, f_n \right\rangle, \tag{3.27}$$

where ϕ and $(f_n)_{n=0}^{\infty}$ are related according to the Wiener-Itô-Segal isomorphism. Moreover, in that case $f_n \in E_{\mathbb{C}}^{\widehat{\otimes} n}$ and

$$\|\phi\|_p^2 = \sum_{n=0}^{\infty} n!\, |f_n|_p^2 < \infty, \qquad \text{for all } p \in \mathbb{R}.$$

In particular, each $\phi \in (E)$ is a continuous function on E^*. (Thus, we do not use the exclusive symbol $\tilde{\phi}$ hereafter.)

To be sure, we recall Wiener-Itô expansion of $\phi \in (L^2)$ and $\Phi \in (E)^*$. As was discussed in Theorem 2.3.5, each $\phi \in (L^2)$ admits an expression:

$$\phi(x) = \sum_{n=0}^{\infty} \left\langle :x^{\otimes n}:, f_n \right\rangle, \tag{3.28}$$

where $f_n \in H_{\mathbb{C}}^{\widehat{\otimes} n}$ and

$$\|\phi\|_0^2 = \sum_{n=0}^{\infty} n!\, |f_n|_0^2 < \infty,$$

namely, $(f_n)_{n=0}^{\infty} \in \Gamma(H_{\mathbb{C}})$. However, $\langle :x^{\otimes n}:, f_n\rangle$ is not a pointwisely defined function but is defined as L^2-function of $x \in E^*$. Moreover, the infinite series in (3.28) converges only in the L^2-sense.

As we agreed in §3.1, we express each $\Phi \in (E)^*$ also in the following form:

$$\Phi(x) = \sum_{n=0}^{\infty} \left\langle :x^{\otimes n}:, F_n \right\rangle, \tag{3.29}$$

where $F_n \in (E_{\mathbb{C}}^{\otimes n})^*_{\mathrm{sym}}$ and

$$\|\Phi\|_{-p}^2 = \sum_{n=0}^{\infty} n!\, |F_n|_{-p}^2 < \infty \qquad \text{for some } p \geq 0.$$

Recall that (3.29) is understood only through the canonical bilinear form. We have thus three ways of understanding expressions (3.27), (3.28) and (3.29) which are seemingly the same; however, there will occur no danger of confusion because they are easily distinguished by the context. Thus we agree that the Wiener-Itô expansion is a general term for such expansions.

For $x \in E^*$ define a linear function δ_x on (E) by

$$\delta_x : \phi \mapsto \phi(x), \qquad \phi \in (E).$$

This is called a *white noise delta function*. The next result has been already established in (3.15) during the proof of Proposition 3.2.3.

Theorem 3.2.12 *$\delta_x \in (E)^*$ for all $x \in E^*$. Moreover,*

$$\|\delta_x\|_{-p} \leq \left\{1 - \left(|\tau|_{-p}^{1/2} + |x|_{-p}\right)^2\right\}^{-1/2},$$

whenever $|\tau|_{-p}^{1/2} + |x|_{-p} < 1$.

Theorem 3.2.13 *The map $x \mapsto \delta_x \in (E)^*$, $x \in E^*$, is continuous.*

PROOF. Since $(E)^*$ is constructed in a similar way to E^*, the topology of $(E)^*$ is again defined by the Hilbertian seminorms:

$$\Phi \mapsto \sup\{|\langle\!\langle \Phi, \phi \rangle\!\rangle| \,;\, \|\phi\|_C \leq 1\}, \qquad \Phi \in (E)^*,$$

where C runs over $\mathcal{C}$. While, it follows from Proposition 3.2.11 that

$$\lim_{y \to x} \sup \{|\langle\!\langle \delta_x - \delta_y, \phi \rangle\!\rangle| \,;\, \|\phi\|_C \leq 1\} = 0,$$

for all $C = (C_p)_{p=0}^{\infty} \in \mathcal{C}$ satisfying the conditions (3.23). It is therefore sufficient to show that such C's constitute a set of defining seminorms of E^*. Note that $|x|_C \leq |x|_{C'}$ for any $x \in E^*$ if $C' \leq C$, namely, if $C'_p \leq C_p$ for all $p = 0, 1, 2, \cdots$. Thus it is sufficient to show that for a given $C \in \mathcal{C}$ there is $C' \in \mathcal{C}$ with (3.23) such that $C' \leq C$. Choose $q \geq 0$ such that $|\tau|_{-q} < e^{-1}(1 - \rho^2)$. Define $C' = (C'_p)_{p=0}^{\infty} \in \mathcal{C}$ by

$$\begin{cases} 0 \quad < C'_0 = \cdots = C'_q < \min\left\{|\tau|_{-q}, C_q, \sqrt{1-\rho^2}\right\}, \\ C'_p \quad = \min\left\{C'_{p-1}, C_p\right\}, \qquad p > q. \end{cases}$$

Then, by construction, $C' \leq C$ and $C_0'^2 < 1 - \rho^2$. Moreover, since $|\tau|_{C'} \leq C_p'^{-2} |\tau|_{-p}$ for all $p \geq 0$ by Lemma 3.2.7, we have

$$C_0'^2 \, |\tau|_{C'} \leq C_0'^2 C_q'^{-2} |\tau|_{-q} = |\tau|_{-q} \leq \frac{1-\rho^2}{e}.$$

This completes the proof. qed

Combining Theorems 3.2.1, 3.2.12 and 3.2.13, we come to the following

Corollary 3.2.14 *The space (E) of white noise test functionals satisfies hypotheses (H1)–(H3) introduced in §1.4.*

As an immediate application of Proposition 1.4.1, we obtain

Corollary 3.2.15 *If $\phi_n \in (E)$ converges to ϕ in (E), then $\lim_{n\to\infty} \phi_n(x) = \phi(x)$ for all $x \in E^*$. Moreover, the convergence is uniform on every compact subset of E^*.*

Finally, from Proposition 1.4.3 and Corollary 3.2.14 we can deduce

Corollary 3.2.16 *Let $n \geq 1$. Each $\phi \in (E)^{\otimes n}$ admits a unique continuous function $\widetilde{\phi}$ on $E^* \times \cdots \times E^*$ (n-times) such that $\phi(x_1, \cdots, x_n) = \widetilde{\phi}(x_1, \cdots, x_n)$ for $\mu \times \cdots \times \mu$-a.e. $(x_1, \cdots, x_n) \in E^* \times \cdots \times E^*$. Moreover, the linear functional:*

$$\delta_{(x_1,\cdots,x_n)} : \phi \mapsto \widetilde{\phi}(x_1, \cdots, x_n), \qquad \phi \in (E)^{\otimes n},$$

belongs to $((E)^{\otimes n})^*$ and the map

$$(x_1, \cdots, x_n) \mapsto \delta_{(x_1,\cdots,x_n)}, \qquad (x_1, \cdots, x_n) \in E^* \times \cdots \times E^*,$$

is continuous.

3.3 S-transform

We repeat the definition of an exponential vector introduced in §2.3.

Definition 3.3.1 For $\xi \in H_{\mathbb{C}}$ the function $\phi_\xi \in (L^2)$ defined by Wiener-Itô expansion

$$\phi_\xi(x) = \sum_{n=0}^{\infty} \left\langle :x^{\otimes n}:, \frac{\xi^{\otimes n}}{n!} \right\rangle, \qquad x \in E^*, \tag{3.30}$$

is called an *exponential vector* or a *coherent state*.

Lemma 3.3.2 *For $\xi, \eta \in H_{\mathbb{C}}$ it holds that*

$$\langle\langle \phi_\xi, \phi_\eta \rangle\rangle = e^{\langle \xi, \eta \rangle} \qquad \textit{and} \qquad \|\phi_\xi\|_0 = e^{|\xi|_0^2/2}.$$

In particular, $\phi_\xi \in (L^2)$ for $\xi \in H_{\mathbb{C}}$ and $\phi_\xi \in L^2(E^*, \mu; \mathbb{R})$ for $\xi \in H$.

PROOF. By definition,

$$\langle\langle \phi_\xi, \phi_\eta \rangle\rangle = \sum_{n=0}^{\infty} n! \left\langle \frac{\xi^{\otimes n}}{n!}, \frac{\eta^{\otimes n}}{n!} \right\rangle = \sum_{n=0}^{\infty} \frac{\langle \xi, \eta \rangle^n}{n!} = e^{\langle \xi, \eta \rangle}.$$

This proves the first identity. The second follows from $\overline{\phi_\xi} = \phi_{\overline{\xi}}$. qed

Lemma 3.3.3 *Let $\xi \in H_{\mathbf{C}}$. Then $\phi_\xi \in (E)$ if and only if $\xi \in E_{\mathbf{C}}$. In that case,*

$$\|\phi_\xi\|_p = \exp\left(\frac{1}{2}|\xi|_p^2\right), \qquad p \in \mathbf{R}.$$

PROOF. In fact, for any $p \in \mathbf{R}$ we have

$$\|\phi_\xi\|_p^2 = \sum_{n=0}^{\infty} n! \left|\frac{\xi^{\otimes n}}{n!}\right|_p^2 = e^{|\xi|_p^2}.$$

Hence $\phi_\xi \in (E)$ if and only if $\xi \in E_{\mathbf{C}}$. qed

Definition 3.3.4 The *S-transfrom* of $\Phi \in (E)^*$ is a function on $E_{\mathbf{C}}$ defined by

$$S\Phi(\xi) = \langle\!\langle \Phi, \phi_\xi \rangle\!\rangle, \qquad \xi \in E_{\mathbf{C}}.$$

Lemma 3.3.5 *Let $\Phi \in (E)^*$ be given with Wiener-Itô expansion:*

$$\Phi(x) = \sum_{n=0}^{\infty} \left\langle :x^{\otimes n}:, F_n \right\rangle.$$

Then,

$$S\Phi(\xi) = \sum_{n=0}^{\infty} \left\langle F_n, \xi^{\otimes n} \right\rangle, \qquad \xi \in E_{\mathbf{C}},$$

where the right hand side converges absolutely.

PROOF. By definition

$$S\Phi(\xi) = \langle\!\langle \Phi, \phi_\xi \rangle\!\rangle = \sum_{n=0}^{\infty} n! \left\langle F_n, \frac{\xi^{\otimes n}}{n!} \right\rangle = \sum_{n=0}^{\infty} \left\langle F_n, \xi^{\otimes n} \right\rangle.$$

Suppose $\|\Phi\|_{-p} < \infty$. Then,

$$\begin{aligned}
\sum_{n=0}^{\infty} |\langle F_n, \xi^{\otimes n} \rangle| &\le \sum_{n=0}^{\infty} |F_n|_{-p} |\xi|_p^n \\
&\le \left(\sum_{n=0}^{\infty} n! |F_n|_{-p}^2\right)^{1/2} \left(\sum_{n=0}^{\infty} \frac{|\xi|_p^{2n}}{n!}\right)^{1/2} \\
&= \|\Phi\|_{-p} \exp\left(\frac{1}{2}|\xi|_p^2\right) < \infty.
\end{aligned}$$

Hence, the series converges absolutely. qed

During the above proof we have established the following

Lemma 3.3.6 *Let $\Phi \in (E)^*$. If $\|\Phi\|_{-p} < \infty$ for $p \in \mathbf{R}$, then*

$$|S\Phi(\xi)| \le \|\Phi\|_{-p} \exp\left(\frac{1}{2}|\xi|_p^2\right), \qquad \xi \in E_{\mathbf{C}}.$$

Theorem 3.3.7 *Let $\xi, \eta \in E_{\mathbb{C}}$ and $\Phi \in (E)^*$. Then*

$$z \mapsto S\Phi(z\xi + \eta), \qquad z \in \mathbb{C},$$

is entire holomorphic.

PROOF. Suppose that the Wiener-Itô expansion of Φ is given as

$$\Phi(x) = \sum_{n=0}^{\infty} \left\langle :x^{\otimes n}:, F_n \right\rangle,$$

where $F_n \in (E^{\otimes n})^*_{\mathrm{sym}}$ and

$$\|\Phi\|_{-p}^2 = \sum_{n=0}^{\infty} n! \, |F_n|_{-p}^2 < \infty$$

for some $p \geq 0$. By Lemma 3.3.5 we see that

$$\begin{aligned} S\Phi(z\xi + \eta) &= \sum_{n=0}^{\infty} \left\langle F_n, (z\xi + \eta)^{\otimes n} \right\rangle \\ &= \sum_{n=0}^{\infty} \sum_{k=0}^{n} \binom{n}{k} z^k \left\langle F_n, \xi^{\otimes k} \otimes \eta^{\otimes (n-k)} \right\rangle \\ &= \sum_{k=0}^{\infty} \left(\sum_{n=0}^{\infty} \binom{n+k}{k} \left\langle F_{n+k}, \xi^{\otimes k} \otimes \eta^{\otimes n} \right\rangle \right) z^k. \end{aligned}$$

For the assertion we need only to prove

$$\frac{1}{R} = \limsup_{k \to \infty} \left| \sum_{n=0}^{\infty} \binom{n+k}{k} \left\langle F_{n+k}, \xi^{\otimes k} \otimes \eta^{\otimes n} \right\rangle \right|^{1/k} = 0.$$

We now observe

$$\begin{aligned} &\left| \sum_{n=0}^{\infty} \binom{n+k}{k} \left\langle F_{n+k}, \xi^{\otimes k} \otimes \eta^{\otimes n} \right\rangle \right| \\ &\quad \leq \sum_{n=0}^{\infty} \frac{(n+k)!}{n!k!} |F_{n+k}|_{-p} |\xi|_p^k |\eta|_p^n \\ &\quad \leq \frac{|\xi|_p^k}{k!} \left(\sum_{n=0}^{\infty} (n+k)! \, |F_{n+k}|_{-p}^2 \right)^{1/2} \left(\sum_{n=0}^{\infty} \frac{(n+k)!}{n!n!} |\eta|_p^{2n} \right)^{1/2} \\ &\quad = \|\Phi\|_{-p} \frac{|\xi|_p^k}{k!} \left(\sum_{n=0}^{\infty} \frac{(n+k)!}{n!n!} |\eta|_p^{2n} \right)^{1/2}. \end{aligned}$$

In view of Lemma 3.2.9, we have

$$\left| \sum_{n=0}^{\infty} \binom{n+k}{k} \left\langle F_{n+k}, \xi^{\otimes k} \otimes \eta^{\otimes n} \right\rangle \right| \leq \|\Phi\|_{-p} \frac{|\xi|_p^k}{k!} \left(|\eta|_p^2 + k \right)^{k/2} \exp\left(\frac{1}{2} |\eta|_p^2 \right).$$

Hence, using an obvious inequality

$$\frac{k^k}{k!} \le e^k,$$

we obtain

$$\begin{aligned}\frac{1}{R} &\le |\xi|_p \limsup_{k\to\infty} \left\{ \frac{(|\eta|_p^2 + k)^{k/2}}{k!} \right\}^{1/k} \\ &\le |\xi|_p \limsup_{k\to\infty} \left\{ \frac{e^k(|\eta|_p^2 + k)^{k/2}}{k^k} \right\}^{1/k} \\ &\le e\,|\xi|_p \limsup_{k\to\infty} \frac{(|\zeta_2|_p^2 + k)^{1/2}}{k} = 0.\end{aligned}$$

This completes the proof. qed

Corollary 3.3.8 *If $\alpha \in \mathbf{C}$, $\alpha \neq 0$, then $\{\phi_{\alpha\xi}; \xi \in E\}$ spans a dense subspace of (E).*

PROOF. We need only to show that $\{\phi_{\alpha\xi}; \xi \in E\}$ spans a dense subspace of $(E)_p$, where $(E)_p$ is the completion of (E) with respect to the norm $\|\cdot\|_p$. Suppose that $\phi \in (E)_p$ is orthogonal to $\phi_{\alpha\xi}$ in $(E)_p$ for all $\xi \in E$. Note that $(E)_p \subset (L^2)$ and we put $\Phi = \Gamma(A)^{2p}\phi \in (E)^*$ for simplicity. Then, for $t \in \mathbf{R}$ and $\xi \in E$,

$$\begin{aligned}S\Phi(t\overline{\alpha}\xi) &= \langle\!\langle \Phi, \phi_{t\overline{\alpha}\xi} \rangle\!\rangle \\ &= \langle\!\langle \Gamma(A)^{2p}\phi, \phi_{t\overline{\alpha}\xi} \rangle\!\rangle \\ &= \left\langle\!\!\left\langle \Gamma(A)^p\phi, \Gamma(A)^p\overline{\phi_{t\alpha\xi}} \right\rangle\!\!\right\rangle = 0 \qquad (3.31)\end{aligned}$$

Let

$$\phi(x) = \sum_{n=0}^{\infty} \left\langle :x^{\otimes n}:, f_n \right\rangle,$$

be the Wiener-Itô expansion of ϕ. Then, for any $z \in \mathbf{C}$ we have

$$S\Phi(z\xi) = \langle\!\langle \Phi, \phi_{z\xi} \rangle\!\rangle = \langle\!\langle \phi, \Gamma(A)^{2p}\phi_{z\xi} \rangle\!\rangle = \sum_{n=0}^{\infty} \left\langle f_n, (A^{2p}\xi)^{\otimes n} \right\rangle z^n. \qquad (3.32)$$

From Theorem 3.3.7 we see that (3.32) is the Taylor expansion of an entire analytic function. But by assumption (3.31) this is identically zero on $\{z = t\overline{\alpha};\ t \in \mathbf{R}\}$. Hence every coefficient in the Taylor series (3.32) vanishes. Namely,

$$\left\langle f_n, (A^{2p}\xi)^{\otimes n} \right\rangle = 0, \qquad \text{for any } \xi \in E,$$

and hence, by the polarization formula,

$$\left\langle f_n, (A^{2p}\xi_1) \otimes \cdots \otimes (A^{2p}\xi_n) \right\rangle = 0$$

for any $\xi_1, \cdots, \xi_n \in E$ and $n = 0, 1, 2, \cdots$ From this we conclude that

$$\langle f_n, e_{i_1} \otimes \cdots \otimes e_{i_n} \rangle = 0,$$

namely, $f_n = 0$ for all $n = 0, 1, 2, \cdots$. Therefore $\phi = 0$. qed

Corollary 3.3.9 *Let $\alpha \in \mathbb{C}$ with $\alpha \neq 0$ and let $\Phi \in (E)^*$. If $S\Phi(\alpha\xi) = 0$ for all $\xi \in E$, then $\Phi = 0$.*

PROOF. By assumption,

$$S\Phi(\alpha\xi) = \langle\langle \Phi, \phi_{\alpha\xi} \rangle\rangle = 0, \qquad \text{for all } \xi \in E.$$

Since $\{\phi_{\alpha\xi}; \xi \in E\}$ spans a dense subspace of (E) by Corollary 3.3.8, we conclude that $\Phi = 0$. qed

As an application of S-transform we prove the following

Proposition 3.3.10 *For any $\xi \in H_{\mathbb{C}}$,*

$$\phi_\xi(x) = \sum_{n=0}^{\infty} \left\langle :x^{\otimes n}:, \frac{\xi^{\otimes n}}{n!} \right\rangle = e^{\langle x, \xi\rangle - \langle \xi, \xi \rangle/2}.$$

In particular,

$$\phi_\xi(x) = \frac{\mu(dx - \xi)}{\mu(dx)}, \qquad x \in E^*, \quad \xi \in H.$$

PROOF. We first prove that for $\xi \in E$,

$$\phi_\xi(x) = e^{\langle x, \xi\rangle - \langle \xi, \xi \rangle/2}, \qquad x \in E^*. \tag{3.33}$$

This is obvious if $\xi = 0$. Suppose $\xi \neq 0$. It follows from Lemma 2.2.7 that

$$\phi_\xi(x) = \sum_{n=0}^{\infty} \left\langle :x^{\otimes n}:, \frac{\xi^{\otimes n}}{n!} \right\rangle = \sum_{n=0}^{\infty} \frac{1}{n!} \left(\frac{|\xi|_0}{\sqrt{2}} \right)^n H_n \left(\frac{\langle x, \xi \rangle}{\sqrt{2}\,|\xi|_0} \right).$$

The last series is the generating function of the Hermite polynomials (B.1) and we obtain

$$\phi_\xi(x) = \exp\left(2\frac{\langle x, \xi\rangle}{\sqrt{2}\,|\xi|_0} \frac{|\xi|_0}{\sqrt{2}} - \left(\frac{|\xi|_0}{\sqrt{2}} \right)^2 \right) = \exp\left(\langle x, \xi \rangle - \frac{1}{2} \langle \xi, \xi \rangle \right),$$

which proves (3.33).

Now for $\xi \in H_{\mathbb{C}}$ we consider the S-transform of ϕ_ξ. It follows from Lemma 3.3.2 that

$$S\phi_\xi(\eta) = \langle\langle \phi_\xi, \phi_\eta \rangle\rangle = e^{\langle \xi, \eta \rangle}, \qquad \eta \in E_{\mathbb{C}}. \tag{3.34}$$

On the other hand, consider the S-transform of the function:

$$\phi(x) = e^{\langle x, \xi\rangle - \langle \xi, \xi \rangle/2}.$$

In view of (3.33), for $\eta \in E$ we have

$$\begin{aligned} S\phi(\eta) &= \langle\langle \phi, \phi_\eta \rangle\rangle \\ &= \int_{E^*} \phi(x)\phi_\eta(x)\mu(dx) \\ &= \int_{E^*} e^{\langle x, \xi\rangle - \langle \xi, \xi\rangle/2} e^{\langle x, \eta \rangle - \langle \eta, \eta\rangle/2} \mu(dx). \end{aligned}$$

Using Lemma 2.1.6, one may immediately obtain

$$S\phi(\xi) = e^{-\langle \xi, \xi\rangle/2 - \langle \eta, \eta\rangle/2} e^{\langle \xi+\eta, \xi+\eta\rangle/2} = e^{\langle \xi, \eta\rangle}, \qquad \eta \in E. \tag{3.35}$$

We see from (3.34) and (3.35) that

$$S\phi_\xi(\eta) = S\phi(\eta), \qquad \eta \in E.$$

Consequently, by Corollary 3.3.9 we conclude that $\phi_\xi = \phi$. The rest follows from Proposition 2.1.6. qed

The second half of Proposition 3.3.10 implies

Corollary 3.3.11 *If $\phi \in (L^2)$, then*

$$S\phi(\xi) = \int_{E^*} \phi(x+\xi)\, d\mu(x), \qquad \xi \in E.$$

3.4 Contraction of tensor products

Recall that $\{e_j\}_{j=0}^\infty$ is a complete orthonormal basis of $H_{\mathbb{C}}$ such that $e_j \in E_{\mathbb{C}}$ and $Ae_j = \lambda_j e_j$. Employing a simplified notation

$$e(\mathbf{j}) = e_{j_1} \otimes \cdots \otimes e_{j_m}, \qquad \mathbf{j} = (j_1, \cdots, j_m),$$

one has

$$|f|_p^2 = \sum_{\mathbf{j}} |\langle f, e(\mathbf{j})\rangle|^2 |e(\mathbf{j})|_p^2, \qquad f \in E_{\mathbb{C}}^{\otimes m},$$

where the sum is taken over all possible $\mathbf{j} = (j_1, \cdots, j_m)$. In fact,

$$\begin{aligned}
|f|_p^2 &= |A^p f|_0^2 = \sum_{\mathbf{j}} |\langle (A^{\otimes m})^p f, e_{j_1} \otimes \cdots \otimes e_{j_m}\rangle|^2 \\
&= \sum_{\mathbf{j}} |\langle f, A^p e_{j_1} \otimes \cdots \otimes A^p e_{j_m}\rangle|^2 \\
&= \sum_{\mathbf{j}} |\langle f, e_{j_1} \otimes \cdots \otimes e_{j_m}\rangle|^2 \lambda_{j_1}^{2p} \cdots \lambda_{j_m}^{2p} \\
&= \sum_{\mathbf{j}} |\langle f, e(\mathbf{j})\rangle|^2 |e(\mathbf{j})|_p^2 .
\end{aligned}$$

We now generalize this situation. Using a similar notation:

$$e(\mathbf{i}) = e_{i_1} \otimes \cdots \otimes e_{i_l}, \qquad \mathbf{i} = (i_1, \cdots, i_l),$$

for $f \in (E_{\mathbb{C}}^{\otimes(l+m)})^*$ we put

$$|f|_{l,m;p,q} = \left(\sum_{\mathbf{i},\mathbf{j}} |\langle f, e(\mathbf{i}) \otimes e(\mathbf{j})\rangle|^2 |e(\mathbf{i})|_p^2 |e(\mathbf{j})|_q^2 \right)^{1/2}, \qquad p, q \in \mathbb{R}. \tag{3.36}$$

This is possibly infinite, however, is always finite for $f \in E_{\mathbb{C}}^{\otimes(l+m)}$. As is immediately seen from definition,

$$|f|_p = |f|_{l,m;p,p}, \qquad p \in \mathbb{R},$$

and

$$|f|_{l,m;p,q} \le \rho^{lr+ms} |f|_{l,m;p+r,q+s}, \qquad p,q \in \mathbb{R}, \quad r,s \ge 0.$$

In fact, for $p,q \in \mathbb{R}$ and $r,s \ge 0$,

$$\begin{aligned} |f|^2_{l,m;p,q} &= \sum_{\mathbf{i},\mathbf{j}} |\langle f, e(\mathbf{i}) \otimes e(\mathbf{j})\rangle|^2 |e(\mathbf{i})|_p^2 |e(\mathbf{j})|_q^2 \\ &\le \sum_{\mathbf{i},\mathbf{j}} |\langle f, e(\mathbf{i}) \otimes e(\mathbf{j})\rangle|^2 \rho^{2lr} |e(\mathbf{i})|^2_{p+r} \rho^{2ms} |e(\mathbf{j})|^2_{q+s} \\ &= \rho^{2(lr+ms)} |f|^2_{l,m;p+r,q+s}. \end{aligned}$$

The above relations are used without special notice.

We now consider two elements $f \in E_{\mathbb{C}}^{\otimes(l+m)}$ and $g \in E_{\mathbb{C}}^{\otimes(l+n)}$. Since

$$\sum_{\mathbf{i}} \langle f, e(\mathbf{i}) \otimes e(\mathbf{j})\rangle \langle g, e(\mathbf{i}) \otimes e(\mathbf{k})\rangle$$

converges absolutely, one may consider a new element in $E_{\mathbb{C}}^{\otimes(m+n)}$ defined by

$$f \otimes^l g = \sum_{\mathbf{j},\mathbf{k}} \left(\sum_{\mathbf{i}} \langle f, e(\mathbf{i}) \otimes e(\mathbf{j})\rangle \langle g, e(\mathbf{i}) \otimes e(\mathbf{k})\rangle \right) e(\mathbf{j}) \otimes e(\mathbf{k}), \tag{3.37}$$

where

$$e(\mathbf{k}) = e_{k_1} \otimes \cdots \otimes e_{k_n}, \qquad \mathbf{k} = (k_1, \cdots, k_n).$$

We must show that (3.37) is well defined, namely the series converges. For that purpose it is sufficient to show that

$$\sum_{\mathbf{j},\mathbf{k}} \left| \sum_{\mathbf{i}} \langle f, e(\mathbf{i}) \otimes e(\mathbf{j})\rangle \langle g, e(\mathbf{i}) \otimes e(\mathbf{k})\rangle \right|^2 |e(\mathbf{j}) \otimes e(\mathbf{k})|_p^2 < \infty$$

for all $p \ge 0$. In the following we prove a more general result.

Lemma 3.4.1 *For any $p,q,r \in \mathbb{R}$ it holds that*

$$|f \otimes^l g|_{m,n;p,q} \le |f|_{l,m;r,p} |g|_{l,n;-r,q}, \qquad f \in E_{\mathbb{C}}^{\otimes(l+m)}, \quad g \in E_{\mathbb{C}}^{\otimes(l+n)}. \tag{3.38}$$

PROOF. Since $|e(\mathbf{i})|_{-r} |e(\mathbf{i})|_r = 1$, we have

$$\begin{aligned} &\left| \sum_{\mathbf{i}} \langle f, e(\mathbf{i}) \otimes e(\mathbf{j})\rangle \langle g, e(\mathbf{i}) \otimes e(\mathbf{k})\rangle \right|^2 \\ &\qquad \le \left(\sum_{\mathbf{i}} |\langle f, e(\mathbf{i}) \otimes e(\mathbf{j})\rangle|^2 |e(\mathbf{i})|_r^2 \right) \left(\sum_{\mathbf{i}} |\langle g, e(\mathbf{i}) \otimes e(\mathbf{k})\rangle|^2 |e(\mathbf{i})|_{-r}^2 \right). \end{aligned}$$

Hence, by definition (3.36) and (3.37) we have

$$\begin{aligned}|f \otimes^l g|^2_{m,n;p,q} &= \sum_{\mathbf{j},\mathbf{k}} \left| \sum_{\mathbf{i}} \langle f, e(\mathbf{i}) \otimes e(\mathbf{j}) \rangle \langle g, e(\mathbf{i}) \otimes e(\mathbf{k}) \rangle \right|^2 |e(\mathbf{j})|^2_p \, |e(\mathbf{k})|^2_q \\ &\le \sum_{\mathbf{i},\mathbf{j}} |\langle f, e(\mathbf{i}) \otimes e(\mathbf{j}) \rangle|^2 \, |e(\mathbf{i})|^2_r \, |e(\mathbf{j})|^2_p \\ &\quad \times \sum_{\mathbf{i},\mathbf{k}} |\langle g, e(\mathbf{i}) \otimes e(\mathbf{k}) \rangle|^2 \, |e(\mathbf{i})|^2_{-r} \, |e(\mathbf{k})|^2_q ,\end{aligned}$$

which proves the assertion. qed

Similarly, for $f \in E_{\mathbf{C}}^{\otimes(l+m)}$ and $g \in E_{\mathbf{C}}^{\otimes(l+n)}$,

$$f \otimes_l g = \sum_{\mathbf{j},\mathbf{k}} \left(\sum_{\mathbf{i}} \langle f, e(\mathbf{j}) \otimes e(\mathbf{i}) \rangle \langle g, e(\mathbf{k}) \otimes e(\mathbf{i}) \rangle \right) e(\mathbf{j}) \otimes e(\mathbf{k})$$

becomes an element in $E_{\mathbf{C}}^{\otimes(m+n)}$.

Definition 3.4.2 The above defined $f \otimes^l g \in E_{\mathbf{C}}^{\otimes(m+n)}$ and $f \otimes_l g \in E_{\mathbf{C}}^{\otimes(m+n)}$ are called the *left contraction* and *right contraction*, respectively.

Note also that

$$f \otimes^0 g = f \otimes_0 g = f \otimes g.$$

Since the argument is entirely parallel, we state the assertions mostly for the left contraction.

Proposition 3.4.3 *For $f \in E_{\mathbf{C}}^{\otimes(l+m)}$ and $g \in E_{\mathbf{C}}^{\otimes(l+n)}$ it holds that*

$$|f \otimes^l g|_p \le \rho^{2pl} \, |f|_p \, |g|_p , \qquad p \ge 0. \tag{3.39}$$

Therefore, the map $(f, g) \mapsto f \otimes^l g$ yields a continuous bilinear map:

$$\otimes^l : E_{\mathbf{C}}^{\otimes(l+m)} \times E_{\mathbf{C}}^{\otimes(l+n)} \longrightarrow E_{\mathbf{C}}^{\otimes(m+n)}.$$

PROOF. In view of Lemma 3.4.1, we obtain

$$|f \otimes^l g|_p = |f \otimes^l g|_{m,n;p,p} \le |f|_{l,m;r,p} \, |g|_{l,n;-r,p}$$

for any $r \in \mathbf{R}$. Hence, putting $r = p$ we come to

$$|f \otimes^l g|_p \le |f|_{l,m;p,p} \, |g|_{l,n;-p,p} \le |f|_p \, \rho^{2pl} \, |g|_p .$$

This proves (3.39). qed

We have defined $f \otimes^l g$ by taking a particular basis $\{e_j\}_{j=0}^{\infty}$, however, the definition is independent of the choice of basis. In fact, $\otimes^l$ is characterized as a unique continuous bilinear map from $E_{\mathbf{C}}^{\otimes(l+m)} \times E_{\mathbf{C}}^{\otimes(l+n)}$ into $E_{\mathbf{C}}^{\otimes(m+n)}$ satisfying the condition that for

$f = \xi_1 \otimes \cdots \otimes \xi_{l+m}$ and $g = \eta_1 \otimes \cdots \otimes \eta_{l+n}$, where $\xi_1, \cdots, \xi_{l+m}, \eta_1, \cdots, \eta_{l+n} \in E_{\mathbf{C}}$, it holds that

$$f \otimes^l g = \langle \xi_1 , \eta_1 \rangle \cdots \langle \xi_l , \eta_l \rangle \, \xi_{l+1} \otimes \cdots \otimes \xi_{l+m} \otimes \eta_{l+1} \otimes \cdots \otimes \eta_{l+n}.$$

Using Lemma 3.4.1 we may extend $\otimes^l$ to a bilinear map from $(E_{\mathbf{C}}^{\otimes(l+m)})^* \times E_{\mathbf{C}}^{\otimes(l+n)}$ into $(E_{\mathbf{C}}^{\otimes(m+n)})^*$. In fact, it follows from Lemma 3.4.1 that

$$|f \otimes^l g|_{-p} = |f \otimes^l g|_{m,n;-p,-p} \le |f|_{l,m;r,-p} \, |g|_{l,n;-r,-p} \,, \qquad r \in \mathbf{R}.$$

Taking $r = -p$, we come to

$$|f \otimes^l g|_{-p} \le |f|_{l,m;-p,-p} \, |g|_{l,n;p,-p} \le |f|_{-p} \, \rho^{2pn} \, |g|_{l,n;p,p} = \rho^{2pn} \, |f|_{-p} \, |g|_{p} \,.$$

Therefore $F \otimes^l g$ is defined for $F \in (E_{\mathbf{C}}^{\otimes(l+m)})^*$ and $g \in E_{\mathbf{C}}^{\otimes(l+n)}$. Moreover,

Lemma 3.4.4 *Let $F \in (E_{\mathbf{C}}^{\otimes(l+m)})^*$ and $g \in E_{\mathbf{C}}^{\otimes(l+n)}$. Then,*

$$|F \otimes^l g|_{-p} \le \rho^{2pn} \, |F|_{-p} \, |g|_{p} \,, \qquad p \ge 0. \tag{3.40}$$

In particular, the map $(F, g) \mapsto F \otimes^l g$ becomes a separately continuous bilinear map:

$$\otimes^l : (E_{\mathbf{C}}^{\otimes(l+m)})^* \times (E_{\mathbf{C}}^{\otimes(l+n)}) \to (E_{\mathbf{C}}^{\otimes(m+n)})^*.$$

PROOF. We show the second half of the assertion. Given a bounded subset $S \subset E_{\mathbf{C}}^{\otimes(m+n)}$, we see from (3.39) that

$$\sup_{h \in S} |\langle F \otimes^l g \,, h \rangle| \le \sup_{h \in S} \rho^{2pn} \, |F|_{-p} \, |g|_{p} \, |h|_{p} = M \, |F|_{-p} \, |g|_{p} \,, \tag{3.41}$$

where $M = \rho^{2pn} \sup_{h \in S} |h|_{p}$. Suppose $F \in (E_{\mathbf{C}}^{\otimes(l+m)})^*$ is fixed and choose $p \ge 0$ with $|F|_{-p} < \infty$. Then (3.41) implies that $g \mapsto F \otimes^l g$ is continuous. On the other hand, for a fixed $g \in E_{\mathbf{C}}^{\otimes(l+n)}$, we see from Proposition 1.2.9 that (3.41) implies the continuity of $F \mapsto F \otimes^l g$. qed

The next result is also useful.

Lemma 3.4.5 *For $F \in (E_{\mathbf{C}}^{\otimes(l+m)})^*$ and $g \in E_{\mathbf{C}}^{\otimes(l+n)}$ it holds that*

$$|F \otimes^l g|_{p} \le \rho^{qn} \, |F|_{l,m;-(p+q),p} \, |g|_{p+q} \,, \qquad p \in \mathbf{R}, \quad q \ge 0. \tag{3.42}$$

In particular, for $F \in (E_{\mathbf{C}}^{\otimes l})^*$ and $g \in E_{\mathbf{C}}^{\otimes(l+n)}$ it holds that

$$|F \otimes^l g|_{p} \le \rho^{qn} \, |F|_{-(p+q)} \, |g|_{p+q} \,, \qquad p \in \mathbf{R}, \quad q \ge 0. \tag{3.43}$$

PROOF. In view of Lemma 3.4.1, we have

$$|F \otimes^l g|_{p} = |F \otimes^l g|_{m,n;p,p} \le |F|_{l,m;r,p} \, |g|_{l,n;-r,p} \,, \qquad r \in \mathbf{R}.$$

Taking $r = -(p+q)$, we obtain

$$|F \otimes^l g|_{p} \le |F|_{l,m;-(p+q),p} \, |g|_{l,n;p+q,p} \le |F|_{l,m;-(p+q),p} \, \rho^{qn} \, |g|_{p+q} \,,$$

which proves (3.42). For (3.43) one need only to take $m = 0$ in (3.42). qed

The left and right contractions are related as follows.

Proposition 3.4.6 *For $F \in (E_{\mathbf{C}}^{\otimes(l+m)})^*$, $g \in E_{\mathbf{C}}^{\otimes(l+n)}$ and $h \in E_{\mathbf{C}}^{\otimes(m+n)}$ it holds that*

$$\langle F \otimes^l g, h\rangle = \langle F, g \otimes_n h\rangle. \tag{3.44}$$

PROOF. Suppose F is fixed. It follows from Lemma 3.4.5 that both sides of (3.44) are continuous bilinear forms in g and h. Therefore we need only to check the identity for $g = e(\mathbf{i}) \otimes e(\mathbf{k})$ and $h = e(\mathbf{j}) \otimes e(\mathbf{k})$. But the verification is straightforward from definition. qed

One may employ (3.44) as the definition of $F \otimes^l g$. Given $F \in (E_{\mathbf{C}}^{\otimes(l+m)})^*$ and $g \in E_{\mathbf{C}}^{\otimes(l+n)}$ we consider a linear form:

$$h \mapsto \langle F, g \otimes_n h\rangle, \qquad h \in E_{\mathbf{C}}^{\otimes(m+n)}.$$

This is continuos by Proposition 3.4.3. Therefore there exists a unique element in $(E_{\mathbf{C}}^{\otimes(m+n)})^*$, which in fact coincides with $F \otimes^l g$. We call $F \otimes^l g$ left contraction again. In a similar way, $F \otimes_l g$, $f \otimes^l G$ and $f \otimes_l G$ are defined for $f \in E_{\mathbf{C}}^{\otimes(l+m)}$ and $G \in (E_{\mathbf{C}}^{\otimes(l+n)})^*$.

Finally, we consider the symmetric case. For symmetric elements $F \in (E_{\mathbf{C}}^{\otimes(l+m)})^*_{\mathrm{sym}}$ and $g \in E_{\mathbf{C}}^{\widehat{\otimes}(l+n)}$, obviously it holds that

$$F \otimes^l g = F \otimes_l g.$$

The symmetrization of $F \otimes_l g$ is denoted by $F \widehat{\otimes}_l g$. Using an obvious inequality:

$$|\widehat{f}|_p \le |f|_p, \qquad f \in E_{\mathbf{C}}^{\otimes n}, \quad p \in \mathbf{R},$$

we may restate (3.39), (3.40) and (3.43) for the symmetric case.

Proposition 3.4.7 *For $p, q \ge 0$,*

$$|f \widehat{\otimes}_l g|_p \le \rho^{2pl} |f|_p |g|_p, \qquad f \in E_{\mathbf{C}}^{\widehat{\otimes}(l+m)}, \quad g \in E_{\mathbf{C}}^{\widehat{\otimes}(l+n)}, \tag{3.45}$$

$$|F \widehat{\otimes}_l g|_{-p} \le \rho^{2pn} |F|_{-p} |g|_p, \qquad F \in (E_{\mathbf{C}}^{\otimes(l+m)})^*_{\mathrm{sym}}, \quad g \in E_{\mathbf{C}}^{\widehat{\otimes}(l+n)}, \tag{3.46}$$

and for $p \in \mathbf{R}$ and $q \ge 0$ we have

$$|F \widehat{\otimes}_l g|_p \le \rho^{qn} |F|_{-(p+q)} |g|_{p+q}, \qquad F \in (E_{\mathbf{C}}^{\otimes l})^*, \quad g \in E_{\mathbf{C}}^{\widehat{\otimes}(l+n)}. \tag{3.47}$$

The following assertion shows a rule of associativity.

Proposition 3.4.8 *For $F \in (E_{\mathbf{C}}^{\otimes k})^*_{\mathrm{sym}}$, $G \in (E_{\mathbf{C}}^{\otimes l})^*_{\mathrm{sym}}$ and $h \in E_{\mathbf{C}}^{\widehat{\otimes}(k+l+m)}$, it holds that*

$$(F \widehat{\otimes} G) \widehat{\otimes}_{k+l} h = F \widehat{\otimes}_k (G \widehat{\otimes}_l h). \tag{3.48}$$

PROOF. Note first that both sides of (3.48) are continuous in h for fixed F and G. In fact, from (3.47) we see that for any $p, q, q' \geq 0$,

$$\left|(F\widehat{\otimes}G)\widehat{\otimes}_{k+l}h\right|_p \leq \rho^{qm}|F\widehat{\otimes}G|_{-(p+q)}\,|h|_{p+q}$$

and

$$\left|F\widehat{\otimes}_k(G\widehat{\otimes}_l h)\right|_p \leq \rho^{ql+q'(k+m)}\,|F|_{-(p+q)}\,|G|_{-(p+q+q')}\,|h|_{p+q+q'}\,.$$

Hence, it is sufficient to show identity (3.48) for $h = \xi^{\otimes(k+l+m)}$, $\xi \in E_{\mathbb{C}}$. But, in that case, the verification is straightforward. qed

Proposition 3.4.9 *Let $F \in (E_{\mathbb{C}}^{\otimes l})^*_{\rm sym}$ and $G \in (E_{\mathbb{C}}^{\otimes m})^*_{\rm sym}$. Then, for any $f \in E_{\mathbb{C}}^{\widehat{\otimes}(l+n)}$ and $g \in E_{\mathbb{C}}^{\widehat{\otimes}(m+n)}$ it holds that*

$$\left\langle F\widehat{\otimes}_l f\,, G\widehat{\otimes}_m g\right\rangle = \langle F \otimes G\,, f \otimes_n g\rangle\,. \tag{3.49}$$

PROOF. In view of (3.46) and (3.47) we observe that for any $p \geq 0$,

$$\left|\left\langle F\widehat{\otimes}_l f\,, G\widehat{\otimes}_m g\right\rangle\right| \leq \left|F\widehat{\otimes}_l f\right|_{-p}\left|G\widehat{\otimes}_m g\right|_p \leq \rho^{2pn}\,|F|_{-p}\,|f|_p\,|G|_{-p}\,|g|_p\,.$$

This implies that the left hand side of (3.49) is a continuous bilinear form in f and g. Obviously, the same is true for the right hand side. Therefore, for the proof we need only to show the identity for $f = \xi^{\otimes(l+n)}$ and $g = \eta^{\otimes(m+n)}$, where $\xi, \eta \in E_{\mathbb{C}}$. But, in that case (3.49) is obviously true. qed

Taking $m = 0$ and $G = 1$ in (3.49), we come to the following

Corollary 3.4.10 *For $F \in (E_{\mathbb{C}}^{\otimes l})^*_{\rm sym}$, $f \in E_{\mathbb{C}}^{\otimes(l+n)}$ and $g \in E_{\mathbb{C}}^{\otimes n}$ it holds that*

$$\left\langle F\widehat{\otimes}_l f\,, g\right\rangle = \left\langle F\,, f\widehat{\otimes}_n g\right\rangle.$$

Some norm estimates discussed in this section are collected in Appendix C.

3.5 Wiener product

Lemma 3.5.1 *For $\xi, \eta \in E_{\mathbb{C}}$ and $m, n \geq 0$, we have*

$$\left\langle :x^{\otimes m}:, \xi^{\otimes m}\right\rangle\left\langle :x^{\otimes n}:, \eta^{\otimes n}\right\rangle = \sum_{k=0}^{m\wedge n} k!\binom{m}{k}\binom{n}{k}\left\langle :x^{\otimes(m+n-2k)}:, \xi^{\otimes m}\widehat{\otimes}_k\eta^{\otimes n}\right\rangle$$

PROOF. We note first the obvious identity:

$$e^{\langle x, s\xi\rangle - s^2\langle \xi, \xi\rangle/2}e^{\langle x, t\eta\rangle - t^2\langle \eta, \eta\rangle/2} = e^{\langle x, s\xi+t\eta\rangle - \langle s\xi+t\eta, s\xi+t\eta\rangle/2}e^{st\langle \xi, \eta\rangle}, \qquad s, t \in \mathbb{C}.$$

Then by Proposition 3.3.10 we have

$$\sum_{m=0}^{\infty}\left\langle :x^{\otimes m}:, \xi^{\otimes m}\right\rangle \frac{s^m}{m!}\sum_{n=0}^{\infty}\left\langle :x^{\otimes n}:, \eta^{\otimes n}\right\rangle \frac{t^n}{n!}$$

$$= \sum_{j=0}^{\infty}\left\langle :x^{\otimes j}:, \frac{(s\xi + t\eta)^{\otimes j}}{j!}\right\rangle \sum_{k=0}^{\infty}\langle \xi, \eta\rangle^k \frac{s^k t^k}{k!}. \tag{3.50}$$

The last expression is computed as follows:

$$\sum_{j=0}^{\infty}\frac{1}{j!}\sum_{l=0}^{j}\binom{j}{l}\left\langle :x^{\otimes j}:, \xi^{\otimes l}\otimes\eta^{\otimes(j-l)}\right\rangle s^l t^{j-l}\sum_{k=0}^{\infty}\langle \xi, \eta\rangle^k \frac{s^k t^k}{k!}$$

$$= \sum_{j,k,l=0}^{\infty}\left\langle :x^{\otimes(j+l)}:, \xi^{\otimes l}\otimes\eta^{\otimes j}\right\rangle \langle \xi, \eta\rangle^k \frac{s^{l+k}t^{j+k}}{j!k!l!}$$

$$= \sum_{j,k,l=0}^{\infty}\left\langle :x^{\otimes(j+l)}:, \xi^{\otimes(l+k)}\widehat{\otimes}_k\eta^{\otimes(j+k)}\right\rangle \frac{s^{l+k}t^{j+k}}{j!k!l!}$$

$$= \sum_{m,n=0}^{\infty}\sum_{\substack{l+k=m\\ j+k=n}}\left\langle :x^{\otimes(m+n-2k)}:, \xi^{\otimes m}\widehat{\otimes}_k\eta^{\otimes n}\right\rangle \frac{s^m t^n}{j!k!k!}$$

$$= \sum_{m,n=0}^{\infty}\sum_{k=0}^{m\wedge n}k!\binom{m}{k}\binom{n}{k}\left\langle :x^{\otimes(m+n-2k)}:, \xi^{\otimes m}\widehat{\otimes}_k\eta^{\otimes n}\right\rangle \frac{s^m t^n}{m!n!}.$$

Therefore, (3.50) becomes

$$\sum_{m,n=0}^{\infty}\left\langle :x^{\otimes m}:, \xi^{\otimes m}\right\rangle\left\langle :x^{\otimes n}:, \eta^{\otimes n}\right\rangle \frac{s^m}{m!}\frac{t^n}{n!}$$

$$= \sum_{m,n=0}^{\infty}\sum_{k=0}^{m\wedge n}k!\binom{m}{k}\binom{n}{k}\left\langle :x^{\otimes(m+n-2k)}:, \xi^{\otimes m}\widehat{\otimes}_k\eta^{\otimes n}\right\rangle \frac{s^m t^n}{m!n!}.$$

The assertion follows by comparing the coefficients of $s^m t^n$ in both sides. qed

Theorem 3.5.2 *For any* $f \in E_{\mathbb{C}}^{\widehat{\otimes} m}$ *and* $g \in E_{\mathbb{C}}^{\widehat{\otimes} n}$, *we have*

$$\left\langle :x^{\otimes m}:, f\right\rangle\left\langle :x^{\otimes n}:, g\right\rangle = \sum_{k=0}^{m\wedge n}k!\binom{m}{k}\binom{n}{k}\left\langle :x^{\otimes(m+n-2k)}:, f\widehat{\otimes}_k g\right\rangle.$$

PROOF. By the polarization formula the assertion is true if both f and g are algebraic tensor products by Lemma 3.5.1. Since $\widehat{\otimes}_k$ is a continuous bilinear map by Proposition 3.4.3, the general case follows. qed

The formula stated in Theorem 3.5.2 is known as *Wiener product formula*. A simple application of Theorem 3.5.2 implies the following

Proposition 3.5.3 *Let* $\phi, \psi \in (E)$ *be given as*

$$\phi(x) = \sum_{m=0}^{\infty}\left\langle :x^{\otimes m}:, f_m\right\rangle, \qquad \psi(x) = \sum_{n=0}^{\infty}\left\langle :x^{\otimes n}:, g_n\right\rangle.$$

Then the Wiener-Itô expansion of $\phi\psi$ is given by

$$\phi(x)\psi(x) = \sum_{l=0}^{\infty} \left\langle :x^{\otimes l}:, h_l \right\rangle, \tag{3.51}$$

where

$$h_l = \sum_{m+n=l} \sum_{k=0}^{\infty} k! \binom{m+k}{k} \binom{n+k}{k} f_{m+k} \widehat{\otimes}_k g_{n+k}. \tag{3.52}$$

Lemma 3.5.4 *Let notations and assumptions be the same as in Proposition 3.5.3. Then for any $l = 0, 1, 2, \cdots$ and $\alpha, \beta, p \geq 0$,*

$$l! \, |h_l|_p^2 \leq (l+1)(\rho^{2\alpha} + \rho^{2\beta})^l \, \|\phi\|_{p+\alpha}^2 \, \|\psi\|_{p+\beta}^2 \sum_{k=0}^{\infty} \binom{l+2k}{2k} \rho^{2k(\alpha+\beta+2p)}. \tag{3.53}$$

PROOF. Note first that

$$|f_{m+k} \widehat{\otimes}_k g_{n+k}|_p \leq \rho^{2pk} \, |f_{m+k}|_p \, |g_{n+k}|_p \, .$$

Then (3.52) is estimated as follows.

$$\begin{aligned}
\sqrt{l!}\, |h_l|_p &\leq \sqrt{l!} \sum_{m+n=l} \sum_{k=0}^{\infty} k! \binom{m+k}{k} \binom{n+k}{k} \rho^{2pk} \, |f_{m+k}|_p \, |g_{n+k}|_p \\
&= \sum_{m+n=l} \sum_{k=0}^{\infty} \left\{ \binom{m+n}{m} \binom{m+k}{k} \binom{n+k}{k} \right\}^{1/2} \rho^{2pk} \\
&\qquad \times \sqrt{(m+k)!} \, |f_{m+k}|_p \sqrt{(n+k)!} \, |g_{n+k}|_p \\
&\leq \sum_{m+n=l} \binom{m+n}{m}^{1/2} \rho^{\alpha m + \beta n} \sum_{k=0}^{\infty} \binom{m+n+2k}{2k}^{1/2} \rho^{(\alpha+\beta+2p)k} \\
&\qquad \times \sqrt{(m+k)!} \, |f_{m+k}|_{p+\alpha} \sqrt{(n+k)!} \, |g_{n+k}|_{p+\beta} \\
&= \sum_{m+n=l} \binom{l}{m}^{1/2} \rho^{\alpha m + \beta n} \sum_{k=0}^{\infty} \binom{l+2k}{2k}^{1/2} \rho^{(\alpha+\beta+2p)k} \\
&\qquad \times \sqrt{(m+k)!} \, |f_{m+k}|_{p+\alpha} \sqrt{(n+k)!} \, |g_{n+k}|_{p+\beta} \, .
\end{aligned}$$

For simplicity we put

$$J_{m,n} = \sum_{k=0}^{\infty} \binom{l+2k}{2k}^{1/2} \rho^{(\alpha+\beta+2p)k} \sqrt{(m+k)!} \, |f_{m+k}|_{p+\alpha} \sqrt{(n+k)!} \, |g_{n+k}|_{p+\beta} \, .$$

Then, applying the Schwarz inequality repeatedly, we obtain

$$\begin{aligned}
l! \, |h_l|_p^2 &\leq \left| \sum_{m+n=l} \binom{l}{m}^{1/2} \rho^{\alpha m + \beta n} J_{m,n} \right|^2 \\
&\leq \sum_{m+n=l} \binom{l}{m} \rho^{2\alpha m + 2\beta n} \sum_{m+n=l} J_{m,n}^2 \\
&= (\rho^{2\alpha} + \rho^{2\beta})^l \sum_{m+n=l} J_{m,n}^2 .
\end{aligned} \tag{3.54}$$

We continue the estimate of $J_{m,n}$.

$$\begin{aligned} J_{m,n}^2 &\le \sum_{k=0}^{\infty} \binom{l+2k}{2k} \rho^{2k(\alpha+\beta+2p)} \sum_{k=0}^{\infty} (m+k)!\, |f_{m+k}|_{p+\alpha}^2 \,(n+k)!\, |g_{n+k}|_{p+\beta}^2 \\ &\le \sum_{k=0}^{\infty} \binom{l+2k}{2k} \rho^{2k(\alpha+\beta+2p)} \sum_{k=0}^{\infty} (m+k)!\, |f_{m+k}|_{p+\alpha}^2 \,\|\psi\|_{p+\beta}^2 \\ &\le \sum_{k=0}^{\infty} \binom{l+2k}{2k} \rho^{2k(\alpha+\beta+2p)} \|\phi\|_{p+\alpha}^2 \,\|\psi\|_{p+\beta}^2 , \end{aligned}$$

where the last quantity is independent of m, n. Consequently, (3.54) becomes

$$\begin{aligned} l!\, |h_l|_p^2 &\le (\rho^{2\alpha} + \rho^{2\beta})^l \sum_{m+n=l} J_{m,n}^2 \\ &\le (l+1)(\rho^{2\alpha} + \rho^{2\beta})^l \,\|\phi\|_{p+\alpha}^2 \,\|\psi\|_{p+\beta}^2 \sum_{k=0}^{\infty} \binom{l+2k}{2k} \rho^{2k(\alpha+\beta+2p)}, \end{aligned}$$

which proves (3.53). qed

Lemma 3.5.5 *For $s, t \in \mathbb{C}$ with $|s|, |t|, |s+t| < 1$, it holds that*

$$\sum_{k,l=0}^{\infty} \binom{l+k}{k} (l+1) s^k t^l = \frac{1-s}{(1-s-t)^2}.$$

PROOF. We first note

$$(1-s-t)^{-1} = \sum_{m=0}^{\infty} (s+t)^m = \sum_{m=0}^{\infty} \sum_{k+l=m} \binom{m}{l} s^k t^l = \sum_{k,l=0}^{\infty} \binom{k+l}{l} s^k t^l.$$

Then we need only to apply the differential operator $\dfrac{\partial}{\partial t} - s\dfrac{\partial}{\partial s}$ on both sides. qed

Theorem 3.5.6 *Let $p, \alpha, \beta \ge 0$ satisfy $\rho^{2\alpha} + \rho^{\alpha+\beta+2p} + \rho^{2\beta} < 1$. Then,*

$$\|\phi\psi\|_p \le \frac{\sqrt{1-\rho^{\alpha+\beta+2p}}}{1-\rho^{2\alpha}-\rho^{\alpha+\beta+2p}-\rho^{2\beta}} \|\phi\|_{p+\alpha} \|\psi\|_{p+\beta}, \qquad \phi, \psi \in (E).$$

In particular, (E) is closed under pointwise multiplication and the multiplication is a continuous bilinear map from $(E) \times (E)$ into (E).

PROOF. We retain the notations used in Proposition 3.5.3. By Lemma 3.5.4 we have

$$\begin{aligned} \|\phi\psi\|_p^2 &= \sum_{l=0}^{\infty} l!\, |h_l|_p^2 \\ &\le \sum_{l=0}^{\infty} (l+1)(\rho^{2\alpha} + \rho^{2\beta})^l \,\|\phi\|_{p+\alpha}^2 \,\|\psi\|_{p+\beta}^2 \sum_{k=0}^{\infty} \binom{l+2k}{2k} \rho^{2k(\alpha+\beta+2p)} \\ &\le \sum_{k,l=0}^{\infty} \binom{l+k}{k} (l+1) \rho^{k(\alpha+\beta+2p)} (\rho^{2\alpha} + \rho^{2\beta})^l \,\|\phi\|_{p+\alpha}^2 \,\|\psi\|_{p+\beta}^2 . \end{aligned}$$

Then, applying Lemma 3.5.5, we obtain

$$\|\phi\psi\|_p^2 \le \frac{1-\rho^{\alpha+\beta+2p}}{(1-\rho^{2\alpha}-\rho^{\alpha+\beta+2p}-\rho^{2\beta})^2}\,\|\phi\|_{p+\alpha}^2\,\|\psi\|_{p+\beta}^2,$$

which completes the proof. qed

The pointwise multiplication of two white noise functionals is also called *Wiener product*. The next result is immediate from Theorem 3.5.6.

Corollary 3.5.7 *For any $p \ge 0$ and $\alpha > 0$ there exist $\beta > 0$ and $C = C(\alpha,\beta)$ such that*

$$\|\phi\psi\|_p \le C\,\|\phi\|_{p+\alpha}\,\|\psi\|_{p+\beta}, \qquad \phi,\psi \in (E).$$

In particular, for any $p \ge 0$ there exist $q > 0$ and $C = C(q) > 0$ such that

$$\|\phi\psi\|_p \le C\,\|\phi\|_{p+q}\,\|\psi\|_{p+q}, \qquad \phi,\psi \in (E).$$

In fact, the above q and C are given as

$$C = \frac{\sqrt{1-\rho^{2q}}}{1-3\rho^{2q}}, \qquad \rho^{-q} > \sqrt{3}.$$

We then consider the product of generalized functional $\Phi \in (E)^*$ and a test functional $\phi \in (E)$. By Theorem 3.5.6,

$$\psi \mapsto \langle\langle \Phi, \phi\psi \rangle\rangle, \qquad \psi \in (E),$$

is a continuous linear functional on (E). Hence there exists a unique generalized functional $\Psi = \Psi(\Phi,\phi) \in (E)^*$ such that

$$\langle\langle \Psi, \psi \rangle\rangle = \langle\langle \Phi, \phi\psi \rangle\rangle.$$

Since $\Psi(\phi,\psi) = \phi\psi = \psi\phi$ for all $\phi,\psi \in (E)$, we write $\Psi(\Phi,\phi) = \Phi\phi = \phi\Phi$. We have thus defined the product $\Phi\phi = \phi\Phi$ uniquely by the formula:

$$\langle\langle \Phi\phi, \psi \rangle\rangle = \langle\langle \Phi, \phi\psi \rangle\rangle, \qquad \Phi \in (E)^*, \quad \phi,\psi \in (E). \tag{3.55}$$

Theorem 3.5.8 *If $p,\alpha,\beta \ge 0$ satisfy $\rho^{2\alpha} + \rho^{\alpha+\beta+2p} + \rho^{2\beta} < 1$, then*

$$\|\phi\Phi\|_{-(p+\beta)} \le \frac{\sqrt{1-\rho^{\alpha+\beta+2p}}}{1-\rho^{2\alpha}-\rho^{\alpha+\beta+2p}-\rho^{2\beta}}\,\|\phi\|_{p+\alpha}\,\|\Phi\|_{-p}, \quad \phi \in (E), \quad \Phi \in (E)^*.$$

PROOF. First observe that

$$\begin{aligned}
\|\phi\Phi\|_{-(p+\beta)} &= \sup\{|\langle\langle \phi\Phi, \psi \rangle\rangle| \,;\, \|\psi\|_{p+\beta} \le 1\} \\
&= \sup\{|\langle\langle \Phi, \phi\psi \rangle\rangle| \,;\, \|\psi\|_{p+\beta} \le 1\} \\
&\le \sup\{\|\Phi\|_{-p}\,\|\phi\psi\|_p \,;\, \|\psi\|_{p+\beta} \le 1\}.
\end{aligned}$$

Then, applying Theorem 3.5.6, we obtain the desired result. qed

Corollary 3.5.9 *The bilinear map $(\Phi, \phi) \mapsto \Phi\phi$ becomes a separately continuous bilinear map from $(E)^* \times (E)$ into $(E)^*$. In particular, each $\Phi \in (E)^*$, as multiplication, is a continuous linear operator from (E) into $(E)^*$.*

The Wiener product formula is also valid for this generalized case. By a simple calculation with (3.55) and Theorem 3.5.2 we obtain

Theorem 3.5.10 *Let $\Phi \in (E)^*$ and $\phi \in (E)$ be given as*

$$\Phi(x) = \left\langle :x^{\otimes m}:, F \right\rangle, \qquad \phi(x) = \left\langle :x^{\otimes n}:, f \right\rangle,$$

where $F \in (E_{\mathbb{C}}^{\otimes m})^*_{\mathrm{sym}}$ and $f \in E_{\mathbb{C}}^{\widehat{\otimes} n}$. Then,

$$(\phi\Phi)(x) = \sum_{k=0}^{m \wedge n} k! \binom{m}{k} \binom{n}{k} \left\langle :x^{\otimes(m+n-2k)}:, F \widehat{\otimes}_k f \right\rangle$$

We next prove a simple formula announced at the end of §2.3.

Proposition 3.5.11 *Let $\{\xi_j\}_{j=0}^{\infty}$ be an orthonormal system for H. For $\mathbf{n} = (n_0, n_1, \cdots)$ with $|\mathbf{n}| = n_0 + n_1 + \cdots = n < \infty$ we put*

$$\phi_{\mathbf{n}}(x) = (n_0! n_1! \cdots)^{-1/2} \left\langle :x^{\otimes n}:, \xi_0^{\otimes n_0} \widehat{\otimes} \xi_1^{\otimes n_1} \widehat{\otimes} \cdots \right\rangle.$$

Then,

$$\phi_{\mathbf{n}}(x) = \prod_{j=0}^{\infty} (n_j! 2^{n_j})^{-1/2} H_{n_j} \left(\frac{\langle x, \xi_j \rangle}{\sqrt{2}} \right).$$

PROOF. By Lemma 2.2.7 we have

$$\begin{aligned} \prod_{j=0}^{\infty} (n_j! 2^{n_j})^{-1/2} H_{n_j} \left(\frac{\langle x, \xi_j \rangle}{\sqrt{2}} \right) &= \prod_{j=0}^{\infty} (n_j! 2^{n_j})^{-1/2} 2^{n_j/2} \left\langle :x^{\otimes n_j}:, \xi_j^{\otimes n_j} \right\rangle \\ &= \prod_{j=0}^{\infty} (n_j!)^{-1/2} \left\langle :x^{\otimes n_j}:, \xi_j^{\otimes n_j} \right\rangle. \end{aligned}$$

Noting $\langle \xi_0, \xi_1 \rangle = 0$, we apply Theorem 3.5.2 to get

$$\left\langle :x^{\otimes n_0}:, \xi_0^{\otimes n_0} \right\rangle \left\langle :x^{\otimes n_1}:, \xi_1^{\otimes n_1} \right\rangle = \left\langle :x^{\otimes(n_0+n_1)}:, \xi_0^{\otimes n_0} \widehat{\otimes} \xi_1^{\otimes n_1} \right\rangle.$$

Using this argument repeatedly, we obtain

$$\prod_{j=0}^{\infty} \left\langle :x^{\otimes n_j}:, \xi_j^{\otimes n_j} \right\rangle = \left\langle :x^{\otimes n}:, \xi_0^{\otimes n_0} \widehat{\otimes} \xi_1^{\otimes n_1} \widehat{\otimes} \cdots \right\rangle,$$

where $n = n_0 + n_1 + \cdots$. This completes the proof. qed

Finally, as an application of Theorem 3.5.6 we shall prove the following

Theorem 3.5.12 *Let $f \in \bigcup_{p>1} L^p(E^*, \mu; \mathbb{C})$. Then,*

$$\Phi_f : \phi \mapsto \int_{E^*} f(x)\phi(x)\,\mu(dx), \qquad \phi \in (E),$$

is a continuous linear functional on (E), namely $\Phi_f \in (E)^*$. Moreover,

$$\|\Phi_f\|_{-(n-1)r} \le \frac{\sqrt{1-\rho^{2r}}}{1-3\rho^{2r}} \|f\|_{L^p(E^*)}, \qquad f \in L^p(E^*, \mu; \mathbb{C}),$$

where $r > 0$ and an integer $n \ge 1$ are taken as

$$\rho^{-r} > \sqrt{3}, \qquad 2n \ge \left(1 - \frac{1}{p}\right)^{-1}.$$

PROOF. Take $r > 0$ such that $\rho^{-r} > \sqrt{3}$ and put

$$C = \frac{\sqrt{1-\rho^{2r}}}{1-3\rho^{2r}}.$$

Then, using Corollary 3.5.7 repeatedly, we obtain for any $\phi \in (E)$,

$$\|\phi^n\|_0 \le C^{n-1} \|\phi\|_r \|\phi\|_{2r} \cdots \|\phi\|_{(n-1)r} \|\phi\|_{(n-1)r} \le C^n \|\phi\|_{(n-1)r}^n,$$

where we used obvious facts:

$$C = \frac{\sqrt{1-\rho^{2r}}}{1-3\rho^{2r}} > 1 \qquad \text{and} \qquad \|\phi\|_p \le \|\phi\|_q \quad \text{whenever} \quad p \le q.$$

It then follows that

$$\int_{E^*} |\phi(x)|^{2n} \mu(dx) = \|\phi^n\|_0^2 \le C^{2n} \|\phi\|_{(n-1)r}^{2n},$$

and we obtain

$$\|\phi\|_{L^{2n}(E^*)} \le C \|\phi\|_{(n-1)r}, \qquad n = 1, 2, \cdots. \tag{3.56}$$

Choose $q \ge 1$ and an integer n satisfying

$$\frac{1}{p} + \frac{1}{q} = 1, \qquad 2n \ge q = \left(1 - \frac{1}{p}\right)^{-1}.$$

Then, the Hölder inequality yields

$$\begin{aligned} |\langle\!\langle \Phi_f, \phi \rangle\!\rangle| &\le \int_{E^*} |f(x)\phi(x)|\,\mu(dx) \\ &\le \|f\|_{L^p(E^*)} \|\phi\|_{L^q(E^*)} \\ &\le \|f\|_{L^p(E^*)} \|\phi\|_{L^{2n}(E^*)}. \end{aligned}$$

Consequently, by (3.56) we obtain

$$|\langle\!\langle \Phi_f, \phi \rangle\!\rangle| \le C \|f\|_{L^p(E^*)} \|\phi\|_{(n-1)r}, \qquad f \in L^p(E^*, \mu; \mathbb{C}), \quad \phi \in (E).$$

This completes the proof. qed

It will be proved in Corollary 5.6.15 that $\Phi_f = 0$ implies $f = 0$. Therefore we have a canonical injection:

$$\bigcup_{p>1} L^p(E^*, \mu; \mathbb{C}) \to (E)^*,$$

which is continuous on each $L^p(E^*, \mu; \mathbb{C})$. Therefore, it is continuous with respect to the inductive limit topology on $\bigcup_{p>1} L^p(E^*, \mu; \mathbb{C})$.

3.6 Characterization theorems

In this section we shall prove the characterization theorems for generalized and test functionals in terms of their S-transform.

Theorem 3.6.1 *Let F be a $\mathbf{C}$-valued function on $E_{\mathbf{C}}$. Then $F = S\Phi$ for some $\Phi \in (E)^*$ if and only if*

(i) *for fixed $\xi, \eta \in E_{\mathbf{C}}$, the function $z \mapsto F(z\xi + \eta)$, $z \in \mathbf{C}$, is an entire holomorphic function on $\mathbf{C}$;*

(ii) *there exist $C \geq 0$, $K \geq 0$ and $p \in \mathbf{R}$ such that*

$$|F(\xi)| \leq C \exp\left(K\, |\xi|_p^2\right), \qquad \xi \in E_{\mathbf{C}}.$$

Theorem 3.6.2 *Let F be a $\mathbf{C}$-valued function on $E_{\mathbf{C}}$. Then $F = S\phi$ for some $\phi \in (E)$ if and only if*

(i) *for fixed $\xi, \eta \in E_{\mathbf{C}}$, the function $z \mapsto F(z\xi + \eta)$, $z \in \mathbf{C}$, is an entire holomorphic function on $\mathbf{C}$;*

(ii) *for any $p \geq 0$ and $\epsilon > 0$ there exists $C \geq 0$ such that*

$$|F(\xi)| \leq C \exp\left(\epsilon\, |\xi|_{-p}^2\right), \qquad \xi \in E_{\mathbf{C}}.$$

We recall some general results on Gâteaux differential. Let $\mathfrak{X}$ be a locally convex space over $\mathbf{C}$. The next remark is obvious.

Lemma 3.6.3 *For a $\mathbf{C}$-valued function F on $\mathfrak{X}$ the following conditions are equivalent:*

(i) *$z \mapsto F(z\xi + \eta)$ is holomorphic on $\mathbf{C}$ for any $\xi, \eta \in \mathfrak{X}$;*

(ii) *$(z_1, \cdots, z_n) \mapsto F(z_1\xi_1 + \cdots + z_n\xi_n)$ is holomorphic on $\mathbf{C}^n$ for any $\xi_1, \cdots, \xi_n \in \mathfrak{X}$, $n = 2, 3, \cdots$.*

Let $\mathfrak{A}(\mathfrak{X})$ be the space of functions $F : \mathfrak{X} \to \mathbf{C}$ satisfying the conditions stated in the above lemma. In general, for a $\mathbf{C}$-valued function F on $\mathfrak{X}$ the Gâteaux derivative at $\eta \in \mathfrak{X}$ in the direction along $\xi \in \mathfrak{X}$ is defined by

$$D_\xi F(\eta) = \left.\frac{\partial}{\partial z}\right|_{z=0} F(z\xi + \eta).$$

Then we note the following obvious result.

Lemma 3.6.4 *Every $F \in \mathfrak{A}(\mathfrak{X})$ is Gâteaux differentiable on $\mathfrak{X}$ and $D_\xi F \in \mathfrak{A}(\mathfrak{X})$ for any $\xi \in \mathfrak{X}$.*

Lemma 3.6.5 *Let $F \in \mathfrak{A}(\mathfrak{X})$ and $\eta \in \mathfrak{X}$. Then,*

$$(\xi_1, \cdots, \xi_n) \mapsto D_{\xi_1} \cdots D_{\xi_n} F(\eta), \qquad \xi_1, \cdots, \xi_n \in \mathfrak{X},$$

is a symmetric n-linear form on $\mathfrak{X}$.

PROOF. We first prove that $\xi \mapsto D_\xi F(\eta)$ is a linear form on $\mathfrak{X}$ for a fixed $\eta \in \mathfrak{X}$. It is obvious that

$$D_{z\xi}F(\eta) = zD_\xi F(\eta), \qquad z \in \mathbf{C}, \quad \xi \in \mathfrak{X}.$$

Let $\xi_1, \xi_2 \in \mathfrak{X}$. Since $F \in \mathfrak{A}(\mathfrak{X})$ by assumption,

$$g(z_1, z_2) = F(z_1\xi_1 + z_2\xi_2 + \eta)$$

is holomorphic on $\mathbf{C} \times \mathbf{C}$. We therefore have

$$g(z, z) = g(0,0) + z\frac{\partial g}{\partial z_1}(0,0) + z\frac{\partial g}{\partial z_2}(0,0) + o(z).$$

Hence

$$F(z\xi_1 + z\xi_2 + \eta) = F(\eta) + zD_{\xi_1}F(\eta) + zD_{\xi_2}F(\eta) + o(z),$$

and consequently,

$$\begin{aligned} D_{\xi_1+\xi_2}F(\eta) &= \lim_{z\to 0}\frac{F(z\xi_1 + z\xi_2 + \eta) - F(\eta)}{z} \\ &= D_{\xi_1}F(\eta) + D_{\xi_2}F(\eta) + \lim_{z\to 0}\frac{o(z)}{z} \\ &= D_{\xi_1}F(\eta) + D_{\xi_2}F(\eta). \end{aligned}$$

We have thus proved that $\xi \mapsto D_\xi F(\eta)$ is linear. It follows from Lemma 3.6.4 that $\xi_1 \mapsto D_{\xi_1}D_{\xi_2}\cdots D_{\xi_n}F(\eta)$ is linear. Hence it is sufficient to show the commutativity: $D_{\xi_1}D_{\xi_2}F(\eta) = D_{\xi_2}D_{\xi_1}F(\eta)$. But this is obvious because $(z_1, z_2) \mapsto F(z_1\xi_1 + z_2\xi_2 + \eta)$ is holomorphic on $\mathbf{C} \times \mathbf{C}$. qed

We need an elementary result.

Lemma 3.6.6 *Let f be an entire holomorphic function on $\mathbf{C}$ with Taylor expansion $f(z) = \sum_{n=0}^\infty a_n z^n$. Assume that*

$$|f(z)| \le Ce^{K|z|^2}, \qquad z \in \mathbf{C},$$

for some $C \ge 0$ and $K \ge 0$. Then,

$$|a_n| \le C\left(\frac{2eK}{n}\right)^{n/2}, \qquad n = 0, 1, 2, \cdots. \tag{3.57}$$

PROOF. Let $R > 0$. It follows from the Cauchy integral formula that

$$a_n = \frac{1}{2\pi i}\int_{|z|=R}\frac{f(z)}{z^{n+1}}dz.$$

Hence by assumption we have

$$|a_n| \le \frac{1}{2\pi}\int_{|z|=R}\frac{|f(z)|}{|z|^{n+1}}|dz| \le CR^{-n}e^{KR^2}, \qquad R > 0.$$

On the other hand, one may easily see that

$$\min_{R>0} R^{-n}e^{KR^2} = \left(\frac{2eK}{n}\right)^{n/2}, \tag{3.58}$$

from which (3.57) follows. qed

Lemma 3.6.7 *Assume that a* $\mathbf{C}$*-valued function* F *on* $E_{\mathbf{C}}$ *satisfies the following two conditions:*

(i) *for fixed* $\xi, \eta \in E_{\mathbf{C}}$*, the function* $z \mapsto F(z\xi + \eta)$*,* $z \in \mathbf{C}$*, is an entire holomorphic function on* $\mathbf{C}$*.*

(ii) *there exist* $C \geq 0$*,* $K \geq 0$ *and* $p \in \mathbf{R}$ *such that*

$$|F(\xi)| \leq C \exp\left(K\, |\xi|_p^2\right), \qquad \xi \in E_{\mathbf{C}}.$$

Then

$$F_n(\xi_1, \cdots, \xi_n) = \frac{1}{n!} D_{\xi_1} \cdots D_{\xi_n} F(0)$$

becomes a continuous n*-linear form on* $E_{\mathbf{C}}$ *and*

$$|F_n|_{-(p+1)} \leq \frac{C}{\sqrt{n!}} (2e^2\delta^2 K)^{n/2}.$$

PROOF. It follows from Lemmas 3.6.3 and 3.6.5 that $F_n(\xi_1, \cdots, \xi_n)$ is a symmetric n-form on $E_{\mathbf{C}}$. We put

$$A_n(\xi) = F_n(\xi, \cdots, \xi).$$

Then the Taylor expansion of the entire holomorphic function $z \mapsto F(z\xi)$ is given by

$$F(z\xi) = \sum_{n=0}^{\infty} D_\xi^n F(0) \frac{z^n}{n!} = \sum_{n=0}^{\infty} A_n(\xi) z^n, \qquad z \in \mathbf{C}.$$

In view of Lemma 3.6.6 we get

$$|A_n(\xi)| \leq C \left(\frac{2eK\, |\xi|_p^2}{n} \right)^{n/2} \leq C \left(\frac{2eK}{n} \right)^{n/2} |\xi|_p^n.$$

By (A.3) in Appendix we get

$$\begin{aligned} &\sup \left\{ |F_n(\xi_1, \cdots, \xi_n)|;\ |\xi_1|_p \leq 1, \cdots, |\xi_n|_p \leq 1 \right\} \\ &\leq \frac{n^n}{n!} C \left(\frac{2eK}{n} \right)^{n/2} = \frac{C}{\sqrt{n!}} (2eK)^{n/2} \sqrt{\frac{n^n}{n!}} \leq \frac{C}{\sqrt{n!}} (2e^2K)^{n/2}. \end{aligned} \tag{3.59}$$

On the other hand,

$$\begin{aligned} |F_n|_{-(p+1)}^2 &= \sum_{j_1, \cdots, j_n = 0}^{\infty} |F_n(e_{j_1}, \cdots, e_{j_n})|^2 \lambda_{j_1}^{-2(p+1)} \cdots \lambda_{j_n}^{-2(p+1)} \\ &= \sum_{j_1, \cdots, j_n = 0}^{\infty} |F_n(\lambda_{j_1}^{-p} e_{j_1}, \cdots, \lambda_{j_n}^{-p} e_{j_n})|^2 \lambda_{j_1}^{-2} \cdots \lambda_{j_n}^{-2}. \end{aligned}$$

Since $\{\lambda_j^{-p} e_j\}_{j=0}^{\infty}$ is a complete orthonormal basis for E_p, applying (3.59), we obtain

$$|F_n|_{-(p+1)}^2 \leq \frac{C^2}{n!} (2e^2K)^n \sum_{j_1, \cdots, j_n = 0}^{\infty} \lambda_{j_1}^{-2} \cdots \lambda_{j_n}^{-2} = \frac{C^2}{n!} (2e^2K)^n \delta^{2n} = \frac{C^2}{n!} (2e^2\delta^2 K)^n.$$

This completes the proof. qed

Proof of Theorem 3.6.1. Suppose that F satisfies the conditions (i) and (ii) in Theorem 3.6.1. Those conditions are the same as ones in Lemma 3.6.7. We then put

$$F_n(\xi_1, \cdots, \xi_n) = \frac{1}{n!} D_{\xi_1} \cdots D_{\xi_n} F(0).$$

It then follows from Lemma 3.6.7 that

$$|F_n|_{-(p+1)} \le \frac{C}{\sqrt{n!}} (2e^2\delta^2 K)^{n/2}$$

and therefore for any $q \ge 0$,

$$|F_n|_{-(p+q+1)} \le \rho^{qn} |F_n|_{-(p+1)} \le \frac{C}{\sqrt{n!}} \left(2e^2\delta^2 K \rho^{2q}\right)^{n/2}.$$

Consequently,

$$\sum_{n=0}^{\infty} n! \, |F_n|^2_{-(p+q+1)} \le C^2 \sum_{n=0}^{\infty} \left(2e^2\delta^2 K \rho^{2q}\right)^n.$$

Now take $q \ge 0$ large enough to hold $2e^2\delta^2 K \rho^{2q} < 1$. Then the last series converges and we see from Theorem 3.1.6 that

$$\Phi(x) = \sum_{n=0}^{\infty} \left\langle :x^{\otimes n}: , F_n \right\rangle$$

belongs to $(E)^*$ with $\|\Phi\|_{-(p+q+1)} < \infty$. Finally by Lemma 3.3.5 we see that

$$S\Phi(\xi) = \sum_{n=0}^{\infty} \left\langle F_n , \xi^{\otimes n} \right\rangle = \sum_{n=0}^{\infty} A_n(\xi) = F(\xi).$$

The converse assertion is immediate from Lemma 3.3.6 and Theorem 3.3.7. qed

Proof of Theorem 3.6.2. We first prove the "only if" part. (i) is immediate from Theorem 3.6.1. Put

$$\phi(x) = \sum_{n=0}^{\infty} \left\langle :x^{\otimes n}: , f_n \right\rangle.$$

Then,

$$F(\xi) = S\phi(\xi) = \sum_{n=0}^{\infty} \left\langle f_n , \xi^{\otimes n} \right\rangle.$$

For any $q \ge 0$ we then observe

$$\begin{aligned}
|F(\xi)| &\le \sum_{n=0}^{\infty} |f_n|_p \, |\xi|_{-p}^n \\
&\le \sum_{n=0}^{\infty} \rho^{qn} \, |f_n|_{p+q} \, |\xi|_{-p}^n \\
&\le \left(\sum_{n=0}^{\infty} n! \, |f_n|^2_{p+q} \right)^{1/2} \left(\sum_{n=0}^{\infty} \frac{\rho^{2qn} \, |\xi|_{-p}^{2n}}{n!} \right)^{1/2} \\
&= \|\phi\|_{p+q} \exp\left(\frac{\rho^{2q}}{2} |\xi|^2_{-p} \right).
\end{aligned}$$

Hence for a given $p \geq 0$ and $\epsilon > 0$ we may find $q \geq 0$ such that $\rho^{2q} < 2\epsilon$. Then we obtain

$$|F(\xi)| \leq \|\phi\|_{p+q} \exp\left(\epsilon\, |\xi|_{-p}^2\right).$$

Namely, F satisfies the condition (ii).

Conversely, let us assume (i) and (ii) in Theorem 3.6.2. By Theorem 3.6.1 there exists $\Phi \in (E)^*$ such that $F = S\Phi$, where

$$\Phi(x) = \sum_{n=0}^{\infty} \left\langle :x^{\otimes n}:, F_n \right\rangle,$$

and

$$F_n(\xi_1, \cdots, \xi_n) = \frac{1}{n!} D_{\xi_1} \cdots D_{\xi_n} F(0).$$

Let $p \geq 0$ be arbitrary and assume $\epsilon > 0$ satisfies $2e^2\delta^2\epsilon < 1$. By assumption, there exists $C \geq 0$ such that

$$|F(\xi)| \leq C \exp\left(\epsilon\, |\xi|_{-(p+1)}^2\right).$$

It then follows from Lemma 3.6.7 that

$$|F_n|_p \leq \frac{C}{\sqrt{n!}} (2e^2\delta^2\epsilon)^{n/2}.$$

Hence

$$\|\Phi\|_p^2 = \sum_{n=0}^{\infty} n!\, |F_n|_p^2 \leq C^2 \sum_{n=0}^{\infty} (2e^2\delta^2\epsilon)^n < \infty.$$

This means that $\Phi \in (E)$. qed

Bibliographical Notes

The construction of generalized white noise functionals in §3.1 is essentially due to Kubo-Takenaka [1], [2] where they also discuss the case of T having no topological structure. However, many advantages have been found in our setup and it has become widely accepted. Note also that the notion of standard construction makes the argument very clear.

Theorem 3.2.1 is known as Kubo-Yokoi's continuous version theorem, which was first proved in Kubo-Yokoi [1], see also Yokoi [1]. Our proof, being different from theirs, is extracted from Obata [9]. There are some closely related results: Lee [4], [5] discussed analyticity of $\phi \in (E)$ when regarded as a function on a Hilbert space $E_{-p} \subset E^*$; Kondrat'ev-Samoylenko [2] discussed analyticity of test functionals in a different context which is much in common with Lee's idea, see also Kondrat'ev [1]. The continuous version theorem is effectively applied to the description of positive generalized white noise functionals. In fact, Yokoi [1], [2] gave a complete answer which supplemented the earlier work of Potthoff [1].

The S-transform in §3.3 is an extension of the S-transform originally introduced by Kubo-Takenaka [2]. In some literature e.g., Kuo [3], the S-transform is also called the U-functional. However, similar concepts have appeared in various contexts by many authors, e.g., Gross [2], Hida-Ikeda [1], M. Krée [1], Krée-Rączka [1], Kuo [1] and Wiener [2]. In particular, T-transform due to Hida and Ikeda will be discussed in §5.5.

Contraction of tensor product in §3.4 is a standard notion. The norm estimates obtained there are improvements of those in Kubo-Takenaka [2].

The Wiener product formula discussed in §3.5 is also well-known. The norm estimates (Theorems 3.5.6 and 3.5.8) are, up to now, the best among similar results due to Kubo-Takenaka [2], Hida-Potthoff-Streit [1], Hida-Kuo-Potthoff-Streit [1], Potthoff-Yan [1], Yan [1] and Yokoi [1]. However, our discussion is still not enough to cover so-called hypercontractivity on Gaussian space, see e.g., Lindsay [1].

The characterization theorem for generalized functionals was first obtained by Potthoff-Streit [1] and for test functionals by Kuo-Potthoff-Streit [1]. Yan [1] has sharpened their discussion and Zhang [1] gave another characterization using a variant of the S-transform called Hermite transform.

Some special white noise functionals are discussed in detail. See Kondrat'ev-Streit [2] for positive definite functionals, Kubo-Kuo [1] for cylindrical functionals, Kuo [5] for Donsker's delta functions and Liu-Yan [1] for homogeneous functionals. To develop de Rham-Hodge-Kodaira theory for infinite dimension Arai-Mitoma [1], [2] discussed $\bigwedge^n K$-valued functionals, where $\bigwedge^n K$ is the n-fold antisymmetric tensor product of a Hilbert space K. In the recent work of Obata [10] has been established a general theory of white noise functionals with values in a standard CH-space. This topic will be reviewed briefly in §6.3.

The generalized white noise functionals are well suited to some problems of quantum physics. In fact, an application to physics is one of the most important motivations of the white noise calculus. An application to the Feynman path integral has started in Hida-Streit [1], [2]. The discussion is considerably developed by de Faria-Potthoff-Streit [1] and Khandekar-Streit [1]. Applications to infinite dimensional Dirichlet forms have been discussed in Albeverio-Hida-Potthoff-Röckner-Streit [1], [2], Hida-Potthoff-Streit [1], [2], Razafimanantena [1], and further discussion toward quantum field theory is found in Albeverio-Hida-Potthoff-Streit [1], and Potthoff-Streit [2]. For application to stochastic integration and stochastic differential equations, see e.g., Holden-Lindstrøm-Øksendal-Ubøe-Zhang [1], Kubo [1], Kubo-Takenaka [3], Kuo-Russek [1]. Some problems in Gaussian random fields are also discussed by Hida [5], [7], [8]. These topics being somehow beyond the purpose of these lecture notes, no effort is made in collecting full references. For further information see also the monograph by Hida-Kuo-Potthoff-Streit [1].

Recently, bigger spaces of generalized white noise functionals have been discussed together with similar characterization theorems, see e.g., Huang-Ren [1], Kondrat'ev-Streit [1], Ouerdiane [1], Streit [2], Streit-Westerkamp [1], Yan [6].

There seems to be a close connection between white noise calculus and the theory of infinite dimensional holomorphy through the characterization theorems in §3.6. In fact, Ouerdiane [1] discusses this direction. For a brief introduction to the infinite dimensional holomorphy, see e.g., Colombeau [1] and references cited therein.

Chapter 4

Operator Theory

4.1 Hida's differential operator

We now introduce the most fundamental operator in white noise calculus. As is discussed in §3.2, each $\phi \in (E)$ admits a Wiener-Itô expansion:

$$\phi(x) = \sum_{n=0}^{\infty} \left\langle :x^{\otimes n}:, f_n \right\rangle, \qquad f_n \in E_{\mathbb{C}}^{\widehat{\otimes} n}, \quad x \in E^*, \tag{4.1}$$

where the series is absolutely convergent at any $x \in E^*$. Moreover,

$$\|\phi\|_p^2 = \sum_{n=0}^{\infty} n! \, |f_n|_p^2 < \infty$$

for all $p \in \mathbb{R}$. We first prove

Theorem 4.1.1 *Assume that $\phi \in (E)$ is given as in (4.1). Then, the series*

$$D_y\phi(x) = \sum_{n=0}^{\infty} n \left\langle :x^{\otimes(n-1)}:, y\widehat{\otimes}_1 f_n \right\rangle, \qquad x, y \in E^*, \tag{4.2}$$

converges absolutely. Moreover, $D_y\phi \in (E)$ and, for any $p \geq 0$ and $q > 0$ we have

$$\|D_y\phi\|_p \leq \left(\frac{\rho^{-2q}}{-2qe \log \rho} \right)^{1/2} |y|_{-(p+q)} \|\phi\|_{p+q}, \qquad \phi \in (E). \tag{4.3}$$

In particular, $D_y \in \mathcal{L}((E), (E))$.

PROOF. Note first that

$$\left| y\widehat{\otimes}_1 f_n \right|_p \leq \rho^{q(n-1)} |y|_{-(p+q)} |f_n|_{p+q}$$

whenever $p \geq 0$, $q \geq 0$ and $|y|_{-(p+q)} < \infty$. Hence, if $q > 0$ we have

$$\begin{aligned}
\sum_{n=0}^{\infty} (n-1)! \left| ny\widehat{\otimes}_1 f_n \right|_p^2 &= \sum_{n=0}^{\infty} n \cdot n! \left| y\widehat{\otimes}_1 f_n \right|_p^2 \\
&\leq \sum_{n=0}^{\infty} n \cdot n! \rho^{2q(n-1)} |y|_{-(p+q)}^2 |f_n|_{p+q}^2 \\
&\leq \sup_{n \geq 0} n\rho^{2q(n-1)} |y|_{-(p+q)}^2 \|\phi\|_{p+q}^2 < \infty.
\end{aligned}$$

Therefore, by Proposition 3.1.5 and Lemma 3.2.4 the series (4.2) converges absolutely at each point $x \in E^*$ and $D_y\phi \in (E)$.

Next we prove (4.3). Using the elementary fact that

$$\max_{x\geq 0} xe^{-\beta x} = \frac{1}{\beta e}, \qquad \beta > 0, \tag{4.4}$$

we obtain

$$\sup_{n\geq 0} n\rho^{2q(n-1)} \leq \frac{\rho^{-2q}}{-2qe\log\rho}.$$

Consequently, for any $\phi \in (E)$ it holds that

$$\|D_y\phi\|_p^2 = \sum_{n=0}^{\infty}(n-1)! \left|ny\hat{\otimes}_1 f_n\right|_p^2 \leq \frac{\rho^{-2q}}{-2qe\log\rho} |y|_{-(p+q)}^2 \|\phi\|_{p+q}^2,$$

as desired. (If $|y|_{-(p+q)} = \infty$, the assertion is obvious.) qed

It is proved in Theorem 3.5.6 that (E) is closed under the pointwise multiplication. Our next task is to prove that D_y is a derivation, namely,

$$D_y(\phi\psi) = (D_y\phi)\psi + \phi(D_y\psi), \qquad \phi, \psi \in (E). \tag{4.5}$$

Actually $D_y\phi(x)$ is the Gâteaux differential of ϕ at $x \in E^*$ in the direction along $y \in E^*$, i.e.,

$$D_y\phi(x) = \lim_{\theta\to 0}\frac{\phi(x+\theta y) - \phi(x)}{\theta}, \qquad x \in E^*. \tag{4.6}$$

The proof is, however, postponed to the next section where we shall prove a much stronger result, see Theorem 4.2.4.

Definition 4.1.2 Recall that $\delta_t \in E^*$ is the Dirac function at $t \in T$. Then

$$\partial_t = D_{\delta_t}$$

is called *Hida's differential operator.*

Proposition 4.1.3 *Let $\phi_\xi \in (E)$ be an exponential vector, $\xi \in E_{\mathbb{C}}$. Then,*

$$D_y\phi_\xi = \langle y, \xi\rangle\,\phi_\xi, \qquad y \in E^*.$$

In particular,

$$\partial_t\phi_\xi = \xi(t)\phi_\xi, \qquad t \in T.$$

PROOF. The exponential vector ϕ_ξ is by definition given by

$$\phi_\xi(x) = \sum_{n=0}^{\infty}\left\langle :x^{\otimes n}:, \frac{\xi^{\otimes n}}{n!}\right\rangle.$$

Then, by (4.2) we see that

$$\begin{aligned} D_y\phi_\xi(x) &= \sum_{n=0}^{\infty} n\left\langle :x^{\otimes(n-1)}:, y\widehat{\otimes}_1\left(\frac{\xi^{\otimes n}}{n!}\right)\right\rangle \\ &= \sum_{n=1}^{\infty} \langle y, \xi\rangle \left\langle :x^{\otimes(n-1)}:, \frac{\xi^{\otimes(n-1)}}{(n-1)!}\right\rangle \\ &= \langle y, \xi\rangle\, \phi_\xi(x). \end{aligned}$$

This proves the assertion. qed

Therefore, in usual Fock space language, ∂_t is an *annihilation operator* and its dual operator $\partial_t^* \in \mathcal{L}((E)^*, (E)^*)$ is a *creation operator*. However, since these operators are defined at each point $t \in T$, we do not need to treat them as operator-valued distibutions as usual theory does.

Proposition 4.1.4 *If the Wiener-Itô expansion of* $\Phi \in (E)^*$ *is given as*

$$\Phi(x) = \sum_{n=0}^{\infty} \left\langle :x^{\otimes n}:, F_n\right\rangle,$$

then for $y \in E^*$,

$$D_y^*\Phi(x) = \sum_{n=0}^{\infty} \left\langle :x^{\otimes(n+1)}:, y\widehat{\otimes} F_n\right\rangle. \tag{4.7}$$

PROOF. Let $\phi \in (E)$ be given as in (4.1). Then, by definition,

$$\begin{aligned} \langle\langle D_y^*\Phi, \phi\rangle\rangle &= \langle\langle \Phi, D_y\phi\rangle\rangle \\ &= \sum_{n=0}^{\infty} n! \left\langle F_n, (n+1) y\widehat{\otimes}_1 f_{n+1}\right\rangle \\ &= \sum_{n=0}^{\infty} (n+1)! \left\langle F_n, y\widehat{\otimes}_1 f_{n+1}\right\rangle. \end{aligned}$$

Then the relation:

$$\left\langle F_n, y\widehat{\otimes}_1 f_{n+1}\right\rangle = \left\langle F_n\widehat{\otimes} y, f_{n+1}\right\rangle,$$

which is shown in Corollary 3.4.10, implies that

$$\left\langle\left\langle D_y^*\Phi, \phi\right\rangle\right\rangle = \sum_{n=0}^{\infty} (n+1)! \left\langle F_n\widehat{\otimes} y, f_{n+1}\right\rangle.$$

This implies (4.7). qed

In §3.1 we introduced the white noise coordinate system $\{x(t)\}_{t\in T}$. As is already explained there, for a fixed $t \in T$,

$$x \mapsto x(t) = \langle : x :, \delta_t\rangle$$

is a generalized functional and therefore we may define a product with $\phi \in (E)$. This is simply denoted by $x(t)\phi(x)$, which is an element of $(E)^*$. In this connection we have

Theorem 4.1.5 *Let $t \in T$. Then*

$$(\partial_t + \partial_t^*)\phi(x) = x(t)\phi(x), \qquad \phi \in (E). \tag{4.8}$$

PROOF. It follows from Corollary 3.5.9 that the multiplication by $x(t)$ is a continuous operator from (E) into $(E)^*$. On the other hand, it is obvious that $\partial_t + \partial_t^* \in \mathcal{L}((E),(E)^*)$. It is therefore sufficient to check the identity (4.8) for $\phi \in (E)$ given as $\phi(x) = \langle :x^{\otimes n}:, f_n\rangle$, $f_n \in E_{\mathbb{C}}^{\hat{\otimes} n}$. A simple application of Theorem 3.5.10 yields

$$\begin{aligned} x(t)\phi(x) &= \sum_{k=0}^{1\wedge n} k! \binom{1}{k}\binom{n}{k} \left\langle :x^{\otimes(1+n-2k)}:, \delta_t \hat{\otimes}_k f_n \right\rangle \\ &= \left\langle :x^{\otimes(n+1)}:, \delta_t \hat{\otimes} f_n \right\rangle + n \left\langle :x^{\otimes(n-1)}:, \delta_t \hat{\otimes}_1 f_n \right\rangle. \end{aligned}$$

This is valid also for $n = 0$. On the other hand, by the definition of Hida's differential operator and by Proposition 4.1.4 we obtain

$$\begin{aligned} \partial_t \phi(x) &= n \left\langle :x^{\otimes(n-1)}:, \delta_t \hat{\otimes}_1 f_n \right\rangle, \\ \partial_t^* \phi(x) &= \left\langle :x^{\otimes(n+1)}:, \delta_t \hat{\otimes} f_n \right\rangle. \end{aligned}$$

The identity (4.8) is then immediate. qed

Finally we give an explicit form of a composition of differential operators D_y. For that purpose we prepare an elementary inequality, where $0^0 = 1$ as usual.

Lemma 4.1.6 *For any $\alpha > 0$ and a non-negative integer m we have*

$$\max_{x \geq 0} (x+m)\cdots(x+1)\rho^{\alpha x} \leq \rho^{-\alpha/2} m^m \left(\frac{\rho^{-\alpha/2}}{-\alpha e \log \rho}\right)^m.$$

PROOF. The inequality being obvious for $m = 0$, we suppose $m \geq 1$. For simplicity we put

$$h_m(x) = (x+m)\cdots(x+1)\rho^{\alpha x}.$$

Using the identity:

$$h_m(x) = \rho^{-\alpha(1+2+\cdots+m)/m} \prod_{j=1}^m (x+j)\rho^{\alpha(x+j)/m} = \rho^{-\alpha(m+1)/2} \prod_{j=1}^m (x+j)\rho^{\alpha(x+j)/m},$$

we obtain

$$h_m(x) \leq \rho^{-\alpha(m+1)/2} \prod_{j=1}^m \left(\max_{x\geq 0} (x+j)\rho^{\alpha(x+j)/m}\right) \leq \rho^{-\alpha(m+1)/2} \left(\max_{x \geq 0} x\rho^{\alpha x/m}\right)^m.$$

Taking (4.4) into account, we come to

$$\max_{x\geq 0} h_m(x) \leq \rho^{-\alpha(m+1)/2} \left(\frac{m}{-\alpha e \log \rho}\right)^m \leq \rho^{-\alpha/2} m^m \left(\frac{\rho^{-\alpha/2}}{-\alpha e \log \rho}\right)^m,$$

where $0 < \rho < 1$ was used. qed

Theorem 4.1.7 *Let $\phi \in (E)$ be given with Wiener-Itô expansion*

$$\phi(x) = \sum_{n=0}^{\infty} \left\langle :x^{\otimes n}:, f_n \right\rangle .$$

Then, for $y_1, \cdots, y_m \in E^*$ we have

$$D_{y_1} \cdots D_{y_m} \phi(x) = \sum_{n=0}^{\infty} \frac{(n+m)!}{n!} \left\langle :x^{\otimes n}:, (y_1 \widehat{\otimes} \cdots \widehat{\otimes} y_m) \widehat{\otimes}_m f_{m+n} \right\rangle . \tag{4.9}$$

Moreover, for any $p \geq 0$, $q > 0$ and $\phi \in (E)$ we have

$$\|D_{y_1} \cdots D_{y_m} \phi\|_p \leq \rho^{-q/2} m^{m/2} \left(\frac{\rho^{-q}}{-2qe \log \rho} \right)^{m/2} |y_1|_{-(p+q)} \cdots |y_m|_{-(p+q)} \|\phi\|_{p+q} .$$

PROOF. Identity (4.9) follows from Theorem 4.1.1 simply by induction. For the norm estimate we first note that

$$\|D_{y_1} \cdots D_{y_m} \phi\|_p^2 = \sum_{n=0}^{\infty} n! \left| \frac{(n+m)!}{n!} (y_1 \widehat{\otimes} \cdots \widehat{\otimes} y_m) \widehat{\otimes}_m f_{m+n} \right|_p^2 .$$

Since

$$\left| (y_1 \widehat{\otimes} \cdots \widehat{\otimes} y_m) \widehat{\otimes}_m f_{m+n} \right|_p \leq \rho^{qn} |y_1 \widehat{\otimes} \cdots \widehat{\otimes} y_m|_{-(p+q)} \, |f_{m+n}|_{p+q} ,$$

we obtain

$$\begin{aligned} \|D_{y_1} \cdots D_{y_m} \phi\|_p^2 &\leq \max_{n \geq 0} \frac{(n+m)!}{n!} \rho^{2qn} |y_1 \widehat{\otimes} \cdots \widehat{\otimes} y_m|_{-(p+q)}^2 \sum_{n=0}^{\infty} (n+m)! \, |f_{m+n}|_{p+q}^2 \\ &= \max_{n \geq 0} \frac{(n+m)!}{n!} \rho^{2qn} \, |y_1|_{-(p+q)}^2 \cdots |y_m|_{-(p+q)}^2 \, \|\phi\|_{p+q}^2 . \end{aligned}$$

Then we need only to apply Lemma 4.1.6. qed

Corollary 4.1.8 *For any $y_1, y_2 \in E^*$ it holds that*

$$[D_{y_1}, D_{y_2}] = 0, \qquad [D_{y_1}^*, D_{y_2}^*] = 0.$$

In particular,

$$[\partial_s, \partial_t] = 0, \qquad [\partial_s^*, \partial_t^*] = 0, \qquad s, t \in T. \tag{4.10}$$

Later we shall observe that ∂_s and ∂_t^* satisfies the so-called *canonical commutation relation*. Namely, in addition to (4.10) it holds that

$$[\partial_t, \partial_s^*] = \delta_s(t) I. \tag{4.11}$$

This expression is, however, somehow informal and the precise statement will be given in Proposition 4.3.12.

4.2 Translation operators

Definition 4.2.1 With each $y \in E^*$ we associate the translation operator T_y by

$$T_y\phi(x) = \phi(x+y), \qquad \phi \in (E), \quad x \in E^*. \tag{4.12}$$

We shall soon show that T_y is a continuous operator on (E). To this end we need the following

Lemma 4.2.2

$$:(x+y)^{\otimes n}: = \sum_{k=0}^{n} \binom{n}{k} :x^{\otimes k}: \widehat{\otimes} y^{\otimes(n-k)}, \qquad x, y \in E^*. \tag{4.13}$$

PROOF. By induction on n. Identity (4.13) holds obviously for $n = 0, 1$. Now we assume that (4.13) is valid up to $n-1$ for $n \geq 2$. It follows from Definition 2.2.2 that

$$:(x+y)^{\otimes n}: = (x+y)\widehat{\otimes} :(x+y)^{\otimes(n-1)}: -(n-1)\tau\widehat{\otimes} :(x+y)^{\otimes(n-2)}:$$

and by the assumption of induction we obtain

$$\begin{aligned}
:(x+y)^{\otimes n}: &= (x+y)\widehat{\otimes} \sum_{k=0}^{n-1} \binom{n-1}{k} :x^{\otimes k}: \widehat{\otimes} y^{\otimes(n-1-k)} \\
&\quad -(n-1)\tau\widehat{\otimes} \sum_{k=0}^{n-2} \binom{n-2}{k} :x^{\otimes k}: \widehat{\otimes} y^{\otimes(n-2-k)} \\
&= \sum_{k=0}^{n-1} \binom{n-1}{k} :x^{\otimes k}: \widehat{\otimes} y^{\otimes(n-k)} + x\widehat{\otimes} y^{\otimes(n-1)} \\
&\quad + \sum_{k=1}^{n-1} \binom{n-1}{k} x\widehat{\otimes} :x^{\otimes k}: \widehat{\otimes} y^{\otimes(n-1-k)} \\
&\quad - \sum_{k=0}^{n-2} (n-1) \binom{n-2}{k} \tau\widehat{\otimes} :x^{\otimes k}: \widehat{\otimes} y^{\otimes(n-2-k)}.
\end{aligned} \tag{4.14}$$

The last two terms are computed as follows.

$$\begin{aligned}
&\sum_{k=0}^{n-2} \binom{n-1}{k+1} x\widehat{\otimes} :x^{\otimes k+1}: \widehat{\otimes} y^{\otimes(n-2-k)} \\
&\qquad - \sum_{k=0}^{n-2} \binom{n-1}{k+1} (k+1)\tau\widehat{\otimes} :x^{\otimes k}: \widehat{\otimes} y^{\otimes(n-2-k)} \\
&= \sum_{k=0}^{n-2} \binom{n-1}{k+1} \left(x\widehat{\otimes} :x^{\otimes(k+1)}: -(k+1)\tau\widehat{\otimes} :x^{\otimes k}: \right) \widehat{\otimes} y^{\otimes(n-2-k)} \\
&= \sum_{k=0}^{n-2} \binom{n-1}{k+1} :x^{\otimes(k+2)}: \widehat{\otimes} y^{\otimes(n-2-k)}.
\end{aligned} \tag{4.15}$$

Consequently, from (4.14) and (4.15) we obtain

$$
\begin{aligned}
:(x+y)^{\otimes n}: &= \sum_{k=0}^{n-1} \binom{n-1}{k} :x^{\otimes k}: \widehat{\otimes} y^{\otimes(n-k)} + x \widehat{\otimes} y^{\otimes(n-1)} \\
&\qquad + \sum_{k=0}^{n-2} \binom{n-1}{k+1} :x^{\otimes(k+2)}: \widehat{\otimes} y^{\otimes(n-2-k)} \\
&= \sum_{k=0}^{n} \binom{n}{k} :x^{\otimes k}: \widehat{\otimes} y^{\otimes(n-k)}.
\end{aligned}
$$

This is the desired identity. qed

Theorem 4.2.3 *$T_y \in \mathcal{L}((E),(E))$ for any $y \in E^*$. Furthermore, for any $p \geq 0$, $q > 0$ with $|y|_{-(p+q)} < \infty$, it holds that*

$$
\|T_y \phi\|_p \leq \frac{\|\phi\|_{p+q}}{(1-\rho^{2q})^{1/2}} \exp\left(\frac{|y|_{-(p+q)}^2}{2(1-\rho^{2q})} \right), \qquad \phi \in (E).
$$

PROOF. Put

$$
\phi(x) = \sum_{n=0}^{\infty} \left\langle :x^{\otimes n}:, f_n \right\rangle.
$$

Then, in view of Lemma 4.2.2 we obtain

$$
\begin{aligned}
T_y \phi(x) &= \sum_{n=0}^{\infty} \left\langle :(x+y)^{\otimes n}:, f_n \right\rangle \\
&= \sum_{n=0}^{\infty} \sum_{k=0}^{n} \binom{n}{k} \left\langle :x^{\otimes k}: \widehat{\otimes} y^{\otimes(n-k)}, f_n \right\rangle \\
&= \sum_{n=0}^{\infty} \sum_{k=0}^{n} \binom{n}{k} \left\langle :x^{\otimes k}:, y^{\otimes(n-k)} \widehat{\otimes}_{n-k} f_n \right\rangle \\
&= \sum_{k=0}^{\infty} \sum_{n=0}^{\infty} \binom{n+k}{k} \left\langle :x^{\otimes k}:, y^{\otimes n} \widehat{\otimes}_n f_{n+k} \right\rangle.
\end{aligned}
$$

Using the inequality

$$
|y^{\otimes n} \widehat{\otimes}_n f_{n+k}|_p \leq \rho^{qk} |y|_{-(p+q)}^n |f_{n+k}|_{p+q},
$$

we have

$$
\begin{aligned}
\|T_y \phi\|_p^2 &= \sum_{k=0}^{\infty} k! \left| \sum_{n=0}^{\infty} \binom{n+k}{k} y^{\otimes n} \widehat{\otimes}_n f_{n+k} \right|_p^2 \\
&\leq \sum_{k=0}^{\infty} k! \left(\sum_{n=0}^{\infty} \frac{(n+k)!}{n!k!} \rho^{qk} |y|_{-(p+q)}^n |f_{n+k}|_{p+q} \right)^2 \\
&\leq \sum_{k=0}^{\infty} k! \left(\sum_{n=0}^{\infty} (n+k)! |f_{n+k}|_{p+q}^2 \right) \left(\sum_{n=0}^{\infty} \frac{(n+k)!}{(n!k!)^2} \rho^{2qk} |y|_{-(p+q)}^{2n} \right) \\
&\leq \|\phi\|_{p+q}^2 \sum_{n=0}^{\infty} \sum_{k=0}^{\infty} \frac{(n+k)!}{n!n!k!} \rho^{2qk} |y|_{-(p+q)}^{2n}.
\end{aligned}
$$

Since

$$\sum_{k=0}^{\infty} \frac{(n+k)!}{n!k!} \rho^{2qk} = (1-\rho^{2q})^{-(n+1)},$$

we conclude that

$$\begin{aligned}
\|T_y \phi\|_p^2 &\leq \|\phi\|_{p+q}^2 \sum_{n=0}^{\infty} \frac{1}{n!} |y|_{-(p+q)}^{2n} (1-\rho^{2q})^{-(n+1)} \\
&= \|\phi\|_{p+q}^2 (1-\rho^{2q})^{-1} \exp\left(|y|_{-(p+q)}^2 (1-\rho^{2q})^{-1} \right).
\end{aligned}$$

This completes the proof. qed

Theorem 4.2.4 *Let $\phi \in (E)$ and $y \in E^*$. Then for any $p \geq 0$, $q > 0$ with $|y|_{-(p+q)} < \infty$ and $\theta \in \mathbb{R}$ it holds that*

$$\left\| \frac{T_{\theta y}\phi - \phi}{\theta} - D_y \phi \right\|_p \leq \frac{|\theta| \, \|\phi\|_{p+q} \, |y|_{-(p+q)}^2}{\sqrt{2}(1-\rho^{2q})^{3/2}} \exp\left(\frac{|\theta|^2 \, |y|_{-(p+q)}^2}{2(1-\rho^{2q})} \right).$$

PROOF. Let $\phi \in (E)$ be given with Wiener-Itô expansion:

$$\phi(x) = \sum_{n=0}^{\infty} \left\langle :x^{\otimes n}:, f_n \right\rangle$$

as usual. Then a direct calculation yields

$$\begin{aligned}
\psi(x) &\equiv \frac{\phi(x+\theta y) - \phi(x)}{\theta} - D_y \phi(x) \\
&= \sum_{n=0}^{\infty} \sum_{k=0}^{n-2} \binom{n}{k} \theta^{n-1-k} \left\langle :x^{\otimes k}:, y^{\otimes (n-k)} \widehat{\otimes}_{n-k} f_n \right\rangle \\
&= \sum_{k=0}^{\infty} \sum_{n=2}^{\infty} \binom{n+k}{k} \theta^{n-1} \left\langle :x^{\otimes k}:, y^{\otimes n} \widehat{\otimes}_n f_{n+k} \right\rangle.
\end{aligned}$$

Hence

$$\begin{aligned}
\|\psi\|_p^2 &= \sum_{k=0}^{\infty} k! \left| \sum_{n=2}^{\infty} \binom{n+k}{k} \theta^{n-1} y^{\otimes n} \widehat{\otimes}_n f_{n+k} \right|_p^2 \\
&\leq \sum_{k=0}^{\infty} k! \left(\sum_{n=2}^{\infty} \frac{(n+k)!}{n!k!} |\theta|^{n-1} \rho^{qk} \, |y|_{-(p+q)}^n \, |f_{n+k}|_{p+q} \right)^2 \\
&\leq \sum_{k=0}^{\infty} k! \left(\sum_{n=2}^{\infty} (n+k)! \, |f_{n+k}|_{p+q}^2 \right) \left(\sum_{n=2}^{\infty} \frac{(n+k)!}{(n!k!)^2} |\theta|^{2n-2} \rho^{2qk} \, |y|_{-(p+q)}^{2n} \right) \\
&\leq \|\phi\|_{p+q}^2 \sum_{k=0}^{\infty} \sum_{n=2}^{\infty} \frac{(n+k)!}{n!n!k!} |\theta|^{2n-2} \rho^{2qk} \, |y|_{-(p+q)}^{2n} \\
&= \|\phi\|_{p+q}^2 \sum_{n=2}^{\infty} \frac{1}{n!} |\theta|^{2n-2} \, |y|_{-(p+q)}^{2n} (1-\rho^{2q})^{-(n+1)} \\
&\leq \|\phi\|_{p+q}^2 \, |\theta|^{-2} (1-\rho^{2q})^{-1} \sum_{n=2}^{\infty} \frac{\left(|\theta|^2 \, |y|_{-(p+q)}^2 (1-\rho^{2q})^{-1} \right)^n}{n!}.
\end{aligned}$$

Using the inequality:

$$\sum_{n=2}^{\infty} \frac{t^n}{n!} \le \frac{t^2}{2} e^t, \qquad t \ge 0,$$

we obtain

$$\begin{aligned}\|\psi\|_p^2 &\le \|\phi\|_{p+q}^2 |\theta|^{-2}(1-\rho^{2q})^{-1} \\ &\quad \times \frac{1}{2}\left(|\theta|^2 |y|_{-(p+q)}^2 (1-\rho^{2q})^{-1}\right)^2 \exp\left(|\theta|^2 |y|_{-(p+q)}^2 (1-\rho^{2q})^{-1}\right) \\ &= \frac{|\theta|^2}{2} \|\phi\|_{p+q}^2 (1-\rho^{2q})^{-3} |y|_{-(p+q)}^4 \exp\left(|\theta|^2 |y|_{-(p+q)}^2 (1-\rho^{2q})^{-1}\right).\end{aligned}$$

This completes the proof. qed

Corollary 4.2.5 *For any $y \in E^*$ and $p \ge 0$ there exists $q > 0$ such that*

$$\lim_{\theta \to 0} \sup_{\|\phi\|_{p+q} \le 1} \left\| \frac{T_{\theta y}\phi - \phi}{\theta} - D_y \phi \right\|_p = 0.$$

In particular,

$$\lim_{\theta \to 0} \frac{T_{\theta y}\phi - \phi}{\theta} = D_y \phi, \qquad \phi \in (E),$$

holds with respect to the topology of (E).

Taking Corollary 3.2.15 into accaount, we deduce the following results announced in (4.5) and (4.6).

Corollary 4.2.6 *For any $y \in E^*$ and $\phi \in (E)$ we have*

$$\lim_{\theta \to 0} \frac{\phi(x + \theta y) - \phi(x)}{\theta} = D_y \phi(x), \qquad x \in E^*.$$

Corollary 4.2.7 *D_y is a continuous derivation on (E) for any $y \in E^*$. Namely,*

$$D_y(\phi\psi) = (D_y\phi)\psi + \phi(D_y\psi), \qquad \phi, \psi \in (E).$$

It follows from Corollary 4.2.5 that D_y is the infinitesimal generator of $\{T_{\theta y}\}_{\theta \in \mathbb{R}}$ which is a one-parameter group of transformations on (E). Further systematic discussion in this direction will be made in §5.2.

4.3 Integral kernel operators

Having introduced the differential operator ∂_t in §4.1, we now develop a general theory of operators which are expressed as superposition of ∂_t and ∂_t^*. We begin with

Lemma 4.3.1 *For $\phi, \psi \in (E)$ we put*

$$\eta_{\phi,\psi}(s_1, \cdots .s_l, t_1, \cdots, t_m) = \langle\!\langle \partial_{s_1}^* \cdots \partial_{s_l}^* \partial_{t_1} \cdots \partial_{t_m} \phi \, , \psi \rangle\!\rangle . \tag{4.16}$$

Then for any $p > 0$ we have

$$|\eta_{\phi,\psi}|_p \le \rho^{-p} \left(l^l m^m \right)^{1/2} \left(\frac{\rho^{-p}}{-2pe \log \rho} \right)^{(l+m)/2} \|\phi\|_p \, \|\psi\|_p .$$

In particular, $\eta_{\phi,\psi} \in E_{\mathbf{C}}^{\otimes(l+m)}$.

PROOF. As usual we put

$$\phi(x) = \sum_{n=0}^{\infty} \left\langle :x^{\otimes n}: , f_n \right\rangle , \qquad \psi(x) = \sum_{n=0}^{\infty} \left\langle :x^{\otimes n}: , g_n \right\rangle .$$

It then follows from Theorem 4.1.7 that

$$\partial_{t_1} \cdots \partial_{t_m} \phi(x) = \sum_{n=0}^{\infty} \frac{(n+m)!}{n!} \left\langle :x^{\otimes n}: , (\delta_{t_1} \widehat{\otimes} \cdots \widehat{\otimes} \delta_{t_m}) \widehat{\otimes}_m f_{n+m} \right\rangle$$

and

$$\partial_{s_1} \cdots \partial_{s_l} \psi(x) = \sum_{n=0}^{\infty} \frac{(n+l)!}{n!} \left\langle :x^{\otimes n}: , (\delta_{s_1} \widehat{\otimes} \cdots \widehat{\otimes} \delta_{s_l}) \widehat{\otimes}_l g_{n+l} \right\rangle .$$

Hence,

$$\begin{aligned}
\eta_{\phi,\psi}(s_1, \cdots, s_l, t_1, \cdots, t_m) &= \langle\!\langle \partial_{t_1} \cdots \partial_{t_m} \phi \, , \partial_{s_1} \cdots \partial_{s_l} \psi \rangle\!\rangle \\
&= \sum_{n=0}^{\infty} n! \frac{(m+n)!}{n!} \frac{(l+n)!}{n!} \left\langle (\delta_{t_1} \widehat{\otimes} \cdots \widehat{\otimes} \delta_{t_m}) \widehat{\otimes}_m f_{n+m} \, , (\delta_{s_1} \widehat{\otimes} \cdots \widehat{\otimes} \delta_{s_l}) \widehat{\otimes}_l g_{n+l} \right\rangle .
\end{aligned}$$

On the other hand, in view of Proposition 3.4.9 we obtain

$$\begin{aligned}
&\left\langle (\delta_{t_1} \widehat{\otimes} \cdots \widehat{\otimes} \delta_{t_m}) \widehat{\otimes}_m f_{n+m} \, , (\delta_{s_1} \widehat{\otimes} \cdots \widehat{\otimes} \delta_{s_l}) \widehat{\otimes}_l g_{n+l} \right\rangle \\
&\qquad = \left\langle (\delta_{s_1} \widehat{\otimes} \cdots \widehat{\otimes} \delta_{s_l}) \otimes (\delta_{t_1} \widehat{\otimes} \cdots \widehat{\otimes} \delta_{t_m}) \, , g_{l+n} \otimes_n f_{n+m} \right\rangle \\
&\qquad = g_{l+n} \otimes_n f_{m+n}(s_1, \cdots, s_l, t_1, \cdots, t_m).
\end{aligned}$$

Therefore,

$$\eta_{\phi,\psi} = \sum_{n=0}^{\infty} \frac{(m+n)!(l+n)!}{n!} g_{l+n} \otimes_n f_{m+n}. \tag{4.17}$$

Since

$$|g_{l+n} \otimes_n f_{m+n}|_p \le \rho^{2pn} \, |f_{m+n}|_p \, |g_{l+n}|_p ,$$

we obtain

$$\begin{aligned}
|\eta_{\phi,\psi}|_p &\le \sum_{n=0}^{\infty} \frac{(m+n)!(l+n)!}{n!} |g_{l+n} \otimes_n f_{m+n}|_p \\
&\le \sum_{n=0}^{\infty} \sqrt{\frac{(m+n)!}{n!}} \sqrt{\frac{(l+n)!}{n!}} \rho^{2pn} \sqrt{(m+n)!} \, |f_{m+n}|_p \sqrt{(l+n)!} \, |g_{l+n}|_p \\
&\le M_0 \left(\sum_{n=0}^{\infty} (m+n)! \, |f_{m+n}|_p^2 \right)^{1/2} \left(\sum_{n=0}^{\infty} (l+n)! \, |g_{l+n}|_p^2 \right)^{1/2} \\
&= M_0 \|\phi\|_p \, \|\psi\|_p ,
\end{aligned}$$

where

$$M_0 = \sup_{n\geq 0} \sqrt{\frac{(m+n)!}{n!}}\sqrt{\frac{(l+n)!}{n!}}\,\rho^{2pn} < \infty$$

for $p > 0$. Furthermore, it follows from Lemma 4.1.6 that

$$\sup_{n\geq 0} \frac{(n+m)!}{n!}\rho^{2pn} \leq \rho^{-p} m^m \left(\frac{\rho^{-p}}{-2pe\log\rho}\right)^m,$$

$$\sup_{n\geq 0} \frac{(n+l)!}{n!}\rho^{2pn} \leq \rho^{-p} l^l \left(\frac{\rho^{-p}}{-2pe\log\rho}\right)^l.$$

Therefore,

$$M_0 \leq \rho^{-p}\left(l^l m^m\right)^{1/2}\left(\frac{\rho^{-p}}{-2pe\log\rho}\right)^{(l+m)/2}.$$

This proves the assertion. qed

Theorem 4.3.2 *For any* $\kappa \in (E_{\mathbf{C}}^{\otimes(l+m)})^*$ *there exists a continuous linear operator* $\Xi_{l,m}(\kappa) \in \mathcal{L}((E),(E)^*)$ *such that*

$$\langle\!\langle \Xi_{l,m}(\kappa)\phi\,,\psi\rangle\!\rangle = \langle \kappa\,, \eta_{\phi,\psi}\rangle\,, \qquad \phi,\psi \in (E),$$

where

$$\eta_{\phi,\psi}(s_1,\cdots,s_l,t_1,\cdots,t_m) = \left\langle\!\left\langle \partial_{s_1}^*\cdots\partial_{s_l}^*\partial_{t_1}\cdots\partial_{t_m}\phi\,,\psi\right\rangle\!\right\rangle.$$

Moreover, for any $p > 0$ *with* $|\kappa|_{-p} < \infty$ *it holds that*

$$\|\Xi_{l,m}(\kappa)\phi\|_{-p} \leq \rho^{-p}\left(l^l m^m\right)^{1/2}\left(\frac{\rho^{-p}}{-2pe\log\rho}\right)^{(l+m)/2}|\kappa|_{-p}\,\|\phi\|_p\,. \tag{4.18}$$

PROOF. First note that for any $\kappa \in (E_{\mathbf{C}}^{\otimes(l+m)})^*$,

$$(\phi,\psi) \mapsto \langle \kappa\,,\eta_{\phi,\psi}\rangle\,, \qquad \phi,\psi\in(E),$$

is a continuous bilinear form on (E). In fact, by Lemma 4.3.1 we have

$$|\langle \kappa\,,\eta_{\phi,\psi}\rangle| \leq |\kappa|_{-p}\,|\eta_{\phi,\psi}|_p \leq M\,|\kappa|_{-p}\,\|\phi\|_p\,\|\psi\|_p\,, \tag{4.19}$$

where

$$M = \rho^{-p}\left(l^l m^m\right)^{1/2}\left(\frac{\rho^{-p}}{-2pe\log\rho}\right)^{(l+m)/2}. \tag{4.20}$$

Therefore there is a continuous linear operator $\Xi_{l,m}(\kappa) \in \mathcal{L}((E),(E)^*)$ such that

$$\langle\!\langle \Xi_{l,m}(\kappa)\phi\,,\psi\rangle\!\rangle = \langle \kappa\,,\eta_{\phi,\psi}\rangle\,,$$

see e.g., Proposition 1.3.12. It then follows from (4.19) that

$$\|\Xi_{l,m}(\kappa)\phi\|_{-p} \leq \sup\{|\langle\kappa\,,\eta_{\phi,\psi}\rangle|\,;\|\psi\|_p\leq 1\} \leq M\,|\kappa|_{-p}\,\|\phi\|_p\,.$$

Then (4.18) follows from (4.20). qed

The operator $\Xi_{l,m}(\kappa)$ is thus defined through two canonical bilinear forms:

$$\langle\langle \Xi_{l,m}(\kappa)\phi\,,\psi\rangle\rangle = \left\langle \kappa\,, \left\langle\left\langle \partial_{s_1}^* \cdots \partial_{s_l}^* \partial_{t_1} \cdots \partial_{t_m}\phi\,,\psi\right\rangle\right\rangle\right\rangle, \qquad \phi,\psi \in (E). \tag{4.21}$$

This suggests us to employ a formal integral expression:

$$\Xi_{l,m}(\kappa) =$$
$$= \int_{T^{l+m}} \kappa(s_1,\cdots,s_l,t_1,\cdots,t_m)\partial_{s_1}^* \cdots \partial_{s_l}^* \partial_{t_1} \cdots \partial_{t_m} ds_1 \cdots ds_l dt_1 \cdots dt_m. \tag{4.22}$$

We call $\Xi_{l,m}(\kappa)$ an *integral kernel operator* with *kernel distribution* κ. It is possible to write down the action of $\Xi_{l,m}(\kappa)$ explicitly using the contraction of tensor product.

Proposition 4.3.3 *Let $\phi \in (E)$ be given with Wiener-Itô expansion:*

$$\phi(x) = \sum_{n=0}^{\infty} \left\langle :x^{\otimes n}:, f_n \right\rangle.$$

Then, for $\kappa \in (E_{\mathbf{C}}^{\otimes(l+m)})^*$ we have

$$\Xi_{l,m}(\kappa)\phi(x) = \sum_{n=0}^{\infty} \frac{(n+m)!}{n!} \left\langle :x^{\otimes(l+n)}:, \kappa \otimes_m f_{n+m} \right\rangle. \tag{4.23}$$

In particular, for $\xi \in E_{\mathbf{C}}$ we have

$$\Xi_{l,m}(\kappa)\phi_\xi(x) = \sum_{n=0}^{\infty} \frac{1}{n!} \left\langle :x^{\otimes(l+n)}:, (\kappa \otimes_m \xi^{\otimes m}) \otimes \xi^{\otimes n} \right\rangle. \tag{4.24}$$

PROOF. Let $\psi \in (E)$ be given as

$$\psi(x) = \sum_{n=0}^{\infty} \left\langle :x^{\otimes n}:, g_n \right\rangle.$$

Then, by definition,

$$\begin{aligned} \langle\langle \Xi_{l,m}(\kappa)\phi\,,\psi\rangle\rangle &= \sum_{n=0}^{\infty} \frac{(m+n)!(l+n)!}{n!} \langle \kappa\,, g_{l+n} \otimes_n f_{m+n} \rangle \\ &= \sum_{n=0}^{\infty} \frac{(m+n)!(l+n)!}{n!} \langle \kappa \otimes_m f_{n+m}\,, g_{l+n} \rangle, \end{aligned}$$

from which (4.23) follows. The proof of (4.24) is then immediate. qed

During the above discussion we have obtained a linear map

$$\kappa \mapsto \Xi_{l,m}(\kappa) \in \mathcal{L}((E),(E)^*), \qquad \kappa \in (E_{\mathbf{C}}^{\otimes(l+m)})^*.$$

But this is not injective, namely, the kernel distribution is not uniquely determined. For the uniqueness we need a "partially symmetrized" kernel. We put

$$s_{l,m}(\kappa) = \frac{1}{l!m!} \sum_{\sigma \in \mathfrak{S}_l \times \mathfrak{S}_m} \kappa^\sigma, \qquad \kappa \in (E_{\mathbf{C}}^{\otimes(l+m)})^*, \tag{4.25}$$

where κ^σ is defined in §1.6. We first note the following

Proposition 4.3.4 *$\Xi_{l,m}(\kappa) = \Xi_{l,m}(s_{l,m}(\kappa))$ for $\kappa \in (E_{\mathbf{C}}^{\otimes(l+m)})^*$.*

PROOF. Since $\Xi_{l,m}(\kappa)$ is defined uniquely by (4.21), the assertion follows immediately from the fact that $[\partial_s, \partial_t] = 0$ and $[\partial_s^*, \partial_t^*] = 0$, $s, t \in T$, which is shown in Corollary 4.1.8. qed

We can now claim the uniqueness of the kernel distribution. Put

$$(E_{\mathbf{C}}^{\otimes(l+m)})^*_{\mathrm{sym}(l,m)} = \left\{\kappa \in (E_{\mathbf{C}}^{\otimes(l+m)})^* \,;\, s_{l,m}(\kappa) = \kappa\right\}.$$

With these notation,

Proposition 4.3.5 *Let $\kappa \in (E_{\mathbf{C}}^{\otimes(l+m)})^*$. Then $\Xi_{l,m}(\kappa) = 0$ if and only if $s_{l,m}(\kappa) = 0$. In other words, the map*

$$\kappa \mapsto \Xi_{l,m}(\kappa) \in \mathcal{L}((E), (E)^*), \qquad \kappa \in (E_{\mathbf{C}}^{\otimes(l+m)})^*_{\mathrm{sym}(l,m)},$$

is injective.

PROOF. Suppose that $s_{l,m}(\kappa) = 0$. Then we see from Proposition 4.3.4 that $\Xi_{l,m}(\kappa) = \Xi_{l,m}(s_{l,m}(\kappa)) = 0$. Conversely, we suppose that $\Xi_{l,m}(\kappa) = 0$. Consider particular functions:

$$\phi_m(x) = \left\langle :x^{\otimes m}:, \xi^{\otimes m}\right\rangle, \qquad \psi_l(x) = \left\langle :x^{\otimes l}:, \eta^{\otimes l}\right\rangle, \qquad \xi, \eta \in E_{\mathbf{C}}. \tag{4.26}$$

Then, it follows from the definition that

$$\langle\langle \Xi_{l,m}(\kappa)\phi_m, \psi_l \rangle\rangle = l!m! \left\langle \kappa, \eta^{\otimes l} \otimes \xi^{\otimes m}\right\rangle.$$

Since $\Xi_{l,m}(\kappa) = 0$ by assumption,

$$0 = \left\langle \kappa, \eta^{\otimes l} \otimes \xi^{\otimes m}\right\rangle = \left\langle s_{l,m}(\kappa), \eta^{\otimes l} \otimes \xi^{\otimes m}\right\rangle.$$

This being true for all $\xi, \eta \in E_{\mathbf{C}}$, we conclude that $s_{l,m}(\kappa) = 0$. qed

Proposition 4.3.6 *Let $\kappa \in (E_{\mathbf{C}}^{\otimes(l+m)})^*$ and $\kappa' \in (E_{\mathbf{C}}^{\otimes(l'+m')})^*$. If $\Xi_{l,m}(\kappa) = \Xi_{l',m'}(\kappa') \neq 0$, then $l = l'$, $m = m'$ and $s_{l,m}(\kappa) = s_{l,m}(\kappa')$.*

PROOF. Suppose that $\Xi_{l,m}(\kappa) = \Xi_{l',m'}(\kappa') \neq 0$. In particular, since $\Xi_{l,m}(\kappa) \neq 0$, it follows from Proposition 4.3.5 that $s_{l,m}(\kappa) \neq 0$ and therefore, there exist $\xi, \eta \in E_{\mathbf{C}}$ such that

$$\left\langle \kappa, \eta^{\otimes l} \otimes \xi^{\otimes m}\right\rangle \neq 0.$$

Let ϕ_m and ψ_l be the same as in (4.26). Then we have

$$\langle\langle \Xi_{l,m}(\kappa)\phi_m, \psi_l \rangle\rangle \neq 0. \tag{4.27}$$

On the other hand, unless $m' \le m$ and $l' \le l$,

$$\langle\langle \Xi_{l',m'}(\kappa')\phi_m, \psi_l \rangle\rangle = \left\langle \kappa', \left\langle\!\left\langle \partial_{t_1} \cdots \partial_{t_{m'}} \phi_m, \partial_{s_1} \cdots \partial_{s_{l'}} \psi_l \right\rangle\!\right\rangle \right\rangle = 0.$$

Therefore, in order to have (4.27) it is necessary that $m' \le m$ and $l' \le l$. A similar argument starting with $\Xi_{l',m'}(\kappa') \neq 0$ implies that $m \le m'$ and $l \le l'$. Hence $l = l'$ and $m = m'$. We then see from Proposition 4.3.5 that $s_{l,m}(\kappa) = s_{l,m}(\kappa')$. qed

With each $(\kappa_{l,m})_{l,m=0}^{\infty} \in \bigoplus_{l,m=0}^{\infty}(E_{\mathbf{C}}^{\otimes(l+m)})^*_{\mathrm{sym}(l,m)}$ (algebraic direct sum) we may associate an operator

$$\sum_{l,m=0}^{\infty} \Xi_{l,m}(\kappa_{l,m}) \in \mathcal{L}((E),(E)^*).$$

It then follows from Propositions 4.3.5 and 4.3.6 that we have a linear injection:

$$\bigoplus_{l,m=0}^{\infty} (E_{\mathbf{C}}^{\otimes(l+m)})^*_{\mathrm{sym}(l,m)} \to \mathcal{L}((E),(E)^*). \tag{4.28}$$

This map will be extended to cover all operators in $\mathcal{L}((E),(E)^*)$ in §4.5 where we discuss an infinite sum of integral kernel operators.

Since (E) is reflexive, $\Xi_{l,m}(\kappa)^*$ is again a continuous linear operator from (E) into $(E)^*$. In fact, $\Xi_{l,m}(\kappa)^*$ is again an integral kernel operator as is seen below. For $\kappa \in (E_{\mathbf{C}}^{\otimes(l+m)})^*$ we define $t_{m,l}(\kappa)$ by

$$\langle t_{m,l}(\kappa), \eta \otimes \zeta \rangle = \langle \kappa, \zeta \otimes \eta \rangle, \qquad \eta \in E_{\mathbf{C}}^{\otimes m}, \quad \zeta \in E_{\mathbf{C}}^{\otimes l}. \tag{4.29}$$

Then, by a straightforward computation we come to

Proposition 4.3.7 *$\Xi_{l,m}(\kappa)^* = \Xi_{m,l}(t_{m,l}(\kappa))$ for $\kappa \in (E_{\mathbf{C}}^{\otimes(l+m)})^*$. In particular, if $\Xi_{l,m}(\kappa)$ is given as in (4.22), then*

$$\Xi_{l,m}(\kappa)^* = \int_{T^{l+m}} \kappa(s_1, \cdots, s_l, t_1, \cdots, t_m)\partial_{t_1}^* \cdots \partial_{t_m}^* \partial_{s_1} \cdots \partial_{s_l} ds_1 \cdots ds_l dt_1 \cdots dt_m.$$

We have so far discussed operators from (E) into $(E)^*$. In some application operators in $\mathcal{L}((E),(E))$ are also important and we shall discuss integral kernel operators in $\mathcal{L}((E),(E))$. For that purpose we use the norms introduced in §3.4. Recall that for $p, q \in \mathbf{R}$ we defined

$$|\kappa|_{l,m;p,q}^2 = \sum_{\mathbf{i},\mathbf{j}} |\langle \kappa, e(\mathbf{i}) \otimes e(\mathbf{j}) \rangle|^2 \, |e(\mathbf{i})|_p^2 \, |e(\mathbf{j})|_q^2, \qquad \kappa \in (E_{\mathbf{C}}^{\otimes(l+m)})^*,$$

where $e(\mathbf{i}) = e_{i_1} \otimes \cdots \otimes e_{i_l}$ with $\mathbf{i} = (i_1, \cdots, i_l)$ and $e(\mathbf{j}) = e_{j_1} \otimes \cdots \otimes e_{j_m}$ with $\mathbf{j} = (j_1, \cdots, j_m)$.

Suppose that $\kappa \in (E_{\mathbf{C}}^{\otimes(l+m)})^*$ and $K \in \mathcal{L}(E_{\mathbf{C}}^{\otimes m}, (E_{\mathbf{C}}^{\otimes l})^*)$ are related according to the canonical isomorphism (Proposition 1.3.12). Then

$$\langle Kf, g \rangle = \langle \kappa, g \otimes f \rangle = \langle \kappa \otimes_m f, g \rangle, \qquad g \in E_{\mathbf{C}}^{\otimes l}, \quad f \in E_{\mathbf{C}}^{\otimes m}. \tag{4.30}$$

In other words, the correspondence between κ and K is given simply as

$$Kf = \kappa \otimes_m f, \qquad f \in E_{\mathbf{C}}^{\otimes m}.$$

The next result will be crucial.

Lemma 4.3.8 *Let $\kappa \in (E_{\mathbf{C}}^{\otimes(l+m)})^*$ and $K \in \mathcal{L}(E_{\mathbf{C}}^{\otimes m}, (E_{\mathbf{C}}^{\otimes l})^*)$ be related as above. Then, the following four conditions are equivalent:*
(i) $\kappa \in (E_{\mathbf{C}}^{\otimes l}) \otimes (E_{\mathbf{C}}^{\otimes m})^*$;
(ii) $K \in \mathcal{L}(E_{\mathbf{C}}^{\otimes m}, E_{\mathbf{C}}^{\otimes l})$;
(iii) *for any $p \geq 0$ there exist $C \geq 0$ and $q \geq 0$ such that $|\langle \kappa, g \otimes f \rangle| \leq C |g|_{-p} |f|_{p+q}$ for $g \in E_{\mathbf{C}}^{\otimes l}$ and $f \in E_{\mathbf{C}}^{\otimes m}$;*
(iv) *for any $p \geq 0$ there exists $q \geq 0$ such that $|\kappa|_{l,m;p,-(p+q)} < \infty$.*

PROOF. (i) $\Longleftrightarrow$ (ii) follows immediately from $\mathcal{L}(E_{\mathbf{C}}^{\otimes m}, E_{\mathbf{C}}^{\otimes l}) \cong (E_{\mathbf{C}}^{\otimes l}) \otimes (E_{\mathbf{C}}^{\otimes m})^*$, which is a part of the kernel theorem (Theorem 1.3.10).

(ii) $\Longrightarrow$ (iii). By continuity of K, for a given $p \geq 0$ there exist $C \geq 0$ and $q \geq 0$ such that

$$|Kf|_p \leq C |f|_{p+q}, \qquad f \in E_{\mathbf{C}}^{\otimes m}.$$

Then in view of (4.30) we have

$$|\langle \kappa, g \otimes f \rangle| = |\langle Kf, g \rangle| \leq C |f|_{p+q} |g|_{-p}, \qquad f \in E_{\mathbf{C}}^{\otimes m}, \quad g \in E_{\mathbf{C}}^{\otimes l}.$$

(iii) $\Longrightarrow$ (iv). Let $p \geq 0$ be given. By assumption there exist $C \geq 0$ and $q \geq 0$ such that

$$|\langle \kappa, g \otimes f \rangle| \leq C |g|_{-(p+1)} |f|_{p+q}, \qquad g \in E_{\mathbf{C}}^{\otimes l}, \quad f \in E_{\mathbf{C}}^{\otimes m}.$$

Therefore,

$$\begin{aligned} |\kappa|_{l,m;p,-(p+q+1)}^2 &= \sum_{\mathbf{i},\mathbf{j}} |\langle \kappa, e(\mathbf{i}) \otimes e(\mathbf{j}) \rangle|^2 |e(\mathbf{i})|_p^2 |e(\mathbf{j})|_{-(p+q+1)}^2 \\ &\leq \sum_{\mathbf{i},\mathbf{j}} C^2 |e(\mathbf{i})|_{-(p+1)}^2 |e(\mathbf{j})|_{p+q}^2 |e(\mathbf{i})|_p^2 |e(\mathbf{j})|_{-(p+q+1)}^2 \\ &= C^2 \sum_{\mathbf{i},\mathbf{j}} |e(\mathbf{i})|_{-1}^2 |e(\mathbf{j})|_{-1}^2 \\ &= C^2 \delta^{2(l+m)} < \infty. \end{aligned}$$

(iv) $\Longrightarrow$ (ii). Recall that for $p \in \mathbf{R}$ and $q \geq 0$,

$$|\kappa \otimes_m f|_p \leq \rho^{qn} |\kappa|_{l,m;p,-(p+q)} |f|_{p+q}, \qquad \kappa \in (E_{\mathbf{C}}^{\otimes(l+m)})^*, \quad f \in E_{\mathbf{C}}^{\otimes(n+m)}, \tag{4.31}$$

which is a simple modification of Lemma 3.4.5. Then the assertion is obvious. qed

Note also that if $|\kappa|_{l,m;p,-(p+q)} < \infty$ for some $p, q \in \mathbf{R}$, then $|\kappa|_{l,m;p,-(p+q+r)} < \infty$ for any $r \geq 0$. This is immediate from definition.

We are now in a position to prove the following

Theorem 4.3.9 *Let $\kappa \in (E_{\mathbf{C}}^{\otimes(l+m)})^*$. Then $\Xi_{l,m}(\kappa) \in \mathcal{L}((E),(E))$ if and only if $\kappa \in (E_{\mathbf{C}}^{\otimes l}) \otimes (E_{\mathbf{C}}^{\otimes m})^*$. In that case, for any $p \in \mathbf{R}$, $q > 0$ and $\alpha, \beta > 0$ with $\alpha + \beta \leq 2q$, it holds that*

$$\begin{aligned} &\|\Xi_{l,m}(\kappa)\phi\|_p \\ &\leq \rho^{-q/2} \left(l^l m^m\right)^{1/2} \left(\frac{\rho^{-\alpha/2}}{-\alpha e \log \rho}\right)^{l/2} \left(\frac{\rho^{-\beta/2}}{-\beta e \log \rho}\right)^{m/2} |\kappa|_{l,m;p,-(p+q)} \|\phi\|_{p+q}, \end{aligned} \tag{4.32}$$

for all $\phi \in (E)$.

PROOF. First suppose that $\kappa \in (E_{\mathbb{C}}^{\otimes l}) \otimes (E_{\mathbb{C}}^{\otimes m})^*$. Let $\phi \in (E)$ be given with Wiener-Itô expansion:

$$\phi(x) = \sum_{n=0}^{\infty} \left\langle :x^{\otimes n}:, f_n \right\rangle .$$

It then follows from Proposition 4.3.3 that

$$\Xi_{l,m}(\kappa)\phi(x) = \sum_{n=0}^{\infty} \frac{(n+m)!}{n!} \left\langle :x^{\otimes(l+n)}:, \kappa \otimes_m f_{n+m} \right\rangle ,$$

and therefore,

$$\|\Xi_{l,m}(\kappa)\phi\|_p^2 = \sum_{n=0}^{\infty} (l+n)! \left(\frac{(n+m)!}{n!} \right)^2 |\kappa \otimes_m f_{n+m}|_p^2 .$$

In view of (4.31), we obtain

$$\begin{aligned} \|\Xi_{l,m}(\kappa)\phi\|_p^2 &\le \sum_{n=0}^{\infty} (l+n)! \left(\frac{(n+m)!}{n!} \right)^2 \rho^{2qn} |\kappa|_{l,m;p,-(p+q)}^2 |f_{n+m}|_{p+q}^2 \\ &= |\kappa|_{l,m;p,-(p+q)}^2 \sum_{n=0}^{\infty} (n+m)! \, |f_{n+m}|_{p+q}^2 \frac{(n+l)!(n+m)!}{n!n!} \rho^{2qn} \\ &\le M_1 |\kappa|_{l,m;p,-(p+q)}^2 \|\phi\|_{p+q}^2 , \end{aligned} \tag{4.33}$$

where

$$M_1 = \sup_{n \ge 0} \frac{(n+l)!(n+m)!}{n!n!} \rho^{2qn} .$$

Take $\alpha, \beta > 0$ with $\alpha + \beta \le 2q$. Then,

$$M_1 \le \sup_{n\ge0} \frac{(n+l)!(n+m)!}{n!n!} \rho^{(\alpha+\beta)n} \le \sup_{n\ge0} \frac{(n+l)!}{n!} \rho^{\alpha n} \times \sup_{n\ge0} \frac{(n+m)!}{n!} \rho^{\beta n} .$$

In view of Lemma 4.1.6 we obtain

$$\begin{aligned} M_1 &\le \rho^{-\alpha/2} l^l \left(\frac{\rho^{-\alpha/2}}{-\alpha e \log \rho} \right)^l \times \rho^{-\beta/2} m^m \left(\frac{\rho^{-\beta/2}}{-\beta e \log \rho} \right)^m \\ &\le \rho^{-q} l^l m^m \left(\frac{\rho^{-\alpha/2}}{-\alpha e \log \rho} \right)^l \left(\frac{\rho^{-\beta/2}}{-\beta e \log \rho} \right)^m . \end{aligned} \tag{4.34}$$

Then (4.32) follows from (4.33) and (4.34). In particular, $\Xi_{l,m}(\kappa)$ is a continuous linear operator on (E) by Lemma 4.3.8.

Conversely, suppose that $\Xi_{l,m}(\kappa)$ is a continuous operator on (E). Then, for any $p \ge 0$ there exist $C \ge 0$ and $q \ge 0$ such that

$$\|\Xi_{l,m}(\kappa)\phi\|_p \le C \|\phi\|_{p+q} , \qquad \phi \in (E).$$

Now consider

$$\phi(x) = \left\langle :x^{\otimes m}:, f \right\rangle , \quad \psi(x) = \left\langle :x^{\otimes l}:, g \right\rangle ,$$

where $f \in E_{\mathbb{C}}^{\otimes m}$ and $g \in E_{\mathbb{C}}^{\otimes l}$. By definition we have

$$\langle\langle \Xi_{l,m}(\kappa)\phi\,,\psi\rangle\rangle = l!m!\,\langle \kappa\,, g\otimes f\rangle\,,$$

and therefore,

$$|\,\langle \kappa\,, g\otimes f\rangle\,| \le \frac{C}{l!m!}\,\|\phi\|_{p+q}\,\|\psi\|_{-p} = \frac{C}{\sqrt{l!m!}}\,|g|_{-p}\,|f|_{p+q}\,.$$

It then follows from Lemma 4.3.8 that $\kappa \in (E_{\mathbb{C}}^{\otimes l})\otimes(E_{\mathbb{C}}^{\otimes m})^*$. qed

We here remark that (4.18) follows from (4.32). Indeed, replacing p and q respectively by $-p$ and $2p$ in (4.32), we obtain

$$\|\Xi_{l,m}(\kappa)\phi\|_{-p} \le \rho^{-p}\left(l^l m^m\right)^{1/2}\left(\frac{\rho^{-\alpha/2}}{-\alpha e\log\rho}\right)^{l/2}\left(\frac{\rho^{-\beta/2}}{-\beta e\log\rho}\right)^{m/2}|\kappa|_{l,m;-p,-p}\,\|\phi\|_{p}\,.$$

Then, taking $\alpha=\beta=2p$, we obtain (4.18).

We give the simplest example of integral kernel operators. In §4.1 we have introduced a differential operator D_y for $y\in E^*$. The next assertion is immediate from definition.

Proposition 4.3.10 *For $y\in E^*$ it holds that*

$$D_y = \Xi_{0,1}(y) = \int_T y(t)\partial_t\,dt, \qquad D_y^* = \Xi_{1,0}(y) = \int_T y(s)\partial_s^*\,ds.$$

In particular,

$$\partial_t = \Xi_{0,1}(\delta_t), \qquad \partial_t = \Xi_{1,0}(\delta_t).$$

Moreover, for $y_1,\cdots,y_m\in E^*$ it holds that

$$D_{y_1}\cdots D_{y_m} = \Xi_{0,m}(y_1\otimes\cdots\otimes y_m) = \int_{T^m} y_1(t_1)\cdots y_m(t_m)\partial_{t_1}\cdots\partial_{t_m}dt_1\cdots dt_m.$$

A precise estimate of $\|D_y\phi\|_p$ obtained in Theorem 4.1.1 also follows from (4.32). In fact, taking $l=0$ and $m=1$ in (4.32), we have

$$\|\Xi_{0,1}(y)\phi\|_p \le \rho^{-q/2}\left(\frac{\rho^{-\beta/2}}{-\beta e\log\rho}\right)^{1/2}|y|_{0,1;p,-(p+q)}\,\|\phi\|_{p+q}\,,$$

namely,

$$\|D_y\phi\|_p \le \rho^{-q/2}\left(\frac{\rho^{-\beta/2}}{-\beta e\log\rho}\right)^{1/2}|y|_{-(p+q)}\,\|\phi\|_{p+q}\,.$$

Since $0<\beta<2q$ is arbitrary, letting $\beta\to 2q$, we obtain (4.3) as desired. Similarly, the estimate of $\|D_{y_1}\cdots D_{y_m}\phi\|_p$ in Theorem 4.1.7 follows from (4.32).

With the help of Theorem 4.3.9 we can formulate the canonical commutation relation announced at the end of §4.1. Note first that both

$$\Xi_{0,1}(y) = \int_T y(t)\partial_t\,dt \qquad\text{and}\qquad \Xi_{1,0}(\xi) = \int_T \xi(s)\partial_s^*\,ds$$

belong to $\mathcal{L}((E),(E))$ if $y\in E_{\mathbb{C}}^*$ and $\xi\in E_{\mathbb{C}}$. In that case the product is meaningful and by a simple computation we obtain

Proposition 4.3.11 *For $\xi \in E_{\mathbf{C}}$ and $y \in E_{\mathbf{C}}^*$ it holds that*

$$[\Xi_{0,1}(y), \Xi_{1,0}(\xi)] = \langle y, \xi \rangle I,$$

where I is the identity operator on (E).

Recall that if Ξ is a continuous linear operator on (E), its adjoint Ξ^* becomes a continuous operator on $(E)^*$. Then, in view of Proposition 4.3.7 and Theorem 4.3.9 we obtain

Theorem 4.3.12 *If $\kappa \in (E_{\mathbf{C}}^{\otimes l}) \otimes (E_{\mathbf{C}}^{\otimes m})^*$, then*

$$\Xi_{m,l}(t_{m,l}(\kappa)) = \int_{T^{m+l}} \kappa(s_1, \cdots, s_l, t_1, \cdots, t_m) \partial_{t_1}^* \cdots \partial_{t_m}^* \partial_{s_1} \cdots \partial_{s_l} dt_1 \cdots dt_m ds_1 \cdots ds_l$$

is extended to a continuous linear operator from $(E)^*$ into itself.

If a continuous linear operator $\Xi \in \mathcal{L}((E),(E)^*)$ can be extended to a continuous linear operator from $(E)^*$ into itself, the extension is denoted by $\tilde{\Xi}$. For example, for $\xi \in E$ we may define a continuous operator $\widetilde{D}_\xi$ acting on $(E)^*$.

4.4 Symbols of operators

Since the exponential vectors $\{\phi_\xi;\ \xi \in E_{\mathbf{C}}\}$ span a dense subspace of (E), the behavior of an operator $\Xi \in \mathcal{L}((E),(E)^*)$ on them is worthwhile to study.

Definition 4.4.1 For $\Xi \in \mathcal{L}((E),(E)^*)$ a function on $E_{\mathbf{C}} \times E_{\mathbf{C}}$ defined by

$$\hat{\Xi}(\xi, \eta) = \langle\langle \Xi\phi_\xi, \phi_\eta \rangle\rangle, \qquad \xi, \eta \in E_{\mathbf{C}},$$

is called the *symbol* of Ξ.

This is an operator-verion of S-transform. In fact, since $\langle\langle \Xi\phi_\xi, \phi_\eta \rangle\rangle = \langle\langle \phi_\xi, \Xi^*\phi_\eta \rangle\rangle$, it holds that

$$\hat{\Xi}(\xi, \eta) = S(\Xi\phi_\xi)(\eta) = S(\Xi^*\phi_\eta)(\xi).$$

The symbol of an integral kernel operator is given in the following

Proposition 4.4.2 *Let $\kappa \in (E_{\mathbf{C}}^{\otimes(l+m)})^*$. Then*

$$\widehat{\Xi_{l,m}(\kappa)}(\xi, \eta) = \langle\langle \Xi_{l,m}(\kappa)\phi_\xi, \phi_\eta \rangle\rangle = \left\langle \kappa, \eta^{\otimes l} \otimes \xi^{\otimes m} \right\rangle e^{\langle \xi, \eta \rangle}, \qquad \xi, \eta \in E_{\mathbf{C}}.$$

PROOF. By Proposition 4.3.3 we have

$$\begin{aligned} \Xi_{l,m}(\kappa)\phi_\xi(x) &= \sum_{n=0}^{\infty} \frac{1}{n!} \left\langle :x^{\otimes(l+n)}:, (\kappa \otimes_m \xi^{\otimes m}) \otimes \xi^{\otimes n} \right\rangle \\ &= \sum_{n=l}^{\infty} \frac{1}{(n-l)!} \left\langle :x^{\otimes n}:, (\kappa \otimes_m \xi^{\otimes m}) \otimes \xi^{\otimes(n-l)} \right\rangle. \end{aligned}$$

On the other hand,

$$\phi_\eta(x) = \sum_{n=0}^{\infty} \left\langle :x^{\otimes n}:, \frac{\eta^{\otimes n}}{n!} \right\rangle.$$

Hence,

$$\begin{aligned}
\langle\!\langle \Xi_{l,m}(\kappa)\phi_\xi, \phi_\eta \rangle\!\rangle &= \sum_{n=l}^{\infty} n! \frac{1}{n!} \frac{1}{(n-l)!} \left\langle (\kappa \otimes_m \xi^{\otimes m}) \otimes \xi^{\otimes(n-l)}, \eta^{\otimes n} \right\rangle \\
&= \sum_{n=l}^{\infty} \frac{1}{(n-l)!} \left\langle \kappa \otimes_m \xi^{\otimes m}, \eta^{\otimes l} \right\rangle \langle \xi, \eta \rangle^{n-l} \\
&= \left\langle \kappa, \eta^{\otimes l} \otimes \xi^{\otimes m} \right\rangle e^{\langle \xi, \eta \rangle}.
\end{aligned}$$

This completes the proof. qed

We now discuss analytic properties of the symbol of an operator Ξ in $\mathcal{L}((E),(E)^*)$ and $\mathcal{L}((E),(E))$.

Proposition 4.4.3 *If $\Xi \in \mathcal{L}((E),(E)^*)$, then for any $\xi, \xi_1, \eta, \eta_1 \in E_{\mathbf{C}}$, the function*

$$z, w \mapsto \widehat{\Xi}(z\xi + \xi_1, w\eta + \eta_1), \qquad z, w \in \mathbf{C},$$

is an entire holomorphic function on $\mathbf{C} \times \mathbf{C}$.

PROOF. Note first that

$$\widehat{\Xi}(z\xi + \xi_1, w\eta + \eta_1) = S(\Xi\phi_{z\xi+\xi_1})(w\eta + \eta_1) = S(\Xi^*\phi_{w\eta+\eta_1})(z\xi + \xi_1).$$

It then follows from Theorem 3.3.7 that the function is entire holomorphic in two variables z and w. qed

Proposition 4.4.4 *Let $\Xi \in \mathcal{L}((E),(E)^*)$. Then, there exist constants $C \ge 0$, $K \ge 0$ and $p \in \mathbf{R}$ such that*

$$|\widehat{\Xi}(\xi, \eta)| \le C \exp K \left(|\xi|_p^2 + |\eta|_p^2 \right), \qquad \xi, \eta \in E_{\mathbf{C}}.$$

PROOF. First note that there exist $C \ge 0$ and $p \ge 0$ such that

$$\|\Xi\phi\|_{-p} \le C \|\phi\|_p, \qquad \phi \in (E). \tag{4.35}$$

In fact, in view of Proposition 1.3.12 we see that $(\phi, \psi) \mapsto \langle\!\langle \Xi\phi, \psi \rangle\!\rangle$ is a continuous bilinear form. Therefore there exist $C \ge 0$ and $p \ge 0$ such that

$$|\langle\!\langle \Xi\phi, \psi \rangle\!\rangle| \le C \|\phi\|_p \|\psi\|_p,$$

from which (4.35) follows. We then consider the operator symbol of Ξ.

$$|\widehat{\Xi}(\xi, \eta)| = |\langle\!\langle \Xi\phi_\xi, \phi_\eta \rangle\!\rangle| \le \|\Xi\phi_\xi\|_{-p} \|\phi_\eta\|_p \le C \|\phi_\xi\|_p \|\phi_\eta\|_p.$$

Since $\|\phi_\xi\|_p = \exp(|\xi|_p^2/2)$ by Lemma 3.3.3, we have

$$|\widehat{\Xi}(\xi, \eta)| \le C \exp \frac{1}{2} \left(|\xi|_p^2 + |\eta|_p^2 \right),$$

as desired. qed

Proposition 4.4.5 *Let $\Xi \in \mathcal{L}((E),(E))$. Then, for any $p \geq 0$ and $\epsilon > 0$ there exist $C \geq 0$ and $q \geq 0$ such that*

$$|\widehat{\Xi}(\xi,\eta)| \leq C \exp \epsilon \left(|\xi|_{p+q}^2 + |\eta|_{-p}^2 \right), \qquad \xi,\eta \in E_{\mathbf{C}}. \tag{4.36}$$

PROOF. Let $p \geq 0$ and $\epsilon > 0$ be given. Choose $r > 0$ such that $\rho^{2r} < 2\epsilon$. Since $\Xi \in \mathcal{L}((E),(E))$, there exist $C \geq 0$ and $s \geq 0$ such that

$$\|\Xi\phi\|_{p+r} \leq C \|\phi\|_{p+r+s} .$$

On the other hand,

$$|\widehat{\Xi}(\xi,\eta)| = |\langle\!\langle \Xi\phi_\xi , \phi_\eta \rangle\!\rangle| \leq \|\Xi\phi_\xi\|_{p+r} \|\phi_\eta\|_{-(p+r)} .$$

Therefore we come to

$$|\widehat{\Xi}(\xi,\eta)| \leq C \|\phi_\xi\|_{p+r+s} \|\phi_\eta\|_{-(p+r)} = C \exp \frac{1}{2} \left(|\xi|_{p+r+s}^2 + |\eta|_{-(p+r)}^2 \right) .$$

Since $\rho^{2r} < 2\epsilon$, we have

$$\begin{aligned} |\widehat{\Xi}(\xi,\eta)| &\leq C \exp \frac{1}{2} \left(\rho^{2r} |\xi|_{p+2r+s}^2 + \rho^{2r} |\eta|_{-p}^2 \right) \\ &\leq C \exp \epsilon \left(|\xi|_{p+2r+s}^2 + |\eta|_{-p}^2 \right) . \end{aligned}$$

Taking $q = 2r + s$, we obtain (4.36). qed

More important is that the above listed properties (Propositions 4.4.3–4.4.5) reproduce the operator on white noise functionals.

Theorem 4.4.6 *Assume that a $\mathbf{C}$-valued function Θ on $E_{\mathbf{C}} \times E_{\mathbf{C}}$ satisfies the following two conditions:*

(i) *for any $\xi, \xi_1, \eta, \eta_1 \in E_{\mathbf{C}}$, the function $z, w \mapsto \Theta(z\xi + \xi_1, w\eta + \eta_1)$ is an entire holomorphic function on $\mathbf{C} \times \mathbf{C}$;*

(ii) *there exist constants $C \geq 0$, $K \geq 0$ and $p \in \mathbf{R}$ such that*

$$|\Theta(\xi,\eta)| \leq C \exp K \left(|\xi|_p^2 + |\eta|_p^2 \right) \qquad \xi,\eta \in E_{\mathbf{C}}.$$

Then, there exists a unique family of kernel distributions $\{\kappa_{l,m}\}_{l,m=0}^{\infty}$, where $\kappa_{l,m} \in (E_{\mathbf{C}}^{\otimes(l+m)})_{\mathrm{sym}(l,m)}^*$, such that

$$|\kappa_{l,m}|_{-(p+1)} \leq C \left(l^l m^m \right)^{-1/2} (2e^3\delta^2)^{(l+m)/2} \left(\frac{\rho^{2p}}{2} + K \right)^{(l+m)/2} \tag{4.37}$$

and

$$\Theta(\xi,\eta) = \sum_{l,m=0}^{\infty} \langle\!\langle \Xi_{l,m}(\kappa_{l,m})\phi_\xi , \phi_\eta \rangle\!\rangle , \qquad \xi,\eta \in E_{\mathbf{C}}. \tag{4.38}$$

Moreover, the series

$$\Xi\phi = \sum_{l,m=0}^{\infty} \Xi_{l,m}(\kappa_{l,m})\phi, \qquad \phi \in (E), \tag{4.39}$$

converges in $(E)^*$, $\Xi \in \mathcal{L}((E),(E)^*)$ *and* $\widehat{\Xi} = \Theta$. *In that case*

$$\|\Xi\phi\|_{-(p+q+1)} \le CM(K,p,q)\,\|\phi\|_{p+q+1}, \qquad \phi \in (E), \tag{4.40}$$

where $M(K,p,q)$ *is a (finite) constant for all* $q > q_0(K,p) > 0$.

Theorem 4.4.7 *Assume that a* $\mathbf{C}$*-valued function* Θ *on* $E_{\mathbf{C}} \times E_{\mathbf{C}}$ *satisfies the following two conditions:*

(i) *for any* $\xi, \xi_1, \eta, \eta_1 \in E_{\mathbf{C}}$, *the function* $z, w \mapsto \Theta(z\xi + \xi_1, w\eta + \eta_1)$ *is an entire holomorphic function on* $\mathbf{C} \times \mathbf{C}$;

(ii) *for any* $p \ge 0$ *and* $\epsilon > 0$ *there exist* $C \ge 0$ *and* $q \ge 0$ *such that*

$$|\Theta(\xi,\eta)| \le C \exp \epsilon \left(|\xi|_{p+q}^2 + |\eta|_{-p}^2 \right), \qquad \xi, \eta \in E_{\mathbf{C}}.$$

Let $\{\kappa_{l,m}\}_{l,m=0}^{\infty}$ *be the family of kernel distributions as in Theorem 4.4.6. Then,* $\kappa_{l,m} \in ((E_{\mathbf{C}}^{\otimes l}) \otimes (E_{\mathbf{C}}^{\otimes m})^*)_{\mathrm{sym}(l,m)} = (E_{\mathbf{C}}^{\widehat{\otimes} l}) \otimes (E_{\mathbf{C}}^{\otimes m})^*_{\mathrm{sym}}$ *and*

$$|\kappa_{l,m}|_{l,m;p-1,-(p+q+1)} \le C(2e^3\delta^2)^{(l+m)/2} \left(l^l m^m\right)^{-1/2} \left(\frac{\rho^{2q}}{2\gamma^2} + \epsilon\right)^{l/2} \left(\frac{\gamma^2}{2} + \epsilon\right)^{m/2}, \tag{4.41}$$

for an arbitrary $\gamma > 0$. *Moreover, the series (4.39) converges in* (E) *and the operator* Ξ *defined there belongs to* $\mathcal{L}((E),(E))$. *In that case*

$$\|\Xi\phi\|_{p-1} \le CM(\epsilon,q,r)\,\|\phi\|_{p+q+r+1}, \qquad \phi \in (E), \tag{4.42}$$

where $M(\epsilon,q,r)$ *is a (finite) constant for* $\epsilon < (2e^3\delta^2)^{-1}$, $r \ge r_0(q) \ge 0$.

For the proofs we need some technical results.

Lemma 4.4.8 *Let* f *be an entire holomorphic function on* $\mathbf{C} \times \mathbf{C}$ *with the Taylor expansion* $f(z,w) = \sum_{m,n=0}^{\infty} a_{mn} z^m w^n$. *Assume that*

$$|f(z,w)| \le C \exp\left(K_1|z|^2 + K_2|w|^2\right), \qquad z, w \in \mathbf{C},$$

for some $C \ge 0$, $K_1 \ge 0$ *and* $K_2 \ge 0$. *Then*

$$|a_{mn}| \le C \left(\frac{2eK_1}{m}\right)^{m/2} \left(\frac{2eK_2}{n}\right)^{n/2}.$$

PROOF. Similar to the proof of Lemma 3.6.6. Since

$$a_{mn} = \left(\frac{1}{2\pi i}\right)^2 \int_{|z|=R_1} \int_{|w|=R_2} \frac{f(z,w)}{z^{m+1}w^{n+1}} dz dw, \qquad R_1 > 0, \quad R_2 > 0,$$

we have

$$\begin{aligned} |a_{mn}| &\le \sup\left\{ |f(z,w)| R_1^{-m} R_2^{-n};\ |z| = R_1, |w| = R_2 \right\} \\ &\le C R_1^{-m} \exp(K_1 R_1^2) \cdot R_2^{-n} \exp(K_2 R_2^2). \end{aligned}$$

Minimalizing the last quantity, we obtain the desired estimate, see also (3.58). qed

Lemma 4.4.9 *Let Θ be a $\mathbf{C}$-valued function on $E_{\mathbf{C}} \times E_{\mathbf{C}}$ and assume that $(z,w) \mapsto \Theta(z\xi+\xi_1, w\eta+\eta_1)$ is an entire holomorphic function on $\mathbf{C}\times\mathbf{C}$ for any $\xi, \xi_1, \eta, \eta_1 \in E_{\mathbf{C}}$. Assume in addition that there exist $C \ge 0$, $K \ge 0$, $p \in \mathbf{R}$ and $q \ge 0$ such that*

$$|\Theta(\xi,\eta)| \le C \exp K \left(|\xi|_{p+q}^2 + |\eta|_{-p}^2 \right), \qquad \xi, \eta \in E_{\mathbf{C}}. \tag{4.43}$$

Put

$$\Psi(\xi,\eta) = e^{-\langle \xi, \eta \rangle} \Theta(\xi,\eta), \qquad \xi,\eta \in E_{\mathbf{C}},$$

and

$$\kappa_{l,m}(\eta_1, \cdots, \eta_l, \xi_1, \cdots, \xi_m) = \frac{1}{l!m!} D_{\xi_1}^{(1)} \cdots D_{\xi_m}^{(1)} D_{\eta_1}^{(2)} \cdots D_{\eta_l}^{(2)} \Psi(0,0), \tag{4.44}$$

where

$$D_{\xi_1}^{(1)} \Psi(\xi,\eta) = \frac{d}{dz}\bigg|_{z=0} \Psi(z\xi_1 + \xi, \eta), \qquad D_{\eta_1}^{(2)} \Psi(\xi,\eta) = \frac{d}{dw}\bigg|_{w=0} \Psi(\xi, w\eta_1 + \eta).$$

Then, $\kappa_{l,m}$ is a continuous $(l+m)$-linear form on $E_{\mathbf{C}}$ with $s_{l,m}(\kappa_{l,m}) = \kappa_{l,m}$. Moreover, for any $\gamma > 0$ it holds that

$$\begin{aligned} &|\kappa_{l,m}|_{l,m;p-1,-(p+q+1)} \\ &\le C(2e^3\delta^2)^{(l+m)/2} \left(l^l m^m\right)^{-1/2} \left(\frac{\rho^{2q}}{2\gamma^2} + K\right)^{l/2} \left(\frac{\gamma^2}{2} + K\right)^{m/2}. \end{aligned} \tag{4.45}$$

PROOF. It follows from Lemma 3.6.7 that $\kappa_{l,m}$ is a $\mathbf{C}$-valued $(l+m)$-linear form on $E_{\mathbf{C}}$. We now put

$$A_{l,m}(\xi,\eta) = \kappa_{l,m}(\underbrace{\eta, \cdots, \eta}_{l \text{ times}}, \underbrace{\xi, \cdots, \xi}_{m \text{ times}}).$$

Then we have the Taylor expansion:

$$\begin{aligned} \Psi(z\xi, w\eta) &= \sum_{l,m=0}^{\infty} \frac{\partial^{l+m}}{\partial w^l \partial z^m} \Psi(z\xi, w\eta)\bigg|_{z=w=0} \frac{w^l z^m}{l!m!} \\ &= \sum_{l,m=0}^{\infty} A_{l,m}(\xi,\eta) z^m w^l. \end{aligned} \tag{4.46}$$

For $\gamma > 0$ we have

$$\begin{aligned}|\langle z\xi\,,w\eta\rangle| &\le |z\xi|_{p+q}\,|w\eta|_{-(p+q)}\\ &\le \frac{\gamma^2}{2}|z|^2\,|\xi|_{p+q}^2 + \frac{1}{2\gamma^2}|w|^2\,|\eta|_{-(p+q)}^2\\ &\le \frac{\gamma^2}{2}|z|^2\,|\xi|_{p+q}^2 + \frac{\rho^{2q}}{2\gamma^2}|w|^2\,|\eta|_{-p}^2\,.\end{aligned}$$

Then by assumption (4.43) we obtain

$$\begin{aligned}|\Psi(z\xi,w\eta)| &= |e^{-\langle z\xi\,,w\eta\rangle}||\Theta(z\xi,w\eta)|\\ &\le C\exp\left(\left(\frac{\gamma^2}{2}+K\right)|z|^2\,|\xi|_{p+q}^2 + \left(\frac{\rho^{2q}}{2\gamma^2}+K\right)|w|^2\,|\eta|_{-p}^2\right).\end{aligned}$$

It then follows from Lemma 4.4.8 that

$$\begin{aligned}|A_{l,m}(\xi,\eta)| &\le C\left(\frac{2e}{m}\left(\frac{\gamma^2}{2}+K\right)|\xi|_{p+q}^2\right)^{m/2}\left(\frac{2e}{l}\left(\frac{\rho^{2q}}{2\gamma^2}+K\right)|\eta|_{-p}^2\right)^{l/2}\\ &= C\left(l^l m^m\right)^{-1/2}(2e)^{(l+m)/2}\left(\frac{\rho^{2q}}{2\gamma^2}+K\right)^{l/2}\left(\frac{\gamma^2}{2}+K\right)^{m/2}|\eta|_{-p}^l\,|\xi|_{p+q}^m\,.\end{aligned}$$

By virtue of the polarization formula (A.4) we obtain

$$\begin{aligned}&\sup\left\{|\kappa_{l,m}(\eta_1,\cdots,\eta_l,\xi_1,\cdots,\xi_m)|\,;\ \begin{matrix}|\xi_i|_{p+q}\le 1, & 1\le i\le l\\ |\eta_j|_{-p}\le 1, & 1\le j\le m\end{matrix}\right\}\\ &\qquad\le e^l e^m C\left(l^l m^m\right)^{-1/2}(2e)^{(l+m)/2}\left(\frac{\rho^{2q}}{2\gamma^2}+K\right)^{l/2}\left(\frac{\gamma^2}{2}+K\right)^{m/2}\\ &\qquad\le C\left(l^l m^m\right)^{-1/2}(2e^3)^{(l+m)/2}\left(\frac{\rho^{2q}}{2\gamma^2}+K\right)^{l/2}\left(\frac{\gamma^2}{2}+K\right)^{m/2}.\end{aligned}$$

Then we have

$$\begin{aligned}&|\langle\kappa_{lm}\,,e(\mathbf{i})\otimes e(\mathbf{j})\rangle|^2\\ &\qquad\le C^2\left(l^l m^m\right)^{-1}(2e^3)^{(l+m)}\left(\frac{\rho^{2q}}{2\gamma^2}+K\right)^l\left(\frac{\gamma^2}{2}+K\right)^m|e(\mathbf{i})|_{-p}^2\,|e(\mathbf{j})|_{p+q}^2\end{aligned}$$

and, by definition

$$\begin{aligned}&|\kappa_{l,m}|_{l,m;p-1,-(p+q+1)}^2 = \sum_{\mathbf{i},\mathbf{j}}|\langle\kappa_{l,m}\,,e(\mathbf{i})\otimes e(\mathbf{j})\rangle|^2\,|e(\mathbf{i})|_{p-1}^2\,|e(\mathbf{j})|_{-(p+q+1)}^2\\ &\qquad\le C^2\left(l^l m^m\right)^{-1}(2e^3)^{(l+m)}\left(\frac{\rho^{2q}}{2\gamma^2}+K\right)^l\left(\frac{\gamma^2}{2}+K\right)^m\sum_{\mathbf{i},\mathbf{j}}|e(\mathbf{i})|_{-1}^2\,|e(\mathbf{j})|_{-1}^2\\ &\qquad\le C^2\left(l^l m^m\right)^{-1}(2e^3\delta^2)^{(l+m)}\left(\frac{\rho^{2q}}{2\gamma^2}+K\right)^l\left(\frac{\gamma^2}{2}+K\right)^m\end{aligned}$$

This completes the proof of (4.45). In particular, $\kappa_{l,m} \in (E_{\mathbb{C}}^{\otimes(l+m)})^*$. It is obvious from Lemma 3.6.5 that $s_{l,m}(\kappa_{l,m}) = \kappa_{l,m}$, namely, $\kappa_{l,m} \in (E_{\mathbb{C}}^{\otimes(l+m)})^*_{\mathrm{sym}(l,m)}$. qed

PROOF OF THEOREM 4.4.6. Without loss of generality we may assume that $p \geq 0$. If we replace p and q respectively by $-p$ and $2p$ in (4.43), we obtain the same expression as in (ii). Let $\kappa_{l,m} \in (E_{\mathbb{C}}^{\otimes(l+m)})^*$ be defined as in (4.44). Then, by (4.45) we obtain

$$|\kappa_{l,m}|_{l,m;-(p+1),-(p+1)} \leq C\left(l^l m^m\right)^{-1/2} (2e^3\delta^2)^{(l+m)/2} \left(\frac{\rho^{4p}}{2\gamma^2} + K\right)^{l/2} \left(\frac{\gamma^2}{2} + K\right)^{m/2}.$$

Putting $\gamma = \rho^p$, we obtain

$$|\kappa_{l,m}|_{-(p+1)} \leq C\left(l^l m^m\right)^{-1/2} (2e^3\delta^2)^{(l+m)/2} \left(\frac{\rho^{2p}}{2} + K\right)^{(l+m)/2}, \tag{4.47}$$

which is (4.37).

We next prove identity (4.39). It follows from Proposition 4.4.2 that

$$\langle\langle \Xi_{l,m}(\kappa_{l,m})\phi_\xi\,, \phi_\eta\rangle\rangle = \left\langle \kappa_{l,m}\,, \eta^{\otimes l} \otimes \xi^{\otimes m}\right\rangle e^{\langle \xi, \eta\rangle} = A_{l,m}(\xi,\eta)e^{\langle \xi,\eta\rangle}.$$

Hence, in view of (4.46),

$$e^{-\langle \xi,\eta\rangle} \sum_{l,m=0}^{\infty} \langle\langle \Xi_{l,m}(\kappa_{l,m})\phi_\xi\,, \phi_\eta\rangle\rangle = \sum_{l,m=0}^{\infty} A_{l,m}(\xi,\eta) = \Psi(\xi,\eta),$$

and therefore,

$$\sum_{l,m=0}^{\infty} \langle\langle \Xi_{l,m}(\kappa_{l,m})\phi_\xi\,, \phi_\eta\rangle\rangle = e^{\langle \xi,\eta\rangle}\Psi(\xi,\eta) = \Theta(\xi,\eta).$$

It follows from the uniqueness of the Taylor coefficients that $\{A_{l,m}(\xi,\eta)\}_{l,m=0}^{\infty}$ is unique, and therefore so is $\{\kappa_{l,m}\}_{l,m=0}^{\infty}$ under the condition that $s_{l,m}(\kappa_{l,m}) = \kappa_{l,m}$.

We then prove that $\sum_{l,m=0}^{\infty} \Xi_{l,m}(\kappa_{l,m})\phi$ converges in $(E)^*$ for any $\phi \in (E)$. It follows from Theorem 4.3.2 that

$$\begin{aligned} &\|\Xi_{l,m}(\kappa_{l,m})\phi\|_{-(p+q+1)} \\ &\leq \rho^{-(p+q+1)}\left(l^l m^m\right)^{1/2} \left(\frac{\rho^{-(p+q+1)}}{-2(p+q+1)e\log\rho}\right)^{(l+m)/2} |\kappa_{l,m}|_{-(p+q+1)} \|\phi\|_{p+q+1} \\ &\leq \rho^{-(p+q+1)}\left(l^l m^m\right)^{1/2} \left(\frac{\rho^{-(p+q+1)}\rho^{2q}}{-2(p+q+1)e\log\rho}\right)^{(l+m)/2} |\kappa_{l,m}|_{-(p+1)} \|\phi\|_{p+q+1}. \end{aligned}$$

In view of (4.47) we obtain

$$\begin{aligned} &\|\Xi_{l,m}(\kappa_{l,m})\phi\|_{-(p+q+1)} \\ &\leq C\rho^{-(p+q+1)} \left\{\left(\frac{e^2\delta^2\rho^{-(p+1)}\rho^q}{-(p+q+1)\log\rho}\right)\left(\frac{\rho^{2p}}{2} + K\right)\right\}^{(l+m)/2} \|\phi\|_{p+q+1}. \end{aligned}$$

Hence the series $\sum_{l,m=0}^{\infty} \|\Xi_{l,m}(\kappa_{l,m})\phi\|_{-(p+q+1)}$ converges for any $\phi \in (E)$ if

$$L = L(q) \equiv \left(\frac{e^2\delta^2 \rho^{-(p+1)} \rho^q}{-(p+q+1)\log \rho} \right) \left(\frac{\rho^{2p}}{2} + K \right) < 1.$$

In fact, we may choose $q > 0$ such that $L < 1$ holds. Then,

$$\sum_{l,m=0}^{\infty} \|\Xi_{l,m}(\kappa_{l,m})\phi\|_{-(p+q+1)} \le C\rho^{-(p+q+1)} \|\phi\|_{p+q+1} \sum_{l,m=0}^{\infty} L^{(l+m)/2}.$$

This means that the series (4.39) converges in $(E)_{-(p+q+1)}$ and Ξ becomes an operator in $\mathcal{L}((E)_{p+q+1}, (E)_{-(p+q+1)})$. In particular, $\Xi \in \mathcal{L}((E), (E)^*)$ and (4.39) converges in $(E)^*$. As for (4.40) we need only to take $q_0 = q_0(K, p)$ as $L(q_0) = 1$ and

$$M(K, p, q) = \rho^{-(p+q+1)} \sum_{l,m=0}^{\infty} L^{(l+m)/2},$$

which is finite for all $q > q_0$. Finally we see that

$$\widehat{\Xi}(\xi, \eta) = \langle\langle \Xi\phi_\xi , \phi_\eta \rangle\rangle = \sum_{l,m=0}^{\infty} \langle\langle \Xi_{l,m}(\kappa_{l,m})\phi_\xi , \phi_\eta \rangle\rangle = \Theta(\xi, \eta)$$

for all $\xi, \eta \in E_{\mathbb{C}}$ as desired. qed

PROOF OF THEOREM 4.4.7. Taking Theorem 4.4.6 into account, we need only to show that $\kappa_{l,m} \in (E_{\mathbb{C}}^{\otimes l}) \otimes (E_{\mathbb{C}}^{\otimes m})^*$, that the series converges in (E) and that the operator Ξ defined there belongs to $\mathcal{L}((E), (E))$.

Let $r \ge 0$. We first see from assumption (ii) that

$$\begin{aligned} |\Theta(\xi, \eta)| &\le C \exp \epsilon \left(|\xi|_{p+q}^2 + |\eta|_{-p}^2 \right) \\ &\le C \exp \epsilon \left(\rho^{2r} |\xi|_{p+q+r}^2 + |\eta|_{-p}^2 \right) \\ &\le C \exp \epsilon \left(|\xi|_{p+q+r}^2 + |\eta|_{-p}^2 \right) \end{aligned}$$

and by Lemma 4.4.9 we obtain

$$\begin{aligned} &|\kappa_{l,m}|_{l,m;p-1,-(p+q+r+1)} \\ &\le C(2e^3\delta^2)^{(l+m)/2} \left(l^l m^m \right)^{-1/2} \left(\frac{\rho^{2(q+r)}}{2\gamma^2} + \epsilon \right)^{l/2} \left(\frac{\gamma^2}{2} + \epsilon \right)^{m/2}, \end{aligned} \tag{4.48}$$

where $\gamma > 0$ is arbitrary. Obviously (4.41) is obtained from (4.48) by taking $r = 0$. Then $\kappa_{l,m} \in (E_{\mathbb{C}}^{\otimes l}) \otimes (E_{\mathbb{C}}^{\otimes m})^*$ follows from Lemma 4.3.8.

Furthermore it follows from Theorem 4.3.9 that for $\phi \in (E)$,

$$\begin{aligned} &\|\Xi_{l,m}(\kappa_{l,m})\phi\|_{p-1} \le \rho^{-(q+r+2)/2} \left(l^l m^m \right)^{1/2} \\ &\times \left(\frac{\rho^{-\alpha/2}}{-\alpha e \log \rho} \right)^{l/2} \left(\frac{\rho^{-\beta/2}}{-\beta e \log \rho} \right)^{m/2} |\kappa_{l,m}|_{l,m;p-1,-(p+q+r+1)} \|\phi\|_{p+q+r+1}, \end{aligned}$$

where $\alpha, \beta > 0$ with $\alpha + \beta \le 2(q+r+2)$. Then, in view of (4.48) we obtain

$$\|\Xi_{l,m}(\kappa_{l,m})\phi\|_{p-1} \le C\rho^{-(q+r+2)/2}(2e^3\delta^2)^{(l+m)/2}\,\|\phi\|_{p+q+r+1} \times \left\{\left(\frac{\rho^{2(q+r)}}{2\gamma^2}+\epsilon\right)\left(\frac{\rho^{-\alpha/2}}{-\alpha e\log\rho}\right)\right\}^{l/2}\left\{\left(\frac{\gamma^2}{2}+\epsilon\right)\left(\frac{\rho^{-\beta/2}}{-\beta e\log\rho}\right)\right\}^{m/2}. \tag{4.49}$$

For simplicity we put

$$A = 2e^3\delta^2\left(\frac{\rho^{2(q+r)}}{2\gamma^2}+\epsilon\right)\left(\frac{\rho^{-\alpha/2}}{-\alpha e\log\rho}\right), \qquad B = 2e^3\delta^2\left(\frac{\gamma^2}{2}+\epsilon\right)\left(\frac{\rho^{-\beta/2}}{-\beta e\log\rho}\right).$$

Then we see from (4.49) that

$$\sum_{l,m=0}^{\infty}\|\Xi_{l,m}(\kappa_{l,m})\phi\|_{p-1} \le C\rho^{-(q+r+2)/2}\,\|\phi\|_{p+q+r+1}\sum_{l,m=0}^{\infty}A^{l/2}B^{m/2}. \tag{4.50}$$

We shall show that $A<1$ and $B<1$ if $\epsilon>0$, $\gamma>0$ and $r\ge 0$ are chosen suitably.

First of all we are given arbitrary $p\ge 0$ and $\epsilon>0$ with $\epsilon<(2e^3\delta^2)^{-1}$. Then take $C\ge 0$ and $q\ge 0$ so as the assumption (ii) is satisfied. We now choose $r\ge 0$ with

$$-\frac{4}{\log\rho} \le 2(q+r+2)$$

and put

$$\alpha = \beta = -\frac{2}{\log\rho}.$$

Then,

$$\frac{\rho^{-\alpha/2}}{-\alpha e\log\rho} = \frac{\rho^{-\beta/2}}{-\beta e\log\rho} = \frac{1}{2}$$

and hence

$$A = e^3\delta^2\left(\frac{\rho^{2(q+r)}}{2\gamma^2}+\epsilon\right), \qquad B = e^3\delta^2\left(\frac{\gamma^2}{2}+\epsilon\right).$$

We take $\gamma>0$ as $e^3\delta^2\gamma^2=1$. Then

$$A = \frac{e^6\delta^4}{2}\rho^{2(q+r)} + \epsilon e^3\delta^2, \qquad B = \frac{1}{2}+\epsilon e^3\delta^2 < 1.$$

Finally, we may take $r\ge 0$ to have $A<1$. In fact, let $r_0=r_0(q)$ be the minimum number satisfying

$$\frac{e^6\delta^4}{2}\rho^{2(q+r_0)} \le \frac{1}{2}, \qquad -\frac{4}{\log\rho}\le 2(q+r_0+2), \qquad r_0\ge 0.$$

Then $A<1$ for all $r\ge r_0$. Consequently, (4.50) becomes

$$\sum_{l,m=0}^{\infty}\|\Xi_{l,m}(\kappa_{l,m})\|_{p-1} \le CM(\epsilon,q,r)\,\|\phi\|_{p+q+r+1}, \qquad \phi\in(E),$$

where

$$M(\epsilon,q,r)=\rho^{-(q+r+2)/2}\sum_{l,m=0}^{\infty}\left(\frac{e^6\delta^4}{2}\rho^{2(q+r)}+\epsilon e^3\delta^2\right)^{l/2}\left(\frac{1}{2}+\epsilon e^3\delta^2\right)^{m/2}.$$

This completes the proof. qed

Simply combining Propositions 4.4.3, 4.4.4 and Theorem 4.4.6, we obtain an operator version of the characterization theorem for generalized white noise functionals (Theorem 3.6.1).

Corollary 4.4.10 *Let Θ be a function on $E_{\mathbf{C}}\times E_{\mathbf{C}}$ with values in $\mathbf{C}$. Then, there exists a continuous operator $\Xi\in\mathcal{L}((E),(E)^*)$ with $\Theta=\widehat{\Xi}$ if and only if*

(i) *for any $\xi,\xi_1,\eta,\eta_1\in E_{\mathbf{C}}$, the function $z,w\mapsto\Theta(z\xi+\xi_1,w\eta+\eta_1)$ is an entire holomorphic function on $\mathbf{C}\times\mathbf{C}$;*

(ii) *there exist constants $C\geq 0$, $K\geq 0$ and $p\in\mathbf{R}$ such that*

$$|\Theta(\xi,\eta)|\leq C\exp K\left(|\xi|_p^2+|\eta|_p^2\right),\qquad \xi,\eta\in E_{\mathbf{C}}.$$

In that case

$$\|\Xi\phi\|_{-(p+q+1)}\leq CM(K,p,q)\,\|\phi\|_{p+q+1},\qquad \phi\in(E),$$

where $M(K,p,q)$ is a (finite) constant for all $q>q_0(K,p)>0$.

Similarly, by Propositions 4.4.3, 4.4.5 and Theorem 4.4.7 we obtain an operator version of the characterization theorem for test white noise functionals (Theorem 3.6.2).

Corollary 4.4.11 *Let Θ be a function on $E_{\mathbf{C}}\times E_{\mathbf{C}}$ with values in $\mathbf{C}$. Then, there exists a continuous operator $\Xi\in\mathcal{L}((E))$ with $\Theta=\widehat{\Xi}$ if and only if*

(i) *for any $\xi,\xi_1,\eta,\eta_1\in E_{\mathbf{C}}$, the function $z,w\mapsto\Theta(z\xi+\xi_1,w\eta+\eta_1)$ is an entire holomorphic function on $\mathbf{C}\times\mathbf{C}$;*

(ii) *for any $p\geq 0$ and $\epsilon>0$ there exist constants $C\geq 0$ and $q\geq 0$ such that*

$$|\Theta(\xi,\eta)|\leq C\exp\epsilon\left(|\xi|_{p+q}^2+|\eta|_{-p}^2\right),\qquad \xi,\eta\in E_{\mathbf{C}}.$$

In that case

$$\|\Xi\phi\|_{p-1}\leq CM(\epsilon,q,r)\,\|\phi\|_{p+q+r+1},\qquad \phi\in(E),$$

where $M(\epsilon,q,r)$ is a (finite) constant for $\epsilon<(2e^3\delta^2)^{-1}$, $r\geq r_0(q)\geq 0$.

In some practical problems operators on Fock space are only defined on the exponential vectors $\{\phi_\xi;\,\xi\in E_{\mathbf{C}}\}$ due to the fact that they are linearly independent (Proposition 2.3.9). The above corollaries then give us criteria when such operators belong to $\mathcal{L}((E),(E)^*)$ or $\mathcal{L}((E),(E))$. A simple application will be illustrated in §4.6.

4.5 Fock expansion

Theorem 4.5.1 *For any $\Xi \in \mathcal{L}((E),(E)^*)$ there exists a unique family of distributions $\{\kappa_{l,m}\}_{l,m=0}^{\infty}$, $\kappa_{l,m} \in (E_{\mathbb{C}}^{\otimes(l+m)})^*_{\mathrm{sym}(l,m)}$, such that*

$$\Xi\phi = \sum_{l,m=0}^{\infty} \Xi_{l,m}(\kappa_{l,m})\phi, \qquad \phi \in (E), \tag{4.51}$$

where the right hand side converges in $(E)^*$. If $\Xi \in \mathcal{L}((E),(E))$, then each kernel distribution $\kappa_{l,m}$ belongs to $((E_{\mathbb{C}}^{\otimes l}) \otimes (E_{\mathbb{C}}^{\otimes m})^*)_{\mathrm{sym}(l,m)} = (E_{\mathbb{C}}^{\hat{\otimes} l}) \otimes (E_{\mathbb{C}}^{\otimes m})^*_{\mathrm{sym}}$ and the right hand side of (4.51) converges in (E).

PROOF. For a given $\Xi \in \mathcal{L}((E),(E)^*)$ we put

$$\Theta(\xi,\eta) = \widehat{\Xi}(\xi,\eta) = \langle\!\langle \Xi\phi_\xi , \phi_\eta \rangle\!\rangle, \qquad \xi,\eta \in E_{\mathbb{C}}. \tag{4.52}$$

Then, by Propositions 4.4.3 and 4.4.4 we see that Θ satisfies the conditions (i) and (ii) in Theorem 4.4.6. Therefore, there exists a unique family of kernels $\{\kappa_{l,m}\}_{l,m=0}^{\infty}$, $\kappa_{l,m} \in (E_{\mathbb{C}}^{\otimes(l+m)})^*_{\mathrm{sym}(l,m)}$, such that

$$\Theta(\xi,\eta) = \sum_{l,m=0}^{\infty} \langle\!\langle \Xi_{l,m}(\kappa_{l,m})\phi_\xi , \phi_\eta \rangle\!\rangle, \qquad \xi,\eta \in E_{\mathbb{C}}.$$

Furthermore, as is stated in Theorem 4.4.6,

$$\Xi'\phi = \sum_{l,m=0}^{\infty} \Xi_{l,m}(\kappa_{l,m})\phi, \qquad \phi \in (E),$$

converges in $(E)^*$, $\Xi' \in \mathcal{L}((E),(E)^*)$ and $\widehat{\Xi'}(\xi,\eta) = \Theta(\xi,\eta)$ for all $\xi,\eta \in E_{\mathbb{C}}$. The last identity and (4.52) yield

$$\langle\!\langle \Xi'\phi_\xi , \phi_\eta \rangle\!\rangle = \langle\!\langle \Xi\phi_\xi , \phi_\eta \rangle\!\rangle, \qquad \xi,\eta \in E_{\mathbb{C}}.$$

Since the exponential vectors span a dense subspace of (E) and both Ξ and Ξ' are continuous, we conclude that $\Xi = \Xi'$.

For the second half of the assertion we need only to employ Proposition 4.4.5 and Theorem 4.4.7. qed

Definition 4.5.2 The unique expression of $\Xi \in \mathcal{L}((E),(E)^*)$ given in (4.51) is called the *Fock expansion* of Ξ and denoted simply by

$$\Xi = \sum_{l,m=0}^{\infty} \Xi_{l,m}(\kappa_{l,m}).$$

Proposition 4.5.3 *Let $\Xi \in \mathcal{L}((E),(E)^*)$ and let $\Xi = \sum_{l,m=0}^{\infty} \Xi_{l,m}(\kappa_{l,m})$ be its Fock expansion. Then,*

$$e^{-\langle\xi,\eta\rangle}\widehat{\Xi}(\xi,\eta) = \sum_{l,m=0}^{\infty} \left\langle \kappa_{l,m} , \eta^{\otimes l} \otimes \xi^{\otimes m} \right\rangle, \qquad \xi,\eta \in E_{\mathbb{C}}. \tag{4.53}$$

PROOF. It follows from Theorem 4.5.1 that

$$\Xi\phi_\xi = \sum_{l,m=0}^{\infty} \Xi_{l,m}(\kappa_{l,m})\phi_\xi,$$

converges in $(E)^*$. Therefore, for $\xi, \eta \in E_{\mathbb{C}}$,

$$\langle\langle \Xi\phi_\xi, \phi_\eta \rangle\rangle = \sum_{l,m=0}^{\infty} \langle\langle \Xi_{l,m}(\kappa_{l,m})\phi_\xi, \phi_\eta \rangle\rangle.$$

Then, in view of Proposition 4.4.2 we obtain (4.53). qed

Thus, in order to obtain the Fock expansion of $\Xi \in \mathcal{L}((E),(E)^*)$ one need only to compute the Taylor expansion of (4.53).

Corollary 4.5.4 *Every bounded operator Ξ on (L^2) admits a Fock expansion.*

PROOF. If Ξ is a bounded operator on (L^2), there exists some $C \geq 0$ such that

$$\|\Xi\phi\|_0 \leq C\,\|\phi\|_0, \qquad \phi \in (E).$$

Hence $\Xi \in \mathcal{L}((E),(E)^*)$ and it admits a Fock expansion. qed

Even though Ξ is a bounded operator on (L^2), the convergence of the Fock expansion can not be discussed within the framework of a Hilbert space. The following result also illustrates this remark.

Proposition 4.5.5 *Let $\kappa \in (E_{\mathbb{C}}^{\otimes(l+m)})^*$. If $\Xi_{l,m}(\kappa)$ admits an extension to a bounded operator on (L^2), then $s_{l,m}(\kappa) = 0$ or $l = m = 0$. Namely, except scalar operators no integral kernel operator admits an extension to a bounded operator on (L^2).*

PROOF. We first observe the action of $\Xi_{l,m}(\kappa)$ to an exponential vector ϕ_ξ. Since

$$\Xi_{l,m}(\kappa)\phi_\xi(x) = \sum_{n=0}^{\infty} \frac{1}{n!}\left\langle :x^{\otimes(n+l)}:, (\kappa \otimes_m \xi^{\otimes m}) \otimes \xi^{\otimes n}\right\rangle$$

by (4.24) in Proposition 4.3.3, we have

$$\begin{aligned} \|\Xi_{l,m}(\kappa)\phi_\xi\|_0^2 &= \sum_{n=0}^{\infty} \frac{(n+l)!}{n!n!}\left|(\kappa \otimes_m \xi^{\otimes m}) \otimes \xi^{\otimes n}\right|_0^2 \\ &= \left|\kappa \otimes_m \xi^{\otimes m}\right|_0^2 \sum_{n=0}^{\infty} \frac{(n+l)!}{n!n!} |\xi|_0^{2n}. \end{aligned} \tag{4.54}$$

Defining $P_l(t)$ by

$$\sum_{n=0}^{\infty} \frac{(n+l)!}{n!n!} t^n = P_l(t)e^t,$$

we can easily see that $P_l(t)$ is a polynomial of degree l. (This P_l appeared also in the proof of Lemma 3.2.9.) Hence (4.54) becomes

$$\|\Xi_{l,m}(\kappa)\phi_\xi\|_0^2 = |\kappa \otimes_m \xi^{\otimes m}|_0^2 P_l\left(|\xi|_0^2\right) e^{|\xi|_0^2}. \tag{4.55}$$

We now suppose that $\Xi_{l,m}(\kappa)$ admits an extension to a bounded operator on (L^2). Then, there exists some $C \geq 0$ such that

$$\|\Xi_{l,m}(\kappa)\phi_\xi\|_0^2 \leq C \|\phi_\xi\|_0^2 = Ce^{|\xi|_0^2}, \qquad \xi \in E_{\mathbf{C}}. \tag{4.56}$$

Combining (4.55) and (4.56) we obtain

$$|\kappa \otimes_m \xi^{\otimes m}|_0^2 P_l\left(|\xi|_0^2\right) \leq C, \qquad \xi \in E_{\mathbf{C}}. \tag{4.57}$$

Suppose that $l \neq 0$. In order that (4.57) is true, we have

$$\kappa \otimes_m \xi^{\otimes m} = 0, \qquad \xi \in E_{\mathbf{C}}.$$

Then, for all $\xi, \eta \in E_{\mathbf{C}}$,

$$0 = \left\langle \kappa \otimes_m \xi^{\otimes m}, \eta^{\otimes l} \right\rangle = \left\langle \kappa, \eta^{\otimes l} \otimes \xi^{\otimes m} \right\rangle = \left\langle s_{l,m}(\kappa), \eta^{\otimes l} \otimes \xi^{\otimes m} \right\rangle.$$

This means that $s_{l,m}(\kappa) = 0$, and therefore $\Xi_{l,m}(\kappa) = 0$. If $m \neq 0$, applying a similar argument to $\Xi_{l,m}(\kappa)^* = \Xi_{m,l}(t_{m,l}(\kappa))$, we come to the same conclusion. Consequently, $\Xi_{l,m}(\kappa) = 0$ unless $l = m = 0$. qed

Therefore, the Fock expansion of a non-scalar bounded operator on (L^2) is always an infinite series of unbounded operators.

4.6 Some examples

We now assemble a few examples of Fock expansions and applications of the characterization theorems for operator symbols.

For $y \in E^*$ the translation operator is defined by

$$T_y\phi(x) = \phi(x + y), \qquad \phi \in (E).$$

It was shown in Theorem 4.2.3 that $T_y \in \mathcal{L}((E), (E))$.

Theorem 4.6.1 *For $y \in E^*$ we have*

$$T_y = \sum_{n=0}^{\infty} \frac{1}{n!} \Xi_{0,n}(y^{\otimes n}). \tag{4.58}$$

PROOF. Let $\xi, \eta \in E_{\mathbf{C}}$. First note that

$$T_y\phi_\xi(x) = \phi_\xi(x + y) = \exp\left(\langle x + y, \xi\rangle - \frac{1}{2}\langle \xi, \xi\rangle\right) = e^{\langle y, \xi\rangle}\phi_\xi(x).$$

Hence

$$\widehat{T}_y(\xi, \eta) = \langle\langle T_y\phi_\xi, \phi_\eta\rangle\rangle = e^{\langle y, \xi\rangle}\langle\langle \phi_\xi, \phi_\eta\rangle\rangle = e^{\langle y, \xi\rangle}e^{\langle \xi, \eta\rangle}.$$

We therefore obtain

$$e^{-\langle \xi, \eta\rangle}\widehat{T}_y(\xi, \eta) = e^{\langle y, \xi\rangle} = \sum_{m=0}^{\infty} \frac{\langle y, \xi\rangle^m}{m!} = \sum_{m=0}^{\infty} \frac{1}{m!}\left\langle y^{\otimes m}, \xi^{\otimes m}\right\rangle. \tag{4.59}$$

On the other hand, if

$$T_y = \sum_{l,m=0}^{\infty} \Xi_{l,m}(\kappa_{l,m})$$

is the Fock expansion, it follows from Proposition 4.5.3 that

$$e^{-\langle \xi,\eta\rangle}\widehat{T}_y(\xi,\eta) = \sum_{l,m=0}^{\infty} \left\langle \kappa_{l,m}, \eta^{\otimes l} \otimes \xi^{\otimes m} \right\rangle. \tag{4.60}$$

Comparing (4.59) and (4.60), we conclude that

$$\begin{aligned} \kappa_{l,m} &= 0 && \text{if } l \geq 1, \\ \kappa_{0,m} &= \frac{1}{m!} y^{\otimes m} && \text{for } m \geq 0. \end{aligned}$$

This completes the proof. qed

The above result yields the Taylor expansion of white noise functional $\phi \in (E)$.

Corollary 4.6.2 *Let $y \in E^*$. Then,*

$$T_y\phi = \sum_{n=0}^{\infty} \frac{1}{n!} D_y^n \phi, \qquad \phi \in (E),$$

where the series converges in (E). In particular,

$$\phi(x+y) = \sum_{n=0}^{\infty} \frac{1}{n!} D_y^n \phi(x), \qquad x \in E^*.$$

PROOF. It follows from Propositions 4.3.10 that

$$\Xi_{0,n}(y^{\otimes n}) = D_y^n, \qquad y \in E^*.$$

Hence the Fock expansion of T_y (Theorem 4.6.1) yields

$$T_y\phi = \sum_{n=0}^{\infty} \frac{1}{n!} D_y^n \phi, \qquad \phi \in (E). \tag{4.61}$$

Since T_y belongs to $\mathcal{L}((E),(E))$, it follows from Theorem 4.5.1 that the series (4.61) converges in (E). qed

We next give an example of a bounded operator on (L^2).

Proposition 4.6.3 *Let π_n be the orthogonal projection from (L^2) onto $\mathcal{H}_n(\mathbb{C})$. Then its Fock expansion is given by*

$$\pi_n = \sum_{m=n}^{\infty} \frac{(-1)^{m-n}}{n!(m-n)!} \Xi_{m,m}(\tau_m),$$

where

$$\tau_m = \sum_{i_1,\cdots,i_m=0}^{\infty} e_{i_1} \otimes \cdots \otimes e_{i_m} \otimes e_{i_1} \otimes \cdots \otimes e_{i_m} \in (E^{\otimes 2m})^*. \tag{4.62}$$

Proof. By definition,

$$\widehat{\pi_n}(\xi,\eta) = \langle\langle \pi_n\phi_\xi , \phi_\eta\rangle\rangle = n!\left\langle \frac{\xi^{\otimes n}}{n!} , \frac{\eta^{\otimes n}}{n!}\right\rangle = \frac{\langle \xi , \eta\rangle^n}{n!}.$$

Hence,

$$\begin{aligned} e^{-\langle\xi,\eta\rangle}\widehat{\pi_n}(\xi,\eta) &= \sum_{m=0}^{\infty} \frac{(-1)^m \langle \xi , \eta\rangle^m}{m!} \frac{\langle \xi , \eta\rangle^n}{n!} \\ &= \sum_{m=n}^{\infty} \frac{(-1)^{m-n} \langle \xi , \eta\rangle^m}{(m-n)!n!} \\ &= \sum_{m=n}^{\infty} \frac{(-1)^{m-n}}{(m-n)!n!} \left\langle \tau_m , \eta^{\otimes m} \otimes \xi^{\otimes m}\right\rangle . \end{aligned}$$

Then the assertion follows immediately from Proposition 4.5.3. qed

Note that $\tau_1 = \tau$. It will be proved in Proposition 5.4.6 that $\Xi_{m,m}(\tau_m)$ is a polynomial in a particular operator called the number operator. While, one may employ the following expression as well.

$$\Xi_{m,m}(\tau_m) = \int_{T^m} \partial_{t_1}^* \cdots \partial_{t_m}^* \partial_{t_1} \cdots \partial_{t_m} dt_1 \cdots dt_m .$$

Recall that any $\Phi \in (E)^*$ gives rise to a continuous operator from (E) into $(E)^*$ by multiplication, see Theorem 3.5.8 and Corollary 3.5.9.

Proposition 4.6.4 *Let $\Phi \in (E)^*$ be given with Wiener-Itô expansion:*

$$\Phi(x) = \sum_{n=0}^{\infty} \left\langle :x^{\otimes n}: , F_n\right\rangle ,$$

where $F_n \in (E^{\otimes n})^*_{\rm sym}$. Then as multiplication operator Φ admits Fock expansion:

$$\Phi = \sum_{l,m=0}^{\infty} \binom{l+m}{m} \Xi_{l,m}(F_{l+m}).$$

Moreover, $\Phi \in \mathcal{L}((E),(E))$ as multiplication operator if and only if $\Phi \in (E)$.

Proof. We first compute the symbol of Φ:

$$\widehat{\Phi}(\xi,\eta) = \langle\langle \Phi\phi_\xi , \phi_\eta\rangle\rangle = \langle\langle \Phi , \phi_\xi\phi_\eta\rangle\rangle .$$

Since $\phi_\xi\phi_\eta = e^{\langle\xi,\eta\rangle}\phi_{\xi+\eta}$ from definition, we obtain

$$\begin{aligned} \widehat{\Phi}(\xi,\eta) &= \langle\langle \Phi , \phi_{\xi+\eta}\rangle\rangle\, e^{\langle\xi,\eta\rangle} \\ &= e^{\langle\xi,\eta\rangle} \sum_{n=0}^{\infty} n! \left\langle F_n , \frac{(\xi+\eta)^{\otimes n}}{n!}\right\rangle \\ &= e^{\langle\xi,\eta\rangle} \sum_{n=0}^{\infty} \left\langle F_n , (\xi+\eta)^{\otimes n}\right\rangle . \end{aligned}$$

Therefore,

$$\begin{aligned} e^{-\langle\xi,\eta\rangle}\widehat{\Phi}(\xi,\eta) &= \sum_{n=0}^{\infty}\left\langle F_n, (\xi+\eta)^{\otimes n}\right\rangle \\ &= \sum_{n=0}^{\infty}\sum_{k=0}^{n}\left\langle F_n, \binom{n}{k}\xi^{\otimes k}\otimes\eta^{\otimes(n-k)}\right\rangle \\ &= \sum_{k=0}^{\infty}\sum_{n=0}^{\infty}\binom{n+k}{k}\left\langle F_{n+k}, \xi^{\otimes k}\otimes\eta^{\otimes n}\right\rangle. \end{aligned}$$

Since F_{n+k} is symmetric, we have

$$e^{-\langle\xi,\eta\rangle}\widehat{\Phi}(\xi,\eta) = \sum_{l,m=0}^{\infty}\binom{l+m}{l}\left\langle F_{l+m}, \eta^{\otimes l}\otimes\xi^{\otimes m}\right\rangle.$$

Then, we need only to apply Proposition 4.5.3.

Suppose $\Phi \in \mathcal{L}((E),(E))$ as multiplication operator. Then it follows from Corollary 4.4.11 that for any $p \ge 0$ and $\epsilon > 0$ there exist some $C \ge 0$ and $q \ge 0$ such that

$$|\widehat{\Phi}(\xi,\eta)| \le C\exp\epsilon\left(|\xi|_{p+q}^2 + |\eta|_{-p}^2\right).$$

Viewing

$$\widehat{\Phi}(\xi,\eta) = \langle\langle\Phi, \phi_{\xi+\eta}\rangle\rangle\, e^{\langle\xi,\eta\rangle} = S\Phi(\xi+\eta)e^{\langle\xi,\eta\rangle}$$

and putting $\xi = 0$, we obtain

$$|S\Phi(\eta)| \le C\exp\left(\epsilon\,|\eta|_{-p}^2\right), \qquad \eta\in E_{\mathbf{C}}.$$

Then it follows from Theorem 3.6.2 that $\Phi \in (E)$. qed

As an immediate consequence we come to a reproduction of Theorem 4.1.5.

Corollary 4.6.5 *For $t \in T$ we have*

$$x(t) = \Xi_{1,0}(\delta_t) + \Xi_{0,1}(\delta_t) = \partial_t + \partial_t^*.$$

Since the exponential vectors are linearly independent (Proposition 2.3.9), a linear operator is uniquely determined on the linear span of the exponential vectors. For example, for each $\lambda \in \mathbf{C}$ a linear operator S_λ is defined by

$$S_\lambda\phi_\xi = e^{(\lambda^2-1)\langle\xi,\xi\rangle/2}\phi_{\lambda\xi}, \qquad \xi\in E_{\mathbf{C}}. \tag{4.63}$$

We then come to a natural question whether or not S_λ is extended to a continuous operator from (E) into itself or into $(E)^*$. This question is easily answered with the help of our characterization theorem for operator symbols. We first prepare a simple but useful result.

Lemma 4.6.6 *For any $K_1, K_2 \ge 0$, $p \ge 0$ and $\epsilon > 0$ there exists $q \ge 0$ such that*

$$K_1|\langle\xi,\xi\rangle| + K_2|\langle\xi,\eta\rangle| \le \epsilon\left(|\xi|_{p+q}^2 + |\eta|_{-p}^2\right), \qquad \xi,\eta\in E_{\mathbf{C}}.$$

PROOF. We first note that for any $q \ge 0$ and $\gamma > 0$,

$$\begin{aligned} K_1|\langle \xi, \xi\rangle| + K_2|\langle \xi, \eta\rangle| &\le K_1 |\xi|_{p+q} |\xi|_{-(p+q)} + K_2 |\xi|_p |\eta|_{-p} \\ &\le K_1\rho^{2(p+q)} |\xi|_{p+q}^2 + \frac{K_2}{2}\left(\gamma^2 |\xi|_p^2 + \frac{1}{\gamma^2}|\eta|_{-p}^2\right) \\ &\le \left(K_1\rho^{2p} + \frac{K_2\gamma^2}{2}\right)\rho^{2q} |\xi|_{p+q}^2 + \frac{K_2}{2\gamma^2}|\eta|_{-p}^2 . \end{aligned}$$

It is possible to choose $\gamma > 0$ and $q \ge 0$ satisfying

$$\frac{K_2}{2\gamma^2} < \epsilon \qquad \text{and} \qquad \left(K_1\rho^{2p} + \frac{K_2\gamma^2}{2}\right)\rho^{2q} < \epsilon.$$

This proves the assertion. qed

Proposition 4.6.7 *For any $\lambda \in \mathbb{C}$ there exists a unique continuous operator $S_\lambda \in \mathcal{L}((E),(E))$ satisfying (4.63). Moreover, for $\lambda \in \mathbb{R}$ it holds that*

$$S_\lambda\phi(x) = \phi(\lambda x), \qquad x \in E^*, \quad \phi \in (E).$$

PROOF. For $\xi, \eta \in E_{\mathbb{C}}$ we set

$$\Theta(\xi,\eta) = \langle\!\langle S_\lambda\phi_\xi, \phi_\eta\rangle\!\rangle = e^{(\lambda^2-1)\langle\xi,\xi\rangle/2}\langle\!\langle \phi_{\lambda\xi}, \phi_\eta\rangle\!\rangle = e^{(\lambda^2-1)\langle\xi,\xi\rangle/2}e^{\lambda\langle\xi,\eta\rangle}.$$

We shall prove that Θ satisfies the conditions (i) and (ii) in Corollary 4.4.11 (or Theorem 4.4.7). In fact, (i) is obvious. To see (ii) let $p \ge 0$ and $\epsilon > 0$ be given arbitrarily. Then, by Lemma 4.6.6 there exists $q \ge 0$ such that

$$|\Theta(\xi,\eta)| \le \exp\epsilon\left(|\xi|_{p+q}^2 + |\eta|_{-p}^2\right), \qquad \xi,\eta \in E_{\mathbb{C}}.$$

Therefore there exists an operator in $\mathcal{L}((E),(E))$ whose operator symbol is Θ. In particular, the action of this operator on exponential vectors is as in (4.63).

Suppose next that $\lambda \in \mathbb{R}$. By definition, for $\xi \in E_{\mathbb{C}}$ and $x \in E^*$,

$$S_\lambda\phi_\xi(x) = e^{(\lambda^2-1)\langle\xi,\xi\rangle/2}\phi_{\lambda\xi}(x) = e^{(\lambda^2-1)\langle\xi,\xi\rangle/2}e^{\langle x,\lambda\xi\rangle - \langle\lambda\xi,\lambda\xi\rangle/2} = \phi_\xi(\lambda x).$$

For a general $\phi \in (E)$ take an approximating sequence ψ_n each of which is a linear combination of exponential vectors. Then $S_\lambda\psi_n \to S_\lambda\phi$ in (E), and therefore pointwisely, i.e., $S_\lambda\psi_n(x) \to S_\lambda\phi(x)$ for any $x \in E^*$. On the other hand, since $S_\lambda\psi_n(x) = \psi_n(\lambda x)$, we conclude that $S_\lambda\phi(x) = \phi(\lambda x)$. qed

Definition 4.6.8 $S_\lambda \in \mathcal{L}((E),(E))$, $\lambda \in \mathbb{C}$, is called a *scaling operator.*

The symbol of S_λ is already obtained during the above proof. We record it as well as the Fock expansion which follows immediately from the Taylor expansion of the operator symbol.

Proposition 4.6.9 *For $\lambda \in \mathbf{C}$ we have*

$$\begin{aligned}\widehat{S_\lambda}(\xi,\eta) &= e^{(\lambda^2-1)\langle\xi,\xi\rangle/2+\lambda\langle\xi,\eta\rangle}, \qquad \xi,\eta \in E_{\mathbf{C}},\\ S_\lambda &= \sum_{l,m=0}^{\infty} \frac{(\lambda-1)^l(\lambda^2-1)^m}{l!\,m!\,2^m}\Xi_{l,l+2m}(\tau_l \otimes \tau^{\otimes m}).\end{aligned}$$

The above method of defining an operator is useful. Here is another example.

Proposition 4.6.10 *For any $T \in \mathcal{L}(E_{\mathbf{C}}, E_{\mathbf{C}})$ there exists a unique operator $\Gamma(T) \in \mathcal{L}((E),(E))$ such that*

$$\Gamma(T)\phi_\xi = \phi_{T\xi}, \qquad \xi \in E_{\mathbf{C}}. \tag{4.64}$$

Moreover, for $\phi \in (E)$ given with Wiener-Itô expansion:

$$\phi(x) = \sum_{n=0}^{\infty} \left\langle :x^{\otimes n}:, f_n\right\rangle$$

it holds that

$$\Gamma(T)\phi(x) = \sum_{n=0}^{\infty} \left\langle :x^{\otimes n}:, T^{\otimes n} f_n\right\rangle.$$

PROOF. We put

$$\Theta(\xi,\eta) = \langle\!\langle \Gamma(T)\phi_\xi\,, \phi_\eta\rangle\!\rangle = \langle\!\langle \phi_{T\xi}\,, \phi_\eta\rangle\!\rangle = e^{\langle T\xi,\eta\rangle}, \qquad \xi,\eta \in E_{\mathbf{C}}.$$

For the first half of the assertion it is sufficient to verify the conditions (i) and (ii) in Corollary 4.4.11. In fact, (i) is immediate. For (ii) let $p \geq 0$ and $\epsilon > 0$ be given. By the continuity of T there exist $C \geq 0$ and $q \geq 0$ such that

$$|T\xi|_p \leq C\,|\xi|_{p+q}\,, \qquad \xi \in E_{\mathbf{C}}.$$

It then follows that

$$\begin{aligned}|\Theta(\xi,\eta)| &\leq \exp(C\,|\xi|_{p+q}\,|\eta|_{-p})\\ &\leq \exp(C\rho^r\,|\xi|_{p+q+r}\,|\eta|_{-p})\\ &\leq \exp\frac{C\rho^r}{2}\left(|\xi|_{p+q+r}^2 + |\eta|_{-p}^2\right)\end{aligned}$$

for any $r \geq 0$. Taking $r \geq 0$ large enough, we have $C\rho^r/2 < \epsilon$. Thus the proof of the first half is completed. The second half is proved in a similar manner as in the proof of Proposition 4.6.7. qed

Recall that we used $\Gamma(A)$ to construct the white noise functionals. This operator is, of course, a special case of the above.

In general, for $T \in \mathcal{L}(E_{\mathbf{C}}, E_{\mathbf{C}})$ we define an operator $d\Gamma(T)$ on (E). Suppose $\phi \in (E)$ is given as

$$\phi(x) = \sum_{n=0}^{\infty} \left\langle :x^{\otimes n}:, f_n\right\rangle, \qquad x \in E^*,$$

as usual. Then we put

$$d\Gamma(T)\phi(x) = \sum_{n=0}^{\infty} \left\langle :x^{\otimes n}:, \gamma_n(T) f_n \right\rangle, \tag{4.65}$$

where

$$\begin{cases} \gamma_n(T) & = \sum_{k=0}^{n-1} I^{\otimes k} \otimes T \otimes I^{\otimes(n-1-k)}, \quad n \geq 1, \\ \gamma_0(T) & = 0. \end{cases} \tag{4.66}$$

Proposition 4.6.11 *$d\Gamma(T) \in \mathcal{L}((E),(E))$ for any $T \in \mathcal{L}(E_{\mathbb{C}}, E_{\mathbb{C}})$.*

PROOF. Let $p \geq 0$. From definition we obtain

$$\|d\Gamma(T)\phi\|_p^2 = \sum_{n=0}^{\infty} n! \, |\gamma_n(T) f_n|_p^2 = \sum_{n=1}^{\infty} n! n^2 \left|(I^{\otimes(n-1)} \otimes T) f_n\right|_p^2. \tag{4.67}$$

Supposing $q \geq 1$, we compute

$$\begin{aligned} &\left|(I^{\otimes(n-1)} \otimes T) f_n\right|_p^2 = \\ &= \sum_{j_1,\cdots,j_n} \left|\left\langle (I^{\otimes(n-1)} \otimes T) f_n, e_{j_1} \otimes \cdots \otimes e_{j_n} \right\rangle\right|^2 |e_{j_1}|_p^2 \cdots |e_{j_n}|_p^2 \\ &= \sum_{j_1,\cdots,j_n} |\langle f_n, (T^* e_{j_1}) \otimes \cdots \otimes e_{j_n} \rangle|^2 \, |e_{j_1}|_p^2 \cdots |e_{j_n}|_p^2 \\ &\leq \sum_{j_1,\cdots,j_n} |f_n|_{p+q}^2 \, |T^* e_{j_1}|_{-(p+q)}^2 \, |e_{j_2}|_{-(p+q)}^2 \cdots |e_{j_n}|_{-(p+q)}^2 \, |e_{j_1}|_p^2 \cdots |e_{j_n}|_p^2 \\ &= |f_n|_{p+q}^2 \sum_{j_1} |T^* e_{j_1}|_{-(p+q)}^2 \, |e_{j_1}|_p^2 \sum_{j_2,\cdots,j_n} |e_{j_2}|_{-q}^2 \cdots |e_{j_n}|_{-q}^2 \\ &\leq |f_n|_{p+q}^2 \, \delta^{2(n-1)} \sum_j |T^* e_j|_{-(p+q)}^2 \, |e_j|_p^2 . \end{aligned} \tag{4.68}$$

On the other hand, since $T \in \mathcal{L}(E_{\mathbb{C}}, E_{\mathbb{C}})$, there exist $q \geq 1$ and $C \geq 0$ such that

$$|T\xi|_{p+1} \leq C \, |\xi|_{p+q} .$$

Then

$$|\langle T^* e_j, \xi \rangle| = |\langle T\xi, e_j \rangle| \leq |T\xi|_{p+1} \, |e_j|_{-(p+1)} \leq C \, |\xi|_{p+q} \, |e_j|_{-(p+1)},$$

and therefore

$$|T^* e_j|_{-(p+q)} \leq C \, |e_j|_{-(p+1)} .$$

Then (4.68) becomes

$$\left|(I^{\otimes(n-1)} \otimes T) f_n\right|_p^2 \leq |f_n|_{p+q}^2 \, \delta^{2(n-1)} \sum_j C^2 \, |e_j|_{-(p+1)}^2 \, |e_j|_p^2 = C^2 \delta^{2n} \, |f_n|_{p+1}^2 .$$

Consequently, (4.67) is estimated as follows.

$$\begin{aligned}\|d\Gamma(T)\phi\|_p^2 &\le \sum_{n=1}^{\infty} n!n^2C^2\delta^{2n}\,|f_n|_{p+1}^2 \\ &\le C^2\sum_{n=1}^{\infty} n!n^2\delta^{2n}\rho^{2nr}\,|f_n|_{p+r+1}^2 \\ &\le C^2\sup_{n\ge 1} n^2(\delta\rho^r)^{2n}\sum_{n=1}^{\infty} n!\,|f_n|_{p+r+1}^2 \\ &\le C^2\,\|\phi\|_{p+r+1}^2\,\sup_{n\ge 0} n^2(\delta\rho^r)^{2n}.\end{aligned}$$

Taking large $r \ge 0$ such that $\delta\rho^r < 1$, we see that $d\Gamma(T) \in \mathcal{L}((E),(E))$. qed

Definition 4.6.12 $\Gamma(T)$ and $d\Gamma(T)$ are called the *second quantization* and *differential second quantization* of T, respectively.

Proposition 4.6.13 *Let $T \in \mathcal{L}(E_{\mathbf{C}}, E_{\mathbf{C}})$. Then for $\Gamma(T)$ we have*

$$\begin{aligned}\widehat{\Gamma(T)}(\xi,\eta) &= e^{\langle T\xi,\eta\rangle}, \qquad \xi,\eta \in E_{\mathbf{C}}, \\ \Gamma(T) &= \sum_{m=0}^{\infty}\frac{1}{m!}\Xi_{m,m}\left((I^{\otimes m}\otimes(T-I)^{\otimes m})^*\tau_m\right),\end{aligned}$$

and for $d\Gamma(T)$,

$$\begin{aligned}\widehat{d\Gamma(T)}(\xi,\eta) &= \langle T\xi,\eta\rangle\, e^{\langle\xi,\eta\rangle}, \qquad \xi,\eta \in E_{\mathbf{C}}, \\ d\Gamma(T) &= \Xi_{11}((I\otimes T)^*\tau).\end{aligned}$$

The verification is a simple computation. Consider a one-parameter group $\{e^{\theta T}\}_{\theta\in\mathbf{R}}$ formally. Since

$$\widehat{\Gamma(e^{\theta T})}(\xi,\eta) = e^{\langle e^{\theta T}\xi,\eta\rangle},$$

differentiating at $\theta = 0$, we obtain

$$\left.\frac{d}{d\theta}\right|_{\theta=0}\widehat{\Gamma(e^{\theta T})}(\xi,\eta) = \langle T\xi,\eta\rangle\, e^{\langle\xi,\eta\rangle} = \widehat{d\Gamma(T)}(\xi,\eta).$$

Hence we expect that

$$\left.\frac{d}{d\theta}\right|_{\theta=0}\Gamma(e^{\theta T}) = d\Gamma(T).$$

A rigorous treatment will be discussed in §5.4.

Bibliographical Notes

We have observed that the idea of ∂_t and ∂_t^* due to Hida [1] is well suited to the Gelfand triple $(E) \subset (L^2) \subset (E)^*$ and plays a principal role in the operator theory on white noise functionals. While, as ∂_t and ∂_t^* are actually field operators at a point $t \in T$

in Fock space, they have been utilized (even formally) in many physical works and, of course, there have been some attempts to define such field operators rigorously. For example, Grossmann [3] defined the creation and annihilation operators at a point using the theory of Gelfand triple, see also Bogolubov-Logunov-Todorov [1], Grossmann [1], [2] and Kristensen-Mejlbo-Thue Poulsen [1]. Those works have a common spirit with our discussion, however, the most remarkable feature of ours may be found in Fock expansion developed in §4.3-4.6. On the other hand, our theory offers a concrete example of an algebra of unbounded operators in quantum field theory. In this section see e.g., Borchers [1], [2] and references cited therein.

A precise norm estimate of translation operators was initiated by Potthoff-Yan [1] and Yan [1]. Theorem 4.2.4 is an improvement of their result and motivates us to introduce the notion of a regular one-parameter subgroup, see §5.2.

The integral kernel operator discussed in §4.3 was first formulated by Hida-Obata-Saitô [1]. However, our expression originates from Kubo-Takenaka [2] where an integral kernel operator involving only creation operators or only annihilation operators is discussed. In particular, the Gross Laplacian belongs to their class, see Kubo-Takenaka [4]. Later on Kuo [6] adopted a similar integral expression for the number operator. On the other hand, the idea of superposition of annihilation and creation operators (with normal-ordering or Wick-ordering) would be a standard notion among theoretical physicists. For mathematical treatment see Berezin [1] and references cited therein. Among others, Krée [1]–[3] and Meyer [1] have studied such operators based on a theory of nuclear spaces, see also Meyer [2].

The notion of symbol of operators in Fock space is due to Berezin [3], Krée [3] and Krée-Rączka [1]. As is mentioned above, it is a big advantage of white noise calculus to have analytic characterization theorems of the symbol of an operator on white noise functionals. The main results in §4.4 are Theorems 4.4.6 and 4.4.7. The former was first proved in Obata [8] and the latter is a new result which completes the characterization.

The Fock expansion in §4.5 would be a well-known idea among physicists, however, they use it in a weaker form (or rather at a formal level). The study of Fock expansion (or a similar expression) in white noise calculus was initiated by Huang [1] and Obata [6]. After some intermediate discussion by Obata [7], [8] we have reached the final result (Theorem 4.5.1) in these lecture notes.

The Taylor expansion of white noise functionals (Corollary 4.6.2) was first proved by Potthoff-Yan [1]. It is now clear that the Taylor expansion is a special case of Fock expansion.

We have hope that our theory of Fock expansion can be applied to quantum stochastic calculus, for its overview see Meyer [2], Parthasarathy [1]. In fact, Huang [1] reformulated the quantum Itô formula due to Hudson-Parthasarathy [1] in terms of white noise calculus and discussed its generalization. It is highly expected that his discussion will be further developed by means of our operator calculus. In some problems of quantum stochastic processes so-called *integral-sum kernel operators* introduced by Maassen [1] play an interesting role. From the viewpoint of operator theory his approach is closely related to ours. In §6.1 we give a quick review on this topic.

Chapter 5

Toward Harmonic Analysis

5.1 First order differential operators

We have observed in the previous chapter that $\{x(t); t \in T\}$ plays a role of coordinate system of white noise space (E^*, μ) and ∂_t the corresponding coordinate differential operator. Hence, for $y \in E^*$,

$$D_y = \int_T y(t)\partial_t dt,$$

which is in fact a directional derivative along the direction y, is justifiably called a *first order differential operator with constant coefficients.* In this section we shall discuss a more general class of first order differential operators and obtain their algebraic characterization.

For notational convenience we put

$$D_j = D_{e_j}, \qquad j = 0, 1, 2, \cdots.$$

The white noise analogue of the *gradient operator* is given by

$$\nabla\phi(t, x) = \partial_t \phi(x), \qquad t \in T, \quad x \in E^*, \quad \phi \in (E). \tag{5.1}$$

For $\omega \in E_{\mathbf{C}} \otimes (E)$ we put

$$\|\omega\|_p = \|(A \otimes \Gamma(A))^p \omega\|_0, \qquad p \in \mathbf{R},$$

where $\|\cdot\|_0$ is the norm of the Hilbert space $H_{\mathbf{C}} \otimes (L^2)$. It is known that these norms determine the topology of $E_{\mathbf{C}} \otimes (E)$.

Lemma 5.1.1 *It holds that*

$$\nabla\phi = \sum_{j=0}^{\infty} e_j \otimes D_j \phi, \qquad \phi \in (E), \tag{5.2}$$

where the series converges in $E_{\mathbf{C}} \otimes (E)$ as well as pointwisely. Moreover, for any $p \geq 0$

$$\|\nabla\phi\|_p^2 = \sum_{j=0}^{\infty} \|e_j \otimes D_j \phi\|_p^2 \leq \left(\frac{\rho^{-2}\delta^2}{-2e \log \rho}\right) \|\phi\|_{p+1}^2, \qquad \phi \in (E). \tag{5.3}$$

In particular, $\nabla \in \mathcal{L}((E), E_{\mathbf{C}} \otimes (E))$.

PROOF. Given $\phi \in (E)$ we put

$$\phi(x) = \sum_{n=0}^{\infty} \left\langle :x^{\otimes n}:, f_n \right\rangle, \qquad f_n \in E_{\mathbb{C}}^{\widehat{\otimes} n},$$

as usual. Then, by definition

$$\nabla\phi(t,x) = \partial_t\phi(x) = \sum_{n=1}^{\infty} n \left\langle :x^{\otimes(n-1)}:, \delta_t \otimes_1 f_n \right\rangle. \tag{5.4}$$

Using the Fourier expansion of f_n in terms of $\{e_j\}_{j=0}^{\infty}$, we obtain

$$\nabla\phi(t,x) = \sum_{j=0}^{\infty}\sum_{n=1}^{\infty} n \left\langle :x^{\otimes(n-1)}:, e_j \otimes_1 f_n \right\rangle e_j(t) = \sum_{j=0}^{\infty} e_j(t) D_j\phi(x). \tag{5.5}$$

As is easily verified, the above infinite series (5.4) and (5.5) converge absolutely at each $t \in T$ and $x \in E^*$.

We next investigate a norm estimate. In view of Theorem 4.1.1 we have

$$\|D_j\phi\|_p \le \left(\frac{\rho^{-2q}}{-2qe\log\rho}\right)^{1/2} |e_j|_{-(p+q)} \|\phi\|_{p+q}, \qquad p \ge 0, \quad q > 0. \tag{5.6}$$

On the other hand, since $\{e_j\}_{j=0}^{\infty}$ is an orthogonal set with respect to every norm $|\cdot|_p$, we see from (5.5) that

$$\|\nabla\phi\|_p^2 = \|(A \otimes \Gamma(A))^p \nabla\phi\|_0^2 = \sum_{j=0}^{\infty} |e_j|_p^2 \|D_j\phi\|_p^2 = \sum_{j=0}^{\infty} \|e_j \otimes D_j\phi\|_p^2, \tag{5.7}$$

which proves the first half of (5.3). Inserting (5.6) into (5.7) we obtain

$$\begin{aligned}
\|\nabla\phi\|_p^2 &\le \sum_{j=0}^{\infty} |e_j|_p^2 \left(\frac{\rho^{-2q}}{-2qe\log\rho}\right) |e_j|_{-(p+q)}^2 \|\phi\|_{p+q}^2 \\
&= \left(\frac{\rho^{-2q}}{-2qe\log\rho}\right) \|\phi\|_{p+q}^2 \sum_{j=0}^{\infty} \lambda_j^{-2q}.
\end{aligned}$$

Thus the second half of (5.3) follows by taking $q = 1$. The rest of the assertion is now immediate. qed

Corollary 5.1.2 *For $y \in E^*$ and $\Phi \in (E)^*$ it holds that*

$$\langle\!\langle y \otimes \Phi, \nabla\phi \rangle\!\rangle = \langle\!\langle \Phi, D_y\phi \rangle\!\rangle, \qquad \phi \in (E).$$

Moreover, the above relation characterizes the gradient operator $\nabla\phi$.

We then introduce first order differential operators in general. Recall that for any $p \ge 0$ there exist $C \ge 0$ and $q > 0$ such that

$$\|\phi\psi\|_p \le C \|\phi\|_{p+q} \|\psi\|_{p+q}, \qquad \phi, \psi \in (E).$$

This result was proved in Theorem 3.5.6.

Lemma 5.1.3 *For $\phi, \psi \in (E)$ put*

$$\omega_{\phi,\psi}(t,x) = (\partial_t\phi)(x) \cdot \psi(x), \qquad t \in T, \quad x \in E^*. \tag{5.8}$$

Then

$$\omega_{\phi,\psi} = \sum_{j=0}^{\infty} e_j \otimes (D_j\phi \cdot \psi)$$

converges in $E_{\mathbb{C}} \otimes (E)$ and, in particular, $\omega_{\phi,\psi} \in E_{\mathbb{C}} \otimes (E)$. Moreover, $\phi, \psi \mapsto \omega_{\phi,\psi}$ is a continuous bilinear map from $(E) \times (E)$ into $E_{\mathbb{C}} \otimes (E)$.

PROOF. For simplicity we write $\omega = \omega_{\phi,\psi}$. It then follows from Lemma 5.1.1 that

$$\omega(t,x) = \nabla\phi(t,x) \cdot \psi(x) = \sum_{j=0}^{\infty} e_j(t) D_j\phi(x)\psi(x), \qquad t \in T, \quad x \in E^*.$$

Given $p \geq 0$ choose $C \geq 0$ and $q > 0$ such that

$$\|D_j\phi \cdot \psi\|_p \leq C \|D_j\phi\|_{p+q} \|\psi\|_{p+q} .$$

Then we obtain

$$\begin{aligned} \|\omega\|_p^2 &= \sum_{j=0}^{\infty} |e_j|_p^2 \|D_j\phi \cdot \psi\|_p^2 \\ &\leq C^2 \|\psi\|_{p+q}^2 \sum_{j=0}^{\infty} |e_j|_p^2 \|D_j\phi\|_{p+q}^2 \\ &\leq C^2\rho^{2q} \|\psi\|_{p+q}^2 \sum_{j=0}^{\infty} |e_j|_{p+q}^2 \|D_j\phi\|_{p+q}^2 \\ &= C^2\rho^{2q} \|\psi\|_{p+q}^2 \|\nabla\phi\|_{p+q}^2 , \end{aligned}$$

and therefore, by (5.3) we come to

$$\|\omega_{\phi,\psi}\|_p \leq M \|\phi\|_{p+q+1} \|\psi\|_{p+q} , \qquad \phi, \psi \in (E), \tag{5.9}$$

where

$$M = \left(\frac{C^2\rho^{2q-2}\delta^2}{-2e \log \rho} \right)^{1/2} .$$

This completes the proof. qed

Proposition 5.1.4 *For each $\tilde{\Phi} \in (E_{\mathbb{C}} \otimes (E))^*$ there exists a unique operator $\Xi \in \mathcal{L}((E),(E)^*)$ such that*

$$\langle\langle \Xi\phi , \psi \rangle\rangle = \langle\langle \tilde{\Phi} , \omega_{\phi,\psi} \rangle\rangle, \qquad \phi, \psi \in (E), \tag{5.10}$$

where $\omega_{\phi,\psi}$ is defined as in (5.8).

PROOF. Choose $p \geq 0$ such as $\|\tilde{\Phi}\|_{-p} < \infty$. Then, by (5.9) we have

$$|\langle\langle \tilde{\Phi}, \omega_{\phi,\psi} \rangle\rangle| \leq \|\tilde{\Phi}\|_{-p} \|\omega_{\phi,\psi}\|_p \leq M \|\tilde{\Phi}\|_{-p} \|\phi\|_{p+q+1} \|\psi\|_{p+q}, \quad \phi, \psi \in (E),$$

for some $q > 0$ and $M \geq 0$. This means that $(\phi, \psi) \mapsto \langle\langle \tilde{\Phi}, \omega_{\phi,\psi} \rangle\rangle$ is a continuous bilinear form on $(E) \times (E)$, and therefore there exists a unique operator $\Xi \in \mathcal{L}((E), (E)^*)$ satisfying (5.10). qed

Definition 5.1.5 The operator Ξ defined as in (5.10) is called a *first order differential operator* with coefficient $\tilde{\Phi} \in (E_{\mathbf{C}} \otimes (E))^*$ and is denoted (somehow formally) by

$$\Xi = \int_T \Phi_t \partial_t dt.$$

Here we put $\Phi_t(x) = \tilde{\Phi}(t, x)$. In fact, $t \mapsto \Phi_t$ is an $(E)^*$-valued distribution on T, namely, an element in $E_{\mathbf{C}}^* \otimes (E)^* \cong (E_{\mathbf{C}} \otimes (E))^*$.

We are now interested in first order differential operators acting from (E) into itself.

Theorem 5.1.6 *Let Ξ be a first order differential operator*

$$\Xi = \int_T \Phi_t \partial_t dt, \tag{5.11}$$

with coefficient $\tilde{\Phi} \in (E_{\mathbf{C}} \otimes (E))^*$, $\Phi_t(x) = \tilde{\Phi}(t, x)$. Then $\Xi \in \mathcal{L}((E), (E))$ if and only if $\tilde{\Phi} \in E_{\mathbf{C}}^* \otimes (E)$.

PROOF. Let $K \in \mathcal{L}(E_{\mathbf{C}}, (E)^*)$ be the operator corresponding to $\tilde{\Phi}$ under the canonical isomorphism $(E_{\mathbf{C}} \otimes (E))^* \cong \mathcal{L}(E_{\mathbf{C}}, (E)^*)$. Then

$$\langle\langle \tilde{\Phi}, \xi \otimes \phi \rangle\rangle = \langle\langle K\xi, \phi \rangle\rangle, \qquad \xi \in E_{\mathbf{C}}, \quad \phi \in (E). \tag{5.12}$$

It follows from the kernel theorem that $\tilde{\Phi} \in E_{\mathbf{C}}^* \otimes (E)$ if and only if $K \in \mathcal{L}(E_{\mathbf{C}}, (E))$.
Suppose that Ξ is given as in (5.11). Then, by definition

$$\langle\langle \Xi\phi, \psi \rangle\rangle = \langle\langle \tilde{\Phi}, \omega_{\phi,\psi} \rangle\rangle, \qquad \phi, \psi \in (E),$$

where $\omega_{\phi,\psi}(t, x) = \partial_t \phi(x) \psi(x)$. On the other hand, we see from Lemma 5.1.3 that

$$\omega_{\phi,\psi} = \sum_{j=0}^{\infty} e_j \otimes (D_j \phi \cdot \psi)$$

converges in $E_{\mathbf{C}} \otimes (E)$. Therefore

$$\begin{aligned} \langle\langle \Xi\phi, \psi \rangle\rangle &= \sum_{j=0}^{\infty} \langle\langle \tilde{\Phi}, e_j \otimes (D_j \phi \cdot \psi) \rangle\rangle \\ &= \sum_{j=0}^{\infty} \langle\langle K e_j, D_j \phi \cdot \psi \rangle\rangle \\ &= \sum_{j=0}^{\infty} \langle\langle D_j \phi \cdot K e_j, \psi \rangle\rangle. \end{aligned}$$

Hence for any $p \geq 0$,

$$|\langle\langle \Xi\phi, \psi\rangle\rangle| \leq \sum_{j=0}^{\infty} \|D_j\phi \cdot Ke_j\|_p \|\psi\|_{-p}, \tag{5.13}$$

though the right hand side is possibly infinite.

We now suppose that $\tilde{\Phi} \in E_{\mathbf{C}}^* \otimes (E)$, i.e., that $K \in \mathcal{L}(E_{\mathbf{C}}, (E))$. Then, $Ke_j \in (E)$ and by Theorem 3.5.6 there exist $q > 0$ and $C_1 \geq 0$ such that

$$\|D_j\phi \cdot Ke_j\|_p \leq C_1 \|D_j\phi\|_{p+q} \|Ke_j\|_{p+q}.$$

Moreover, since $K \in \mathcal{L}(E_{\mathbf{C}}, (E))$ there exist $r \geq 0$ and $C_2 \geq 0$ such that

$$\|Ke_j\|_{p+q} \leq C_2 |e_j|_{p+q+r}.$$

Thus (5.13) becomes

$$\begin{aligned} |\langle\langle \Xi\phi, \psi\rangle\rangle| &\leq C_1C_2 \|\psi\|_{-p} \sum_{j=0}^{\infty} \|D_j\phi\|_{p+q} |e_j|_{p+q+r} \\ &= C_1C_2 \|\psi\|_{-p} \sum_{j=0}^{\infty} \|D_j\phi\|_{p+q} |e_j|_{p+q+r+1} \lambda_j^{-1} \\ &\leq C_1C_2 \|\psi\|_{-p} \left(\sum_{j=0}^{\infty} \|D_j\phi\|_{p+q}^2 |e_j|_{p+q+r+1}^2 \right)^{1/2} \left(\sum_{j=0}^{\infty} \lambda_j^{-2} \right)^{1/2} \\ &\leq C_1C_2\delta \|\psi\|_{-p} \left(\sum_{j=0}^{\infty} \|e_j \otimes D_j\phi\|_{p+q+r+1}^2 \right)^{1/2}. \end{aligned}$$

It then follows from Lemma 5.1.1 that

$$|\langle\langle \Xi\phi, \psi\rangle\rangle| \leq C_1C_2\delta \|\psi\|_{-p} \|\nabla\phi\|_{p+q+r+1},$$

and hence

$$\|\Xi\phi\|_p \leq C_1C_2\delta \|\nabla\phi\|_{p+q+r+1} \leq C_1C_2\delta \left(\frac{\rho^{-2}\delta^2}{-2e\log\rho} \right)^{1/2} \|\phi\|_{p+q+r+2}.$$

Consequently, $\Xi \in \mathcal{L}((E), (E))$.

Conversely, suppose that $\Xi \in \mathcal{L}((E), (E))$. Then, for any $p \geq 0$ there exist $q \geq 0$ and $C \geq 0$ such that

$$\|\Xi\phi\|_p \leq C \|\phi\|_{p+q}, \qquad \phi \in (E). \tag{5.14}$$

Let $\xi \in E_{\mathbf{C}}$ be fixed and consider

$$\phi(x) = \langle x, \xi\rangle, \qquad x \in E^*.$$

As is easily verified, $\omega_{\phi,\psi} = \xi \otimes \psi$ for any $\psi \in (E)$. Hence by (5.12) we obtain

$$\langle\langle \Xi\phi, \psi\rangle\rangle = \langle\langle \tilde{\Phi}, \omega_{\phi,\psi}\rangle\rangle = \langle\langle \tilde{\Phi}, \xi \otimes \psi\rangle\rangle = \langle\langle K\xi, \psi\rangle\rangle.$$

In view of (5.14) we obtain

$$|\langle\langle K\xi\,,\psi\rangle\rangle| = |\langle\langle \Xi\phi\,,\psi\rangle\rangle| \le \|\Xi\phi\|_p\,\|\psi\|_{-p} \le C\,\|\phi\|_{p+q}\,\|\psi\|_{-p},$$

and therefore

$$\|K\xi\|_p \le C\,\|\phi\|_{p+q} = C\,|\xi|_{p+q}, \qquad \xi \in E_{\mathbb{C}}.$$

This implies that $K \in \mathcal{L}(E_{\mathbb{C}}, (E))$. qed

Such an operator Ξ described as in Theorem 5.1.6 is called a *first order differential operator with smooth coefficients.* This would be reasonable because in that case $t \mapsto \Phi_t$ is an (E)-valued distributions on T.

The symbol of a first order differential operator is easily obtained.

Proposition 5.1.7 *Let Ξ be a first order differential operator given as*

$$\Xi = \int_T \Phi_t \partial_t \, dt,$$

where $\Phi_t(x) = \tilde{\Phi}(t,x)$, $\tilde{\Phi} \in (E_{\mathbb{C}} \otimes (E))^*$. Then it holds that

$$e^{-\langle \xi\,,\eta\rangle}\widehat{\Xi}(\xi,\eta) = \langle\langle \tilde{\Phi}\,,\xi\otimes\phi_{\xi+\eta}\rangle\rangle, \qquad \xi,\eta \in E_{\mathbb{C}}. \tag{5.15}$$

PROOF. We use the notation introduced in Lemma 5.1.3. By definition (5.8) we have

$$\omega_{\phi_\xi,\phi_\eta}(t,x) = \partial_t\phi_\xi(x)\phi_\eta(x) = \xi(t)\phi_\xi(x)\phi_\eta(x), \qquad \xi,\eta \in E_{\mathbb{C}},$$

namely,

$$\omega_{\phi_\xi,\phi_\eta} = \xi\otimes(\phi_\xi\phi_\eta) = e^{\langle \xi\,,\eta\rangle}\xi\otimes\phi_{\xi+\eta}.$$

Therefore,

$$\widehat{\Xi}(\xi,\eta) = \langle\langle \Xi\phi_\xi\,,\phi_\eta\rangle\rangle = \langle\langle \tilde{\Phi}\,,\omega_{\phi_\xi,\phi_\eta}\rangle\rangle = e^{\langle \xi\,,\eta\rangle}\langle\langle \tilde{\Phi}\,,\xi\otimes\phi_{\xi+\eta}\rangle\rangle,$$

which shows (5.15). qed

Corollary 5.1.8 *Let $\kappa \in E_{\mathbb{C}}^*$. Then the first order differential operator with coefficient $\kappa\otimes 1 \in E_{\mathbb{C}}^*\otimes(E)$ coincides with $\Xi_{0,1}(\kappa)$. In particular, the first order differential operator with coefficient $y\otimes 1$, $y \in E^*$, coincides with $\Xi_{0,1}(y) = D_y$.*

Such an operator described as in Corollary 5.1.8 is called a *first order differential operator with constant coefficients.*

We are going into an algebraic counterpart. Since (E) is a topological algebra, we have a natural concept of derivations on (E). Moreover, since each $\Phi \in (E)^*$ gives rise to a continuous operator in $\mathcal{L}((E),(E)^*)$ by multiplication, the following definition is adequate.

Definition 5.1.9 A linear operator $\Xi : (E) \to (E)^*$ is called a *derivation* if

$$\Xi(\phi\psi) = \Xi\phi\cdot\psi + \phi\cdot\Xi\psi, \qquad \phi,\psi \in (E). \tag{5.16}$$

It has been already observed in Corollary 4.2.7 that D_y is a derivation for any $y \in E^*$. The rest of this section is devoted to answering a natural question: what are the derivations on white noise functionals? The main assertion is stated in the following

Theorem 5.1.10 *Any continuous derivation in $\mathcal{L}((E),(E)^*)$ is a first order differential operator and vice versa. Furthermore, any continuous derivation in $\mathcal{L}((E),(E))$ is a first order differential operator with smooth coefficients and vice versa.*

The essence of the proof lies in Fock expansion. We begin with a few lemmas.

Lemma 5.1.11 *Let $\Xi \in \mathcal{L}((E),(E)^*)$. Then, it is a derivation if and only if*

$$e^{\langle \xi, \eta \rangle} \widehat{\Xi}(\xi + \eta, \zeta) = e^{\langle \eta, \zeta \rangle} \widehat{\Xi}(\xi, \eta + \zeta) + e^{\langle \xi, \zeta \rangle} \widehat{\Xi}(\eta, \xi + \zeta), \qquad \xi, \eta, \zeta \in E_{\mathbb{C}}. \tag{5.17}$$

PROOF. Recall that the exponential vectors span a dense subspace of (E) and that multiplication (Wiener product) is a continuous bilinear map. Hence Ξ is a derivation if and only if

$$\begin{aligned} \langle\langle \Xi(\phi_\xi \phi_\eta), \phi_\zeta \rangle\rangle &= \langle\langle \Xi\phi_\xi \cdot \phi_\eta, \phi_\zeta \rangle\rangle + \langle\langle \phi_\xi \cdot \Xi\phi_\eta, \phi_\zeta \rangle\rangle \\ &= \langle\langle \Xi\phi_\xi, \phi_\eta\phi_\zeta \rangle\rangle + \langle\langle \Xi\phi_\eta, \phi_\xi\phi_\zeta \rangle\rangle, \qquad \xi, \eta, \zeta \in E_{\mathbb{C}}. \end{aligned} \tag{5.18}$$

Then with an obvious relation $\phi_\xi \phi_\eta = e^{\langle \xi, \eta \rangle} \phi_{\xi+\eta}$, we see that (5.17) and (5.18) are equivalent. qed

Lemma 5.1.12 *Any first order differential operator is a derivation in $\mathcal{L}((E),(E)^*)$.*

PROOF. Let Ξ be a first order differential operator with coefficient $\tilde{\Phi} \in (E_{\mathbb{C}} \otimes (E))^*$. Then, by Proposition 5.1.7 we have

$$e^{-\langle \xi, \eta \rangle} \widehat{\Xi}(\xi, \eta) = \langle\langle \tilde{\Phi}, \xi \otimes \phi_{\xi+\eta} \rangle\rangle, \qquad \xi, \eta \in E_{\mathbb{C}}.$$

Then the verification of condition (5.17) in Lemma 5.1.11 is straightforward. qed

Lemma 5.1.13 *Let $\Xi \in \mathcal{L}((E),(E)^*)$ be a derivation and let*

$$\Xi = \sum_{l,m=0}^{\infty} \Xi_{l,m}(\kappa_{l,m}) \tag{5.19}$$

be the Fock expansion. Then $\kappa_{l,0} = 0$ for all $l \geq 0$ and

$$\left\langle \kappa_{l,m+1}, \eta^{\otimes l} \otimes \xi^{\otimes (m+1)} \right\rangle = \binom{l+m}{l} \left\langle \kappa_{l+m,1}, (\eta^{\otimes l} \otimes \xi^{\otimes m}) \otimes \xi \right\rangle, \quad \xi, \eta \in E_{\mathbb{C}},$$

for all $l, m \geq 0$.

PROOF. By assumption the symbol $\widehat{\Xi}$ satisfies (5.17). On the other hand, we have

$$e^{-\langle \xi, \eta\rangle}\widehat{\Xi}(\xi,\eta) = \sum_{l,m=0}^{\infty} \left\langle \kappa_{l,m}, \eta^{\otimes l} \otimes \xi^{\otimes m}\right\rangle,$$

which is a general result, see Proposition 4.5.3. Then, it is easy to obtain

$$\begin{aligned}\binom{m+n}{n}&\left\langle \kappa_{l,m+n}, \zeta^{\otimes l} \otimes \xi^{\otimes m} \otimes \eta^{\otimes n}\right\rangle \\ &= \binom{l+m}{m}\left\langle \kappa_{l+m,n}, \zeta^{\otimes l} \otimes \xi^{\otimes m} \otimes \eta^{\otimes n}\right\rangle \\ &\quad + \binom{l+n}{n}\left\langle \kappa_{l+n,m}, \zeta^{\otimes l} \otimes \eta^{\otimes n} \otimes \xi^{\otimes m}\right\rangle, \qquad (5.20)\end{aligned}$$

for any $\xi, \eta, \zeta \in E_{\mathbf{C}}$. Then $\kappa_{l,0} = 0$, $l \geq 0$, follows by putting $m = n = 0$ in (5.20). We next put $n = 1$ and $\eta = \xi$ in (5.20) to obtain

$$\begin{aligned}l!(m+1)!&\left\langle \kappa_{l,m+1}, \zeta^{\otimes l} \otimes \xi^{\otimes (m+1)}\right\rangle \\ &= (l+m)!\left\langle \kappa_{l+m,1}, (\zeta^{\otimes l} \otimes \xi^{\otimes m}) \otimes \xi\right\rangle \\ &\quad +(l+1)!m!\left\langle \kappa_{l+1,m}, (\zeta^{\otimes l} \otimes \xi) \otimes \xi^{\otimes m}\right\rangle.\end{aligned}$$

Applying this argument to the second term successively, we come to

$$l!(m+1)!\left\langle \kappa_{l,m+1}, \zeta^{\otimes l} \otimes \xi^{\otimes (m+1)}\right\rangle = (m+1)(l+m)!\left\langle \kappa_{l+m,1}, (\zeta^{\otimes l} \otimes \xi^{\otimes m}) \otimes \xi\right\rangle.$$

This completes the proof. qed

PROOF OF THEOREM 5.1.10. Suppose that we are given a continuous derivation Ξ with Fock expansion

$$\Xi = \sum_{l,m=0}^{\infty} \Xi_{l,m}(\kappa_{l,m}).$$

We first introduce a continuous bilinear form Ω on $E_{\mathbf{C}} \times (E)$:

$$\Omega(\xi,\phi) = \sum_{n=0}^{\infty} n!\,\langle \kappa_{n,1}, f_n \otimes \xi\rangle, \qquad \xi \in E_{\mathbf{C}}, \quad \phi \in (E), \qquad (5.21)$$

where $\phi(x) = \sum_{n=0}^{\infty} \langle :x^{\otimes n}:, f_n\rangle$. We shall prove the convergence of (5.21). In fact, for any $p, q \geq 0$ we have

$$\begin{aligned}\sum_{n=0}^{\infty} &n!\,|\langle \kappa_{n1}, f_n \otimes \xi\rangle| \\ &\leq \sum_{n=0}^{\infty} n!\,|\kappa_{n,1}|_{-(p+q+1)}\,|f_n \otimes \xi|_{p+q+1} \\ &\leq \left(\sum_{n=0}^{\infty} n!\,|\kappa_{n,1}|_{-(p+q+1)}^2\right)^{1/2} \left(\sum_{n=0}^{\infty} n!\,|f_n|_{p+q+1}^2\right)^{1/2} |\xi|_{p+q+1} \\ &= |\xi|_{p+q+1}\,\|\phi\|_{p+q+1} \left(\sum_{n=0}^{\infty} n!\,|\kappa_{n,1}|_{-(p+q+1)}^2\right)^{1/2}. \qquad (5.22)\end{aligned}$$

Since $\Xi \in \mathcal{L}((E),(E)^*)$, there exist $C \geq 0$, $K \geq 0$ and $p \geq 0$ such that

$$\left|\widehat{\Xi}(\xi,\eta)\right| \leq C \exp K \left(|\xi|_p^2 + |\eta|_p^2\right), \qquad \xi,\eta \in E_{\mathbf{C}}.$$

It then follows from Theorem 4.4.6 that the kernel distributions $\kappa_{l,m}$ of Ξ satisfies

$$|\kappa_{l,m}|_{-(p+1)} \leq C \left(l^l m^m\right)^{-1/2} (2e^3\delta^2)^{(l+m)/2} \left(\frac{\rho^{2p}}{2} + K\right)^{(l+m)/2}.$$

In particular,

$$\begin{aligned} |\kappa_{n,1}|_{-(p+q+1)} &\leq \rho^{q(n+1)} |\kappa_{n,1}|_{-(p+1)} \\ &\leq C\rho^{q(n+1)} n^{-n/2} (2e^3\delta^2)^{(n+1)/2} \left(\frac{\rho^{2p}}{2} + K\right)^{(n+1)/2}. \end{aligned}$$

Therefore,

$$\sum_{n=0}^{\infty} n!\, |\kappa_{n,1}|^2_{-(p+q+1)} \leq C^2 \sum_{n=0}^{\infty} \frac{n!}{n^n} \left\{2e^3\delta^2\rho^{2q}\left(\frac{\rho^{2p}}{2} + K\right)\right\}^{n+1} < \infty \tag{5.23}$$

for a sufficiently large $q \geq 0$. In conclusion, we see from (5.22) and (5.23) that

$$\sum_{n=0}^{\infty} n!\, |\langle \kappa_{n,1}, f_n \otimes \xi\rangle| \leq C_1 |\xi|_{p+q+1} \|\phi\|_{p+q+1}, \qquad \xi \in E_{\mathbf{C}}, \quad \phi \in (E),$$

for some $C_1 \geq 0$, $p \geq 0$ and $q \geq 0$. Therefore Ω in (5.21) is well defined on $E_{\mathbf{C}} \times (E)$ and becomes a continuous bilinear form.

Let $\tilde{\Phi} \in (E_{\mathbf{C}} \otimes (E))^*$ be the element corresponding to Ω, namely,

$$\langle\langle \tilde{\Phi}, \xi \otimes \phi \rangle\rangle = \Omega(\xi,\phi), \qquad \xi \in E_{\mathbf{C}}, \quad \phi \in (E).$$

Let Ξ' be the first order differential operator with coefficient $\tilde{\Phi}$. It then follows from Proposition 5.1.7 that

$$e^{-\langle \xi,\eta\rangle} \widehat{\Xi'}(\xi,\eta) = \langle\langle \tilde{\Phi}, \xi \otimes \phi_{\xi+\eta}\rangle\rangle = \Omega(\xi,\phi_{\xi+\eta}).$$

On the other hand, in view of (5.21),

$$\begin{aligned} \Omega(\xi,\phi_{\xi+\eta}) &= \sum_{n=0}^{\infty} n! \left\langle \kappa_{n,1}, \frac{(\xi+\eta)^{\otimes n}}{n!} \otimes \xi \right\rangle \\ &= \sum_{n=0}^{\infty}\sum_{l=0}^{n} \binom{n}{l} \left\langle \kappa_{n,1}, (\eta^{\otimes l} \otimes \xi^{\otimes(n-l)}) \otimes \xi \right\rangle \\ &= \sum_{l,m=0}^{\infty} \binom{l+m}{l} \left\langle \kappa_{l+m,1}, (\eta^{\otimes l} \otimes \xi^{\otimes m}) \otimes \xi \right\rangle. \end{aligned}$$

Since Ξ is a continuous derivation, the kernel distributions $\kappa_{l,m}$ in its Fock expansion satisfy a special relation stated in Lemma 5.1.13. We have thus obtained

$$e^{-\langle \xi,\eta\rangle} \widehat{\Xi'}(\xi,\eta) = \Omega(\xi,\phi_{\xi+\eta}) = \sum_{l,m=0}^{\infty} \left\langle \kappa_{l,m}, \eta^{\otimes l} \otimes \xi^{\otimes m}\right\rangle.$$

The last expression coincides with $e^{-\langle\xi,\eta\rangle}\widehat{\Xi}(\xi,\eta)$ and therefore $\Xi = \Xi'$, i.e., Ξ is the first order differential operator with coefficient $\tilde{\Phi}$. The rest of the assertion is now immediate from Lemma 5.1.12 and Theorem 5.1.6. qed

Note that any derivation maps constant functions into zero.

Corollary 5.1.14 *Any continuous derivation on (E) which maps linear functionals into constants is a first order differential operator with constant coefficients and vice versa.*

In general, a first order differential operator with smooth coefficients defines a smooth vector field on E^*. It seems very interesting to establish relationship between such vector fields and (local) one-parameter trasformation groups on E^*. Some special examples will be discussed later.

5.2 Regular one-parameter transformation group

In this section we discuss some general notion. Let $\mathfrak{X}$ be a nuclear Fréchet space with defining Hilbertian seminorms $\{\|\cdot\|_\alpha\}_{\alpha\in A}$, taking $\mathfrak{X} = E$ or $\mathfrak{X} = (E)$ into consideration. We denote by $GL(\mathfrak{X})$ the group of all linear homeomorphisms from $\mathfrak{X}$ onto itself. Obviously, $GL(\mathfrak{X}) \subset \mathcal{L}(\mathfrak{X},\mathfrak{X})$.

A one-parameter subgroup $\{g_\theta\}_{\theta\in\mathbb{R}} \subset GL(\mathfrak{X})$ is called *differentiable* if

$$\lim_{\theta\to 0}\frac{g_\theta\xi - \xi}{\theta}$$

converges in $\mathfrak{X}$ for any $\xi \in \mathfrak{X}$. In that case a linear operator X from $\mathfrak{X}$ into itself is defined by

$$X\xi = \lim_{\theta\to 0}\frac{g_\theta\xi - \xi}{\theta}, \qquad \xi \in \mathfrak{X}. \tag{5.24}$$

As usual, this operator X is called the *infinitesimal generator* of the differentiable one-parameter subgroup $\{g_\theta\}_{\theta\in\mathbb{R}} \subset GL(\mathfrak{X})$.

Proposition 5.2.1 *Let $\{g_\theta\}_{\theta\in\mathbb{R}} \subset GL(\mathfrak{X})$ be a differentiable one-parameter subgroup. Then its infinitesimal generator X is always continuous, i.e., $X \in \mathcal{L}(\mathfrak{X},\mathfrak{X})$. Moreover, the convergence of (5.24) is uniform on every compact (or equivalently, bounded) subset of $\mathfrak{X}$, namely,*

$$\lim_{\theta\to 0}\sup_{\xi\in K}\left\|\frac{g_\theta\xi - \xi}{\theta} - X\xi\right\|_\alpha = 0 \tag{5.25}$$

for any $\alpha \in A$ and any compact (or bounded) subset $K \subset \mathfrak{X}$.

PROOF. The assertion follows by a simple application of the Banach-Steinhaus theorem. Recall also that a subset of a nuclear space $\mathfrak{X}$ is compact if and only if it is closed and bounded. qed

Let $\{g_\theta\}_{\theta\in\mathbb{R}} \subset GL(\mathfrak{X})$ be a differentiable one-parameter subgroup with infinitesimal generator X. Then, as is easily verified, for any $\theta \in \mathbb{R}$ and any $\xi \in \mathfrak{X}$ we have

$$g_\theta X\xi = Xg_\theta\xi = \lim_{\epsilon\to 0}\frac{g_{\theta+\epsilon}\xi - g_\theta\xi}{\epsilon}. \tag{5.26}$$

Moreover, the convergence is uniform on every compact (or equivalently, bounded) subset in $\mathfrak{X}$. Hence we have

$$\lim_{\epsilon\to 0}\sup_{\xi\in K}\|g_{\theta+\epsilon}\xi - g_\theta\xi\|_\alpha = 0 \tag{5.27}$$

for any $\alpha \in A$ and any compact (or bounded) subset $K \subset \mathfrak{X}$.

A differentiable one-parameter subgroup is uniquely determined by its infinitesimal generator, namely, we have

Proposition 5.2.2 *Let $\{g_\theta\}_{\theta\in\mathbb{R}}$ and $\{h_\theta\}_{\theta\in\mathbb{R}}$ be two differentiable one-parameter subgroups of $GL(\mathfrak{X})$ with the same infinitesimal generator X. Then $g_\theta = h_\theta$ for all $\theta \in \mathbb{R}$.*

PROOF. Let $\xi_0 \in \mathfrak{X}$ be arbitrarily fixed. For simplicity we put $\xi(\theta) = h_{-\theta}\xi_0$. It becomes a differentiable curve in $\mathfrak{X}$ and from (5.26) we see that

$$\frac{d}{d\theta}\xi(\theta) = -Xh_{-\theta}\xi_0 = -X\xi(\theta). \tag{5.28}$$

We next observe that $\{g_\theta\xi(\theta)\}_{\theta\in\mathbb{R}}$ is also a differentiable curve in $\mathfrak{X}$. In fact,

$$\begin{aligned}\frac{g_{\theta+\epsilon}\xi(\theta+\epsilon) - g_\theta\xi(\theta)}{\epsilon} &= \\ = \left\{g_{\theta+\epsilon}\left(\frac{\xi(\theta+\epsilon)-\xi(\theta)}{\epsilon}\right) - g_\theta\left(\frac{\xi(\theta+\epsilon)-\xi(\theta)}{\epsilon}\right)\right\} \\ +g_\theta\left(\frac{\xi(\theta+\epsilon)-\xi(\theta)}{\epsilon}\right) + \frac{g_\epsilon g_\theta\xi(\theta) - g_\theta\xi(\theta)}{\epsilon}.\end{aligned} \tag{5.29}$$

Note that $K = \{(\xi(\theta+\epsilon)-\xi(\theta))/\epsilon; 0 < |\epsilon| \le 1\} \cup \{-X\xi(\theta)\}$ is a bounded subset of $\mathfrak{X}$ for a fixed $\theta \in \mathbb{R}$. Then it follows from (5.27) that

$$\left\|g_{\theta+\epsilon}\left(\frac{\xi(\theta+\epsilon)-\xi(\theta)}{\epsilon}\right) - g_\theta\left(\frac{\xi(\theta+\epsilon)-\xi(\theta)}{\epsilon}\right)\right\|_\alpha \le \sup_{\xi\in K}\|g_{\theta+\epsilon}\xi - g_\theta\xi\|_\alpha \to 0, \tag{5.30}$$

as $\epsilon \to 0$ for any $\alpha \in A$. Thus, by (5.28), (5.29) and (5.30) we obtain

$$\frac{d}{d\theta}(g_\theta\xi(\theta)) = \lim_{\epsilon\to 0}\frac{g_{\theta+\epsilon}\xi(\theta+\epsilon) - g_\theta\xi(\theta)}{\epsilon} = g_\theta(-X\xi(\theta)) + Xg_\theta\xi(\theta) = 0.$$

Hence $g_\theta\xi(\theta) = g_0\xi(0) = \xi_0$ for all $\theta \in \mathbb{R}$, and therefore $g_\theta\xi_0 = h_\theta\xi_0$. Since $\xi_0 \in \mathfrak{X}$ is arbitrary, we conclude that $g_\theta = h_\theta$. qed

In general, not every $X \in \mathcal{L}(\mathfrak{X}, \mathfrak{X})$ can be an infinitesimal generator of a differentiable one-parameter subgroup of $GL(\mathfrak{X})$. We give here a sufficient condition.

Proposition 5.2.3 *Let $X \in \mathcal{L}(\mathfrak{X}, \mathfrak{X})$ and assume that there exists $r > 0$ such that $\{(rX)^n/n!\}_{n=0}^{\infty}$ is equicontinuous, namely, for every $\alpha \in A$ there exist $C = C(\alpha) \geq 0$ and $\beta = \beta(\alpha) \in A$ such that*

$$\sup_{n\geq 0} \frac{1}{n!} \|(rX)^n \xi\|_\alpha \leq C \|\xi\|_\beta, \qquad \xi \in \mathfrak{X}.$$

Then there exists a differentiable one-parameter subgroup $\{g_\theta\}_{\theta\in\mathbf{R}} \subset GL(\mathfrak{X})$ with infinitesimal generator X.

PROOF. By assumption, the series

$$g_\theta \xi = \sum_{n=0}^{\infty} \frac{\theta^n}{n!} X^n \xi, \qquad \xi \in \mathfrak{X}, \quad |\theta| < r, \tag{5.31}$$

is convergent in $\mathfrak{X}$ and $\|g_\theta \xi\|_\alpha \leq C(1 - |\theta|/r)^{-1} \|\xi\|_\beta$, namely, $g_\theta \in \mathcal{L}(\mathfrak{X}, \mathfrak{X})$ for $|\theta| < r$. Furthermore, $g_0 = I$ and $g_{\theta_1+\theta_2} = g_{\theta_1} g_{\theta_2}$ whenever $|\theta_1|, |\theta_2|, |\theta_1 + \theta_2| < r$. We now define g_θ for all $\theta \in \mathbf{R}$. For a given $\theta \in \mathbf{R}$ choose a positive integer n such that $|\theta/n| < r$ and put $g_\theta = (g_{\theta/n})^n$. As is easily seen, this definition is independent of the choice of n, and therefore $g_{\theta_1+\theta_2} = g_{\theta_1} g_{\theta_2}$ for all $\theta_1, \theta_2 \in \mathbf{R}$. Since

$$\begin{aligned} \left\| \frac{g_\theta \xi - \xi}{\theta} - X\xi \right\|_\alpha &\leq \sum_{n=2}^{\infty} \frac{|\theta|^{n-1}}{n!} \|X^n \xi\|_\alpha \\ &\leq |\theta| C r^{-2} \left(1 - \frac{|\theta|}{r}\right)^{-1} \|\xi\|_\beta, \qquad |\theta| < r, \end{aligned} \tag{5.32}$$

$\{g_\theta\}_{\theta\in\mathbf{R}}$ is a differentiable one-parameter subgroup of $GL(\mathfrak{X})$ with infinitesimal generator X.
qed

The condition in Proposition 5.2.3 is satisfied if for any $\alpha \in A$ there exist constant numbers $C \geq 0$, $0 \leq \delta < 1$ and $\beta \in A$ such that

$$\|X^n \xi\|_\alpha \leq C (n!)^\delta \|\xi\|_\beta, \qquad \xi \in \mathfrak{X}.$$

This slightly stronger condition is sometimes useful. Under the above condition g_θ is defined by (5.31) for all $\theta \in \mathbf{R}$.

On the other hand, inequality (5.32) shows that the above constructed $\{g_\theta\}_{\theta\in\mathbf{R}}$ is more than being differentiable, cf. (5.25). This observation leads us to the following

Definition 5.2.4 A differentiable one-parameter subgroup $\{g_\theta\}_{\theta\in\mathbf{R}} \subset GL(\mathfrak{X})$ with infinitesimal generator X is called *regular* if for any $\alpha \in A$ there exists $\beta \in A$ such that

$$\lim_{\theta\to 0} \sup_{\|\xi\|_\beta \leq 1} \left\| \frac{g_\theta \xi - \xi}{\theta} - X\xi \right\|_\alpha = 0.$$

An example appeared already in Corollary 4.2.5 which we shall reformulate below. Another examples will be discussed.

Theorem 5.2.5 *For $y \in E^*$ let D_y and T_y be the differential and translation operators, respectively. Then $\{T_{\theta y}\}_{\theta\in\mathbf{R}}$ is a regular one-parameter subgroup of $GL((E))$ with infinitesimal generator D_y.*

5.3 Infinite dimensional Laplacians

The trace $\tau \in (E \otimes E)^*_{\mathrm{sym}}$ is defined in §2.2. Consider integral kernel operators with τ as kernel distribution:

$$\Delta_G = \Xi_{0,2}(\tau) = \int_{T\times T} \tau(s,t)\partial_s\partial_t dsdt, \tag{5.33}$$

$$N = \Xi_{1,1}(\tau) = \int_{T\times T} \tau(s,t)\partial_s^*\partial_t dsdt. \tag{5.34}$$

Definition 5.3.1 The operators Δ_G and N are called the *Gross Laplacian* and the *number operator*, respectively.

Proposition 5.3.2 *Both Δ_G and N belong to $\mathcal{L}((E),(E))$.*

PROOF. We apply Theorem 4.3.9. Since $\tau \in (E_{\mathbb{C}} \otimes E_{\mathbb{C}})^*$, it is immediate that $\Delta_G \in \mathcal{L}((E),(E))$. Furthermore, note that $\tau \in E_{\mathbb{C}} \otimes E_{\mathbb{C}}^*$. In fact, the definition

$$\langle \tau, \xi \otimes \eta \rangle = \langle \xi, \eta \rangle, \qquad \xi, \eta \in E_{\mathbb{C}},$$

implies that the corresponding operator under the isomorphism $\mathcal{L}(E_{\mathbb{C}}, E_{\mathbb{C}}) \cong E_{\mathbb{C}} \otimes E_{\mathbb{C}}^*$ is the identity. Therefore $N \in \mathcal{L}((E),(E))$. qed

In view of Proposition 4.3.7, we have

$$\Delta_G^* = \Xi_{2,0}(\tau) = \int_{T\times T} \tau(s,t)\partial_s^*\partial_t^* dsdt. \tag{5.35}$$

We shall consider Δ_G and N as natural analogues of a finite dimensional Laplacian.

Lemma 5.3.3 *For $\xi, \eta \in E_{\mathbb{C}}$,*

$$\widehat{N}(\xi,\eta) = \langle \xi, \eta \rangle e^{\langle \xi, \eta \rangle}, \quad \widehat{\Delta_G}(\xi,\eta) = \langle \xi, \xi \rangle e^{\langle \xi, \eta \rangle}, \quad \widehat{\Delta_G^*}(\xi,\eta) = \langle \eta, \eta \rangle e^{\langle \xi, \eta \rangle}.$$

Lemma 5.3.4 *Let $\xi \in E_{\mathbb{C}}$ and put $\psi_n(x) = \langle :x^{\otimes n}:, \xi^{\otimes n} \rangle$, $n = 0, 1, \cdots$. Then,*

$$\begin{aligned} N\psi_n &= n\psi_n, \\ \Delta_G\psi_n &= n(n-1)\langle \xi, \xi \rangle \psi_{n-2}, \\ (\Delta_G^*\psi_n)(x) &= \left\langle :x^{\otimes(n+2)}:, \tau \otimes \xi^{\otimes n} \right\rangle. \end{aligned}$$

In particular, $\mathcal{H}_n(\mathbb{C}) \cap (E)$ is the eigenspace of N with eigenvalue n.

The above results are straightforward with the help of Propositions 4.4.2 and 4.3.3. The reason why N is called the number operator becomes clear now. There are another expressions for the number operator.

Proposition 5.3.5 $N = d\Gamma(I) = \nabla^*\nabla$.

PROOF. $N = d\Gamma(I)$ follows, e.g., by comparing their operator symbols, see Proposition 4.6.13 and Lemma 5.3.3. Since $\nabla \in \mathcal{L}((E), E_{\mathbb{C}}\otimes(E))$ and $E_{\mathbb{C}}\otimes(E) \subset (E_{\mathbb{C}}\otimes(E))^*$, the composition $\nabla^*\nabla$ is meaningful and becomes an operator in $\mathcal{L}((E),(E)^*)$. We shall compute the symbol. Since $\nabla\phi_\xi = \xi \otimes \phi_\xi$, $\xi \in E_{\mathbb{C}}$, we have

$$\langle\langle \nabla^*\nabla\phi_\xi\,,\phi_\eta\rangle\rangle = \langle\langle \nabla\phi_\xi\,,\nabla\phi_\eta\rangle\rangle = \langle \xi\,,\eta\rangle\,\langle\langle \phi_\xi\,,\phi_\eta\rangle\rangle = e^{\langle \xi\,,\eta\rangle}\,\langle \xi\,,\eta\rangle\,, \qquad \xi,\eta \in E_{\mathbb{C}},$$

which coincides with $\widehat{N}(\xi,\eta)$. qed

Following our convention introduced at the end of §4.3, we write $\tilde{\Xi}$ for a continuous extension (if it exists) of $\Xi \in \mathcal{L}((E),(E)^*)$ to an operator from $(E)^*$ into $(E)^*$.

Proposition 5.3.6 *The number operator is symmetric, i.e., $N^* = \widetilde{N}$. Moreover, if $\Phi \in (E)^*$ is given as*

$$\Phi(x) = \sum_{n=0}^{\infty} \left\langle :x^{\otimes n}:, F_n \right\rangle,$$

then

$$\widetilde{N}\Phi(x) = N^*\Phi(x) = \sum_{n=0}^{\infty} n \left\langle :x^{\otimes n}:, F_n \right\rangle. \tag{5.36}$$

PROOF. Note first that $N^* \in \mathcal{L}((E)^*,(E)^*)$ since $N \in \mathcal{L}((E),(E))$. Hence we need only to prove (5.36) but the verification is obvious by Lemma 5.3.4. qed

In §4.6 we introduced an element in $(E^{\otimes 2m})^*$:

$$\tau_m = \sum_{i_1,\cdots,i_m=0}^{\infty} e_{i_1} \otimes \cdots \otimes e_{i_m} \otimes e_{i_1} \otimes \cdots \otimes e_{i_m}.$$

Then $\Xi_{m,m}(\tau_m)$ is a polynomial in the number operator N.

Proposition 5.3.7 *For $m = 1,2,\cdots$ we have*

$$\Xi_{m,m}(\tau_m) = N(N-1)\cdots(N-m+1). \tag{5.37}$$

PROOF. We prove (5.37) by operator symbols. Let $\xi,\eta \in E_{\mathbb{C}}$. Since

$$N\phi_\xi(x) = \sum_{k=0}^{\infty} k \left\langle :x^{\otimes k}:, \frac{\xi^{\otimes k}}{k!} \right\rangle,$$

which is immediate from the definition, we obtain

$$\begin{aligned}\prod_{j=0}^{m-1}(N-j)\phi_\xi(x) &= \sum_{k=0}^{\infty} k(k-1)\cdots(k-(m-1)) \left\langle :x^{\otimes k}:, \frac{\xi^{\otimes k}}{k!} \right\rangle \\ &= \sum_{k=m}^{\infty} \left\langle :x^{\otimes k}:, \frac{\xi^{\otimes k}}{(k-m)!} \right\rangle\end{aligned}$$

and therefore,

$$\left\langle\left\langle \prod_{j=0}^{m-1}(N-j)\phi_\xi\,,\phi_\eta \right\rangle\right\rangle = \sum_{k=m}^{\infty} k! \left\langle \frac{\xi^{\otimes k}}{(k-m)!}, \frac{\eta^{\otimes k}}{k!} \right\rangle = \sum_{k=0}^{\infty} \frac{\langle \xi\,,\eta\rangle^{m+k}}{k!} = \langle \xi\,,\eta\rangle^m\, e^{\langle \xi\,,\eta\rangle}.$$

On the other hand, in view of Proposition 4.4.2 we have

$$\begin{aligned}\langle\langle \Xi_{m,m}(\tau_m)\phi_\xi\,,\phi_\eta\rangle\rangle &= \langle \tau_m\,,\eta^{\otimes m}\otimes\xi^{\otimes m}\rangle e^{\langle\xi\,,\eta\rangle} \\ &= \sum_{j_1,\cdots,j_m=0}^{\infty} \langle e_{j_1}\,,\eta\rangle\cdots\langle e_{j_m}\,,\eta\rangle\,\langle e_{j_1}\,,\xi\rangle\cdots\langle e_{j_m}\,,\xi\rangle\, e^{\langle\xi\,,\eta\rangle} \\ &= \langle\xi\,,\eta\rangle^m\, e^{\langle\xi\,,\eta\rangle}.\end{aligned}$$

Therefore we come to

$$\left\langle\!\!\left\langle \prod_{j=0}^{m-1}(N-j)\phi_\xi\,,\phi_\eta\right\rangle\!\!\right\rangle = \langle\langle \Xi_{m,m}(\tau_m)\phi_\xi\,,\phi_\eta\rangle\rangle\,,$$

from which (5.37) follows. qed

A white noise analogue of Euclidean norm is given by

$$R(x) = \left\langle :x^{\otimes 2}:\,,\tau\right\rangle. \tag{5.38}$$

Obviously, $R\in(E)^*$ and, therefore, belongs to $\mathcal{L}((E),(E)^*)$ as multiplication operator. We here note the following

Proposition 5.3.8

$$R = \Xi_{2,0}(\tau) + 2\Xi_{1,1}(\tau) + \Xi_{0,2}(\tau) = \Delta_G^* + 2N + \Delta_G.$$

PROOF. The Fock expansion of a multiplication operator is known, see Proposition 4.6.4. Applying the formula there, we obtain

$$R = \sum_{l+m=2}\binom{l+m}{m}\Xi_{l,m}(\tau) = \Xi_{2,0}(\tau) + 2\Xi_{1,1}(\tau) + \Xi_{0,2}(\tau),$$

as desired. qed

Similarity between Δ_G, N and a finite dimensional Laplacian is also observed in terms of discrete coordinate. For simplicity we put

$$D_j = D_{e_j}, \qquad j=0,1,2,\cdots,$$

as before.

Proposition 5.3.9 *For $\phi\in(E)$,*

$$\Delta_G\phi = \sum_{j=0}^{\infty} D_j^2\phi, \qquad N\phi = \sum_{j=0}^{\infty} D_j^*D_j\phi,$$

where the right hand sides converge in (E).

PROOF. Let $p \geq 0$. In view of Theorem 4.1.7 (with $q = 1$) we obtain

$$\left\|D_j^2\phi\right\|_p \leq 2\rho^{-1/2}\left(\frac{\rho^{-1}}{-2e\log\rho}\right)|e_j|_{-(p+1)}^2 \|\phi\|_{p+1} = \frac{\rho^{-3/2}}{-e\log\rho}\rho^{2p}\lambda_j^{-2}\|\phi\|_{p+1},$$

for $\phi \in (E)$. Hence

$$\sum_{j=0}^{\infty}\left\|D_j^2\phi\right\|_p \leq \frac{\rho^{2p-3/2}\delta^2}{-e\log\rho}\|\phi\|_{p+1}, \qquad \phi \in (E).$$

Thus, $\sum_{j=0}^{\infty} D_j^2\phi$ converges in (E) and $\phi \mapsto \sum_{j=0}^{\infty} D_j^2\phi$ becomes a continuous operator on (E) which is denoted by $\sum_{j=0}^{\infty} D_j^2$ simply. The action on an exponential vector is easily computed.

$$\sum_{j=0}^{\infty} D_j^2\phi_\xi = \sum_{j=0}^{\infty}\langle\xi, e_j\rangle^2\phi_\xi = \langle\xi,\xi\rangle\phi_\xi, \qquad \xi \in E_{\mathbf{C}}.$$

It then follows from Lemma 5.3.4 that $\sum_{j=0}^{\infty} D_j^2 = \Delta_G$ on exponential vectors and therefore on (E) since both are continuous operator on it.

As for N, noting that $D_j^*D_j = \Xi_{1,1}(e_j \otimes e_j)$, we obtain by Theorem 4.3.9 (putting $\alpha = \beta = q = 2$)

$$\left\|D_j^*D_j\phi\right\|_p \leq \frac{\rho^{-2}}{-2e\log\rho}\lambda_j^{-2}\|\phi\|_{p+2},$$

and therefore

$$\sum_{j=0}^{\infty}\left\|D_j^*D_j\phi\right\|_p \leq \frac{\rho^{-2}\delta^2}{-2e\log\rho}\|\phi\|_{p+2}.$$

The rest is proved similarly as above. qed

Proposition 5.3.10 *$\Delta_G + N$ is a derivation in $\mathcal{L}((E),(E))$.*

PROOF. We put $\Xi = \Delta_G + N$. As an immediate consequence of Lemma 5.3.3,

$$\widehat{\Xi}(\xi,\eta) = \langle\xi,\xi+\eta\rangle e^{\langle\xi,\eta\rangle}, \qquad \xi,\eta \in E_{\mathbf{C}}.$$

Then the condition in Lemma 5.1.11 is easily verified. qed

Therefore $\Delta_G + N$ is a first order differential operator with smooth coefficients (see Theorem 5.1.10). In fact, we observe that

$$\Delta_G + N = \int_{T\times T}\tau(s,t)(\partial_s + \partial_s^*)\partial_t\, dsdt = \int_{T\times T}\tau(s,t)x(s)\partial_t\, dsdt = \int_T x(t)\partial_t\, dt.$$

The corresponding one-parameter group of transformations on E^* is given by scaling operators S_λ introduced in §4.6.

Theorem 5.3.11 *$\{S_{e^\theta}\}_{\theta\in\mathbf{R}}$ is a regular one-parameter subgroup of $GL((E))$ with infinitesimal generator $\Delta_G + N$.*

PROOF. We study the operator symbol of S_{e^θ}. For $\xi, \eta \in E_{\mathbb{C}}$ we put

$$f(\theta) = \widehat{S_{e^\theta}}(\xi,\eta) = \exp\left(\frac{(e^{2\theta}-1)}{2}\langle \xi, \xi\rangle + e^\theta \langle \xi, \eta\rangle\right).$$

Then,

$$\begin{aligned} f'(\theta) &= (e^{2\theta}\langle \xi, \xi\rangle + e^\theta\langle \xi, \eta\rangle) f(\theta), \\ f''(\theta) &= (2e^{2\theta}\langle \xi, \xi\rangle + e^\theta\langle \xi, \eta\rangle) f(\theta) + (e^{2\theta}\langle \xi, \xi\rangle + e^\theta\langle \xi, \eta\rangle)^2 f(\theta). \end{aligned}$$

Let $\theta_0 > 0$ be fixed arbitrarily. Then, whenever $|\theta| \le \theta_0$,

$$\begin{aligned} |f''(\theta)| &\le (e^{2\theta}|\langle \xi, \xi\rangle| + e^\theta|\langle \xi, \eta\rangle| + 1)^2 |f(\theta)| \\ &\le (e^{2\theta_0}|\langle \xi, \xi\rangle| + e^{\theta_0}|\langle \xi, \eta\rangle| + 1)^2 \exp\left(\frac{(e^{2\theta_0}+1)}{2}|\langle \xi, \xi\rangle| + e^{\theta_0}|\langle \xi, \eta\rangle|\right). \end{aligned}$$

As is easily seen, the last quantity is bounded by

$$\le C\exp(K_1|\langle \xi, \xi\rangle| + K_2|\langle \xi, \eta\rangle|)$$

with some constants $C = C(\theta_0)$, $K_1 = K_1(\theta_0)$ and $K_2 = K_2(\theta_0)$. For simplicity we put

$$g_\theta(\xi,\eta) = f(\theta) - f(0) - f'(0)\theta.$$

Then by Taylor expansion, whenever $|\theta| \le \theta_0$ it holds that

$$\begin{aligned} |g_\theta(\xi,\eta)| &\le \frac{|\theta|^2}{2}\max_{|\theta|\le\theta_0}|f''(\theta)| \\ &\le \frac{|\theta|^2}{2} C\exp(K_1|\langle \xi, \xi\rangle| + K_2|\langle \xi, \eta\rangle|), \qquad \xi,\eta \in E_{\mathbb{C}}. \end{aligned}$$

Now suppose we are given $p \ge 0$ and $\epsilon > 0$. Then with the help of Lemma 4.6.6 we may find $q = q(p,\epsilon,\theta_0) \ge 0$ such that

$$|g_\theta(\xi,\eta)| \le \frac{|\theta|^2}{2} C\exp\epsilon\left(|\xi|_{p+q}^2 + |\eta|_{-p}^2\right), \qquad |\theta| \le \theta_0, \quad \xi,\eta \in E_{\mathbb{C}}.$$

It then follows from Theorem 4.4.7 that there exists $\Xi_\theta \in \mathcal{L}((E),(E))$ such that $\widehat{\Xi_\theta} = g_\theta$ and

$$\|\Xi_\theta\phi\|_{p-1} \le \frac{|\theta|^2}{2} CM(\epsilon,q,r)\,\|\phi\|_{p+q+r+1}, \qquad \phi \in (E),$$

for some $r \ge 0$. On the other hand, since

$$\begin{aligned} f(0) &= e^{\langle \xi, \eta\rangle} = \widehat{I}(\xi,\eta), \\ f'(0) &= (\langle \xi, \xi\rangle + \langle \xi, \eta\rangle)e^{\langle \xi, \eta\rangle} = \widehat{\Delta}_G(\xi,\eta) + \widehat{N}(\xi,\eta), \end{aligned}$$

we see that

$$\Xi_\theta = S_{e^\theta} - I - \theta(\Delta_G + N).$$

Consequently we obtain

$$\sup_{\|\phi\|_{p+q+r+1}\le 1}\left\|\frac{S_{e^\theta}\phi - \phi}{\theta} - (\Delta_G + N)\phi\right\|_{p-1} \le \frac{|\theta|}{2} CM(\epsilon,q,r) \to 0,$$

as $\theta \to 0$. This completes the proof. qed

5.4 Infinite dimensional rotation group

Our discussion is based on a Gelfand triple $E \subset H \subset E^*$. We now set

$$O(E;H) = \{g \in GL(E);\ |g\xi|_0 = |\xi|_0 \quad \text{for all } \xi \in E\},$$

which is, obviously, a subgroup of $GL(E)$. In other words, $O(E;H)$ is the group of automorphisms of the Gelfand triple $E \subset H \subset E^*$. While, since each $g \in O(E;H)$ is extended to an orthogonal operator on the Hilbert space H, we may regard $O(E;H)$ as a subgroup of the full orthogonal group $O(H)$. The group $O(E;H)$ is called the *infinite dimensional rotation group* (associated with the Gelfand triple $E \subset H \subset E^*$).

The infinite dimensional rotation group $O(E;H)$ acts on the Gaussian space E^* by means of the adjoint:

$$\langle x, g\xi \rangle = \langle g^*x, \xi \rangle, \qquad x \in E^*, \quad \xi \in E.$$

One of the most fundamental property of $O(E;H)$ from the viewpoint of white noise calculus is the invariance of the Gaussian measure.

Proposition 5.4.1 *The Gaussian measure μ is invariant under $O(E;H)$.*

PROOF. Let $g \in O(E;H)$ and denote by $g^*\mu$ the image measure of μ under $g^* : E^* \to E^*$. Then its characteristic functional is given by

$$\widehat{g^*\mu}(\xi) = \int_{E^*} e^{i\langle g^*x, \xi \rangle} \mu(dx) = \int_{E^*} e^{i\langle x, g\xi \rangle} \mu(dx) = \hat{\mu}(g\xi), \qquad \xi \in E.$$

From the explicit expression $\hat{\mu}(\xi) = e^{-|\xi|_0^2/2}$ it follows that $\hat{\mu}(g\xi) = \hat{\mu}(\xi)$ and hence $\widehat{g^*\mu} = \hat{\mu}$. Therefore, $g^*\mu = \mu$ by the uniqueness of a characteristic functional. qed

Similarly we put

$$U((E);(L^2)) = \{U \in GL((E));\ \|U\phi\|_0 = \|\phi\|_0, \quad \text{for all } \phi \in (E)\}.$$

With each $g \in O(E;H)$ we associate its second quantization $\Gamma(g)$. It follows from Proposition 4.6.10 that $\Gamma(g) \in \mathcal{L}((E),(E))$. Moreover,

Proposition 5.4.2 *$\Gamma(g) \in U((E);(L^2))$ for any $g \in O(E;H)$. Moreover,*

$$\Gamma(g)\phi(x) = \phi(g^*x), \qquad x \in E^*, \quad \phi \in (L^2) \tag{5.39}$$

and

$$\Gamma(g_1g_2) = \Gamma(g_1)\Gamma(g_2), \qquad g_1, g_2 \in O(E;H). \tag{5.40}$$

In particular, $\Gamma(g)$ is a unitary representation of $O(E;H)$ on (L^2).

PROOF. Suppose that $\phi \in (E)$ is given with Wiener-Itô expansion:

$$\phi(x) = \sum_{n=0}^{\infty} \left\langle :x^{\otimes n}:, f_n \right\rangle, \qquad f_n \in E_{\mathbb{C}}, \quad x \in E^*.$$

Then, by Proposition 4.6.10 we have

$$\Gamma(g)\phi(x) = \sum_{n=0}^{\infty} \left\langle :x^{\otimes n}:, g^{\otimes n} f_n \right\rangle.$$

On the other hand, since g is an orthogonal operator,

$$\|\Gamma(g)\phi\|_0^2 = \sum_{n=0}^{\infty} n! \left| g^{\otimes n} f_n \right|_0^2 = \sum_{n=0}^{\infty} n! \, |f_n|_0^2 = \|\phi\|_0^2.$$

It is obvious that $\Gamma(g_1 g_2) = \Gamma(g_1)\Gamma(g_2)$ and $\Gamma(I) = I$. Hence $\Gamma(g) \in U((E), (L^2))$.

Finally we must prove the identity (5.39). It is certainly valid for an exponential vector ϕ_ξ. In fact,

$$\Gamma(g)\phi_\xi(x) = \phi_{g\xi}(x) = e^{\langle x, g\xi \rangle - \langle g\xi, g\xi \rangle/2} = e^{\langle g^* x, \xi \rangle - \langle \xi, \xi \rangle /2} = \phi_\xi(g^* x).$$

Then, by approximation (for instance, modelled after the argument in the proof of Proposition 4.6.7) we obtain (5.39) for an arbitrary $\phi \in (E)$. qed

By definition $\Gamma(g)^* \in \mathcal{L}((E)^*, (E)^*)$. In this connection we note

Proposition 5.4.3 *For any $g \in O(E; H)$, $\Gamma(g)$ admits a continuous extension to an operator in $\mathcal{L}((E)^*, (E)^*)$ which is denoted by $\widetilde{\Gamma(g)}$. In that case,*

$$\widetilde{\Gamma(g)} = \Gamma(g^{-1})^*, \qquad g \in O(E; H).$$

PROOF. Since $\Gamma(g) \in \mathcal{L}((E), (E))$, its adjoint $\Gamma(g)^* \in \mathcal{L}((E)^*, (E)^*)$. Then for any $\phi \in (E)$,

$$\langle\langle \overline{\phi}, \phi \rangle\rangle = \|\phi\|_0^2 = \|\Gamma(g)\phi\|_0^2 = \langle\langle \overline{\Gamma(g)\phi}, \Gamma(g)\phi \rangle\rangle = \langle\langle \overline{\phi}, \overline{\Gamma(g)^*} \Gamma(g)\phi \rangle\rangle.$$

On the other hand, since $\Gamma(g)$ is a real operator, $\overline{\Gamma(g)^*} = \Gamma(g)^*$. Therefore,

$$\Gamma(g)^* \Gamma(g)\phi = \phi, \qquad \phi \in (E).$$

Thus $\Gamma(g)^*$ is the unique continuous extension of $\Gamma(g^{-1})$. qed

If $\{g_\theta\}_{\theta \in \mathbb{R}} \subset O(E; H)$ is a one-parameter subgroup, so is $\{\Gamma(g_\theta)\}_{\theta \in \mathbb{R}} \subset U((E), (L^2))$. However, it is not clear whether $\{\Gamma(g_\theta)\}$ is differentiable. We shall prove the differentiability under stronger condition.

Theorem 5.4.4 *Let $\{g_\theta\}_{\theta \in \mathbb{R}}$ be a regular one-parameter subgroup of $O(E; H)$ with infinitesimal generator X. Then, $\{\Gamma(g_\theta)\}_{\theta \in \mathbb{R}}$ is a regular one-parameter subgroup of $U((E); (L^2))$ with infinitesimal generator $d\Gamma(X)$. Moreover, there exists a skew-symmetric distribution $\kappa \in E \otimes E^*$ such that*

$$d\Gamma(X) = \int_{T \times T} \kappa(s, t)(\partial_s^* \partial_t - \partial_t^* \partial_s) \, ds dt. \tag{5.41}$$

As is easily seen, the infinitesimal generator X of a regular one-parameter subgroup $\{g_\theta\}_{\theta\in\mathbb{R}}$ is skew-symmetric in the sense that

$$\langle X\xi\,,\eta\rangle = -\langle \xi\,,X\eta\rangle\,, \qquad \xi,\eta\in E. \tag{5.42}$$

Taking this remark into account, we devide Theorem 5.4.4 into two propositions below where the statements are slightly more general than necessary.

Before going into the proof we make a small remark. Recall that $\{x(t)\}$ is regarded as white noise coordinate system (see §3.1) and, as multiplication operator,

$$x(t) = \partial_t + \partial_t^*, \qquad t\in T,$$

see Theorem 4.1.5 and Corollary 4.6.5. Hence a white noise analogy of an infinitesimal generator of finite dimensional rotations would be formally given as

$$x(s)\partial_t - x(t)\partial_s = (\partial_s^* + \partial_s)\partial_t - (\partial_t^* + \partial_t)\partial_s = \partial_s^*\partial_t - \partial_t^*\partial_s,$$

which is, in fact, an operator in $\mathcal{L}((E),(E)^*)$. Then expression (5.41) is understood to be a linear combination of $x(s)\partial_t - x(t)\partial_s$.

Proposition 5.4.5 *If $\{B_\theta\}_{\theta\in\mathbb{R}}$ is a regular one-parameter subgroup of $GL(E)$ with infinitesimal generator X, then $\{\Gamma(B_\theta)\}_{\theta\in\mathbb{R}}$ is a regular one-parameter subgroup of $GL((E))$ with infinitesimal generator $d\Gamma(X)$.*

The proof is postponed.

Proposition 5.4.6 *Let $X\in\mathcal{L}(E_{\mathbb{C}},E_{\mathbb{C}})$ be skew-symmetric in the sense of (5.42). Then there exists a skew-symmetric distribution $\kappa\in E_{\mathbb{C}}\otimes E_{\mathbb{C}}^*$ such that*

$$d\Gamma(X) = \int_{T\times T}\kappa(s,t)(\partial_s^*\partial_t - \partial_t^*\partial_s)\,dsdt. \tag{5.43}$$

PROOF. By the canonical isomorphism $E_{\mathbb{C}}\otimes E_{\mathbb{C}}^* \cong \mathcal{L}(E_{\mathbb{C}},E_{\mathbb{C}})$ there exists $\kappa\in E_{\mathbb{C}}\otimes E_{\mathbb{C}}^*$ such that

$$\langle \kappa\,,\eta\otimes\xi\rangle = \frac{1}{2}\langle \eta\,,X\xi\rangle\,, \qquad \xi,\eta\in E_{\mathbb{C}}. \tag{5.44}$$

Since X is skew-symmetric, we have

$$\langle \kappa\,,\eta\otimes\xi\rangle = \frac{1}{2}\langle \eta\,,X\xi\rangle = -\frac{1}{2}\langle \xi\,,X\eta\rangle = -\langle \kappa\,,\xi\otimes\eta\rangle\,.$$

It then follows from Theorem 4.3.9 that the operator

$$\int_{T\times T}\kappa(s,t)(\partial_s^*\partial_t - \partial_t^*\partial_s)dsdt = 2\Xi_{1,1}(\kappa)$$

belongs to $\mathcal{L}((E),(E))$. We now compare the symbols of $d\Gamma(X)$ and $2\Xi_{1,1}(\kappa)$. By Proposition 4.6.13 we have

$$\widehat{d\Gamma(X)}(\xi,\eta) = \langle X\xi\,,\eta\rangle\, e^{\langle \xi,\eta\rangle}.$$

On the other hand, by Proposition 4.4.2 we have

$$2\widehat{\Xi_{1,1}(\kappa)}(\xi,\eta) = 2\langle \kappa, \eta\otimes\xi\rangle e^{\langle \xi,\eta\rangle}.$$

Then, from (5.44) we see that $d\Gamma(X) = 2\Xi_{1,1}(\kappa)$. qed

For the proof of Proposition 5.4.5 we need some technical results.

Lemma 5.4.7 *For $i = 1, 2, \cdots, d$, let $K_i \in \mathcal{L}(E_{\mathbb{C}}^{\otimes m_i}, E_{\mathbb{C}}^{\otimes l_i})$, $m_i \ge 0$, $l_i \ge 0$. Assume that $|K_i\xi_i|_p \le C_i |\xi_i|_{p+q}$, $\xi_i \in E_{\mathbb{C}}^{\otimes m_i}$, for some $p, q \ge 0$ and $C_i \ge 0$. Then,*

$$|(K_1\otimes\cdots\otimes K_d)\omega|_p \le C_1\cdots C_d\delta^m |\omega|_{p+q+1}, \qquad \omega \in E_{\mathbb{C}}^{\otimes m}, \tag{5.45}$$

$$|(I^{\otimes k}\otimes K_1\otimes\cdots\otimes K_d)\omega|_p \le C_1\cdots C_d\rho^{(q+1)k}\delta^m |\omega|_{p+q+1}, \quad \omega \in E_{\mathbb{C}}^{\otimes(k+m)}, \tag{5.46}$$

where $m = m_1 + \cdots + m_d$.

PROOF. For simplicity we put

$$e(\mathbf{j}_i) = e_{j_1}\otimes\cdots\otimes e_{j_{m_i}}, \qquad \mathbf{j}_i = (j_1, \cdots, j_{m_i}).$$

Then every $\omega \in E_{\mathbb{C}}^{\otimes m} = E_{\mathbb{C}}^{\otimes m_1}\otimes\cdots\otimes E_{\mathbb{C}}^{\otimes m_d}$ admits a Fourier series expansion of the form:

$$\omega = \sum_{\mathbf{j}_1,\cdots \mathbf{j}_d} \langle \omega, e(\mathbf{j}_1)\otimes\cdots\otimes e(\mathbf{j}_d)\rangle e(\mathbf{j}_1)\otimes\cdots\otimes e(\mathbf{j}_d).$$

Hence,

$$(K_1\otimes\cdots\otimes K_d)\omega = \sum_{\mathbf{j}_1,\cdots \mathbf{j}_d} \langle \omega, e(\mathbf{j}_1)\otimes\cdots\otimes e(\mathbf{j}_d)\rangle K_1e(\mathbf{j}_1)\otimes\cdots\otimes K_de(\mathbf{j}_d),$$

and therefore, by Schwartz inequality and the assumption we come to

$$\begin{aligned}
&|(K_1\otimes\cdots\otimes K_d)\omega|_p^2 \\
&\quad\le \sum_{\mathbf{j}_1,\cdots \mathbf{j}_d} \langle \omega, e(\mathbf{j}_1)\otimes\cdots\otimes e(\mathbf{j}_d)\rangle^2 |e(\mathbf{j}_1)\otimes\cdots\otimes e(\mathbf{j}_d)|_{p+q+1}^2 \\
&\qquad\times \sum_{\mathbf{j}_1,\cdots \mathbf{j}_d} |K_1e(\mathbf{j}_1)\otimes\cdots\otimes K_de(\mathbf{j}_d)|_p^2 |e(\mathbf{j}_1)\otimes\cdots\otimes e(\mathbf{j}_d)|_{-(p+q+1)}^2 \\
&\quad\le |\omega|_{p+q+1}^2 C_1^2\cdots C_d^2 \sum_{\mathbf{j}_1,\cdots \mathbf{j}_d} |e(\mathbf{j}_1)|_{-1}^2\cdots|e(\mathbf{j}_d)|_{-1}^2 \\
&\quad= |\omega|_{p+q+1}^2 C_1^2\cdots C_d^2 \delta^{2(m_1+\cdots+m_d)}.
\end{aligned}$$

This proves (5.45).

As is easily verified, if $K \in \mathcal{L}(E_{\mathbb{C}}^{\otimes m}, E_{\mathbb{C}}^{\otimes l})$ satisfies $|K\eta|_p \le C|\eta|_{p+q}$, $\eta \in E_{\mathbb{C}}^{\otimes m}$, for some $p, q \ge 0$ and $C \ge 0$, then for any $n \ge 0$,

$$\left|(K\otimes I^{\otimes n})f_{m+n}\right|_p \le C\rho^{qn} |f_{m+n}|_{p+q}, \qquad f_{m+n} \in E_{\mathbb{C}}^{\otimes(m+n)}. \tag{5.47}$$

Then (5.46) follows from (5.45) and (5.47). qed

Lemma 5.4.8 *Let $B \in \mathcal{L}(E_{\mathbb{C}}, E_{\mathbb{C}})$ and assume that $|B\xi|_p \le C_1 |\xi|_{p+q}$ and $|(B-I)\xi|_p \le C_2 |\xi|_{p+q}$ for some $C_1, C_2 \ge 0$. Then,*

$$\left|(B^{\otimes n} - I^{\otimes n})f_n\right|_p \le \delta C_2(\delta C_1 + \rho^{q+1})^{n-1} |f_n|_{p+q+1}, \qquad f_n \in E_{\mathbb{C}}^{\otimes n}.$$

PROOF. We need only simple calculation and (5.46). In fact,

$$\begin{aligned}
\left|(B^{\otimes n} - I^{\otimes n})f_n\right|_p &\le \sum_{k=0}^{n-1} \left|\left(B^{\otimes(n-1-k)} \otimes (B-I) \otimes I^{\otimes k}\right) f_n\right|_p \\
&\le \sum_{k=0}^{n-1} C_1^{n-1-k} C_2 \rho^{(q+1)k} \delta^{n-k} |f_n|_{p+q+1} \\
&\le \delta C_2(\delta C_1 + \rho^{q+1})^{n-1} |f_n|_{p+q+1}.
\end{aligned}$$

This completes the proof. qed

PROOF OF PROPOSITION 5.4.5. Suppose that $p \ge 0$ is given. By the regularity of $\{B_\theta\}_{\theta\in\mathbb{R}}$ there exists $q \ge 0$ such that

$$\lim_{\theta\to 0} \sup_{|\xi|_{p+q}\le 1} \left|\frac{B_\theta\xi - \xi}{\theta} - X\xi\right|_p = 0. \tag{5.48}$$

Moreover, we may assume that with some $C \ge 0$,

$$|X\xi|_p \le C |\xi|_{p+q}, \tag{5.49}$$

$$\delta^2\rho^q + \delta\rho^{q+1} < 1. \tag{5.50}$$

Suppose next that $\epsilon > 0$ is given. In view of (5.48) there exists $\theta_0 > 0$ such that

$$\left|\frac{B_\theta\xi - \xi}{\theta} - X\xi\right|_p \le \epsilon |\xi|_{p+q}, \qquad |\theta| < \theta_0. \tag{5.51}$$

Furthermore, by (5.50) we may assume

$$\delta^2(\epsilon + C)\theta_0 + \delta^2\rho^q + \delta\rho^{q+1} < 1. \tag{5.52}$$

It follows from (5.49) and (5.51) that

$$|B_\theta\xi - \xi|_p \le (\epsilon + C)|\theta| |\xi|_{p+q} \tag{5.53}$$

and

$$|B_\theta\xi|_p \le M_4 |\xi|_{p+q}, \qquad |\theta| < \theta_0, \tag{5.54}$$

where

$$M_4 = (\epsilon + C)\theta_0 + \rho^q.$$

A simple calculation yields

$$\begin{aligned}
\frac{B_\theta^{\otimes n} - I^{\otimes n}}{\theta} - \gamma_n(X) &= \sum_{k=0}^{n-1} \left\{I^{\otimes k} \otimes \left(\frac{B_\theta - I}{\theta} - X\right) \otimes B_\theta^{\otimes(n-1-k)}\right\} \\
&\quad + \sum_{k=0}^{n-1} \left\{I^{\otimes k} \otimes X \otimes (B_\theta^{\otimes(n-1-k)} - I^{\otimes(n-1-k)})\right\},
\end{aligned}$$

and therefore, for $f_n \in E_{\mathbb{C}}^{\otimes n}$ it holds that

$$\left|\frac{B_\theta^{\otimes n} f_n - f_n}{\theta} - \gamma_n(X) f_n\right|_p \le \sum_{k=0}^{n-1}\left|\left(I^{\otimes k}\otimes\left(\frac{B_\theta - I}{\theta} - X\right)\otimes B_\theta^{\otimes(n-1-k)}\right) f_n\right|_p + \sum_{k=0}^{n-1}\left|\left(I^{\otimes k}\otimes X\otimes (B_\theta^{\otimes(n-1-k)} - I^{\otimes(n-1-k)})\right) f_n\right|_p . \tag{5.55}$$

In view of (5.51), (5.54) and (5.46) in Lemma 5.4.7, we obtain

$$\begin{aligned}\left|\left(I^{\otimes k}\otimes\left(\frac{B_\theta - I}{\theta} - X\right)\otimes B_\theta^{\otimes(n-1-k)}\right) f_n\right|_p &\le \epsilon M_4^{n-1-k}\rho^{(q+1)k}\delta^{n-k}\left|f_n\right|_{p+q+1} \\ &= \epsilon\delta(\delta M_4)^{n-1-k}\rho^{(q+1)k}\left|f_n\right|_{p+q+1} .\end{aligned}$$

Hence,

$$\sum_{k=0}^{n-1}\left|\left(I^{\otimes k}\otimes\left(\frac{B_\theta - I}{\theta} - X\right)\otimes B_\theta^{\otimes(n-1-k)}\right) f_n\right|_p \le \epsilon\delta M_5^{n-1}\left|f_n\right|_{p+q+1} \le \epsilon\rho\delta(\rho M_5)^{n-1}\left|f_n\right|_{p+q+2} , \tag{5.56}$$

where

$$M_5 = \delta M_4 + \rho^{q+1} = \delta(\epsilon + C)\theta_0 + \delta\rho^q + \rho^{q+1}.$$

On the other hand, by (5.53), (5.54) and Lemma 5.4.8 we get

$$\left|\left(B_\theta^{\otimes(n-1-k)} - I^{\otimes(n-1-k)}\right)\omega\right|_p \le \delta(\epsilon + C) M_5^{n-2-k}|\theta|\,|\omega|_{p+q+1} . \tag{5.57}$$

In view of (5.49), (5.57) and (5.46), we obtain

$$\begin{aligned}\left|\left(I^{\otimes k}\otimes X\otimes (B_\theta^{\otimes(n-1-k)} - I^{\otimes(n-1-k)})\right) f_n\right|_p &\le C\rho\cdot\delta(\epsilon + C) M_5^{n-2-k}|\theta|\rho^{(q+2)k}\delta^{n-k}\left|f_n\right|_{p+q+2} \\ &= C\rho\delta^2(\epsilon + C) M_5^{-1}(\delta M_5)^{n-1-k}\rho^{(q+2)k}|\theta|\left|f_n\right|_{p+q+2} \\ &\le C\delta^2\rho^{-q}(\epsilon + C)(\delta M_5)^{n-1-k}\rho^{(q+2)k}|\theta|\left|f_n\right|_{p+q+2} ,\end{aligned}$$

where we used $M_5^{-1} < \rho^{-q-1}$. Therefore we have

$$\sum_{k=0}^{n-1}\left|\left(I^{\otimes k}\otimes X\otimes (B_\theta^{\otimes(n-1-k)} - I^{\otimes(n-1-k)})\right) f_n\right|_p \le C\delta^2\rho^{-q}(\epsilon + C)(\delta M_5 + \rho^{q+2})^{n-1}|\theta|\left|f_n\right|_{p+q+2} . \tag{5.58}$$

From (5.55), (5.56) and (5.58) we see that

$$\left|\frac{B_\theta^{\otimes n} f_n - f_n}{\theta} - \gamma_n(X) f_n\right|_p \le \epsilon\rho\delta(\rho M_5)^{n-1}\left|f_n\right|_{p+q+2} + C\delta^2\rho^{-q}(\epsilon + C)(\delta M_5 + \rho^{q+2})^{n-1}|\theta|\left|f_n\right|_{p+q+2} .$$

Since $\rho M_5 < \delta M_5 + \rho^{q+2} < 1$ by (5.52), the last quantity is bounded by

$$\begin{aligned}&\left\{\epsilon\rho\delta(\rho M_5)^{-1} + C\delta^2\rho^{-q}(\epsilon + C)(\delta M_5 + \rho^{q+2})^{-1}|\theta|\right\} |f_n|_{p+q+2} \\ &\le (\epsilon\delta\rho^{-q-1} + C\delta(\epsilon + C)\rho^{-2q-1}|\theta|) |f_n|_{p+q+2},\end{aligned}$$

where $M_5^{-1} < \rho^{-q-1}$ is used again. We thus conclude that

$$\left\| \frac{\Gamma(B_\theta)\phi - \phi}{\theta} - d\Gamma(X)\phi \right\|_p \le (\epsilon\delta\rho^{-q-1} + |\theta|C(\epsilon + C)\delta\rho^{-2q-1}) \|\phi\|_{p+q+2},$$

whenever $|\theta| < \theta_0$. Consequently,

$$\limsup_{\theta\to 0} \sup_{\|\phi\|_{p+q+2}\le 1} \left\| \frac{\Gamma(B_\theta)\phi - \phi}{\theta} - d\Gamma(X)\phi \right\|_p \le \epsilon\delta\rho^{-q-1}.$$

Since $\epsilon > 0$ is arbitrary,

$$\lim_{\theta\to 0} \sup_{\|\phi\|_{p+q+2}\le 1} \left\| \frac{\Gamma(B_\theta)\phi - \phi}{\theta} - d\Gamma(X)\phi \right\|_p = 0.$$

Therefore $\Gamma(B_\theta)$ is a regular one-parameter subgroup of $GL((E))$ with infinitesimal generator $d\Gamma(X)$. qed

5.5 Rotation-invariant operators

In §5.3 we have introduced the Gross Laplacian and the number operator:

$$\begin{aligned} \Delta_G &= \Xi_{0,2}(\tau) = \int_{T\times T} \tau(s,t)\partial_s\partial_t\, dsdt, \\ N &= \Xi_{1,1}(\tau) = \int_{T\times T} \tau(s,t)\partial_s^*\partial_t\, dsdt. \end{aligned}$$

Both operators belong to $\mathcal{L}((E),(E))$. The main purpose of this section is to characterize these operators in terms of their invariance property under $O(E;H)$.

We say that a continuous operator Ξ from (E) into $(E)^*$ is *rotation-invariant* if

$$\Gamma(g)^*\Xi\Gamma(g) = \Xi \qquad \text{for all} \quad g \in O(E;H). \tag{5.59}$$

(Recall that $\Gamma(g) \in U((E);(L^2))$, see Proposition 5.4.2.) By definition, if Ξ is rotation-invariant, so is Ξ^*. By virtue of Proposition 5.4.3, condition (5.59) for $\Xi \in \mathcal{L}((E),(E))$ is equivalent to the following

$$[\Gamma(g), \Xi] = 0 \qquad \text{for all} \quad g \in O(E;H). \tag{5.60}$$

Apparently, (5.60) means the usual rotation-invariance.

It is rather straightforward to see that N and Δ_G are rotation-invariant. (This fact follows from Lemma 5.5.6 and Proposition 5.5.7 below.) Therefore Δ_G^* is also rotation-invariant, while N^* is the extension of N, see Proposition 5.3.6. The goal of this section is the following significant characterization of rotation-invariant operators.

Theorem 5.5.1 *Any rotation-invariant operator in $\mathcal{L}((E),(E)^*)$ is generated by Δ_G^*, Δ_G and N, and any rotation-invariant operator in $\mathcal{L}((E),(E))$ is generated by Δ_G and N.*

The above statement involves somehow vague wording. In fact, we shall prove the following assertions instead.

Theorem 5.5.2 *Let $\Xi \in \mathcal{L}((E),(E)^*)$ and let $\Xi = \sum_{l,m=0}^{\infty} \Xi_{l,m}(\kappa_{l,m})$ be its Fock expansion. Then Ξ is rotation-invariant if and only if all $\Xi_{l,m}(\kappa_{l,m})$ are rotation-invariant.*

Theorem 5.5.3 *Let $\kappa \in (E_{\mathbb{C}}^{\otimes(l+m)})^*_{\cdot}$ and assume that $\Xi_{l,m}(\kappa)$ is rotation-invariant. If $l+m$ is odd, then $\Xi_{l,m}(\kappa) = 0$. If $l+m$ is even, then $\Xi_{l,m}(\kappa)$ is a linear combination of $(\Delta_G^\alpha)^* N^\beta \Delta_G^\gamma$ with α, β, γ being non-negative integers such that $\alpha+\beta+\gamma \le (l+m)/2$.*

Theorem 5.5.4 *Let $\kappa \in (E_{\mathbb{C}}^{\otimes l}) \otimes (E_{\mathbb{C}}^{\otimes m})^*$ and assume that $\Xi_{l,m}(\kappa)$ is rotation-invariant. If $l+m$ is odd, then $\Xi_{l,m}(\kappa) = 0$. If $l+m$ is even, then $\Xi_{l,m}(\kappa)$ is a linear combination of $N^\beta \Delta_G^\gamma$ with β, γ being non-negative integers such that $\beta+\gamma \le (l+m)/2$.*

As is easily verified,

$$[\Delta_G, N] = 2\Delta_G. \tag{5.61}$$

Hence any product of Δ_G^*, Δ_G and N (whenever it is well defined on (E)) may be rearranged as a sum of $(\Delta_G^\alpha)^* N^\beta \Delta_G^\gamma$ with α, β, γ being non-negative integers. This is also related to the normal ordering of creation and annihilation operators.

PROOF OF THEOREM 5.5.2. Let

$$\Xi = \sum_{l,m=0}^{\infty} \Xi_{l,m}(\kappa_{l,m})$$

be the Fock expansion of $\Xi \in \mathcal{L}((E),(E)^*)$, where we my assume without loss of generality that $\kappa_{l,m} \in (E_{\mathbb{C}}^{\otimes(l+m)})^*_{\mathrm{sym}(l,m)}$. Recall that $\Gamma(g)\phi_\xi = \phi_{g\xi}$, $\xi \in E_{\mathbb{C}}$. Then for any $\xi, \eta \in E_{\mathbb{C}}$ we have

$$\begin{aligned}(\Gamma(g)^*\Xi\Gamma(g))\widehat{\ }(\xi,\eta) &= \langle\langle \Xi\Gamma(g)\phi_\xi\,, \Gamma(g)\phi_\eta \rangle\rangle \\ &= \langle\langle \Xi\phi_{g\xi}\,, \phi_{g\eta}\rangle\rangle \\ &= \widehat{\Xi}(g\xi, g\eta).\end{aligned} \tag{5.62}$$

Moreover, from Proposition 4.5.3 we see that

$$\begin{aligned}\widehat{\Xi}(g\xi, g\eta) &= e^{\langle g\xi, g\eta\rangle} \sum_{l,m=0}^{\infty} \left\langle \kappa_{l,m}\,, (g\eta)^{\otimes l} \otimes (g\xi)^{\otimes m} \right\rangle \\ &= e^{\langle \xi,\eta\rangle} \sum_{l,m=0}^{\infty} \left\langle (g^{\otimes(l+m)})^*\kappa\,, \eta^{\otimes l} \otimes \xi^{\otimes m} \right\rangle.\end{aligned} \tag{5.63}$$

From (5.62) and (5.63) one may derive the Fock expansion of $\Gamma(g)^*\Xi\Gamma(g)$:

$$\Gamma(g)^*\Xi\Gamma(g) = \sum_{l,m=0}^{\infty} \Xi_{l,m}\left((g^{\otimes(l+m)})^*\kappa_{l,m}\right). \tag{5.64}$$

In particular,

$$\Gamma(g)^*\Xi_{l,m}(\kappa)\Gamma(g) = \Xi_{l,m}\left((g^{\otimes(l+m)})^*\kappa\right). \tag{5.65}$$

Therefore, by the uniqueness of the Fock expansion, Ξ is rotation-invariant if and only if $\Xi_{l,m}(\kappa_{l,m})$ is rotation-invariant for all $l, m = 0, 1, 2, \cdots$. qed

We say that $F \in (E_{\mathbb{C}}^{\otimes n})^*$ is *rotation-invariant* if $(g^{\otimes n})^* F = F$ for all $g \in O(E; H)$. During the proof of Theorem 5.5.2 we have established the following

Lemma 5.5.5 *Let $\kappa \in (E_{\mathbb{C}}^{\otimes(l+m)})^*_{\mathrm{sym}(l,m)}$. Then $\Xi_{l,m}(\kappa)$ is rotation-invariant if and only if κ is rotation-invariant.*

Thus the proofs of Theorems 5.5.3 and 5.5.4 are essentially reduced to listing up the rotation-invariant distributions. The full list is, in fact, described satisfactorily as below, though the proof is long and deferred at the end of this section. For the definition of F^σ for $F \in (E_{\mathbb{C}}^{\otimes n})^*$ and $\sigma \in \mathfrak{S}_n$, see §1.6.

Proposition 5.5.6 *Assume that $F \in (E_{\mathbb{C}}^{\otimes n})^*$ is rotation-invariant. If n is odd, then $F = 0$. If n is even, say $n = 2m$, then F is a linear combination of $(\tau^{\otimes m})^\sigma$, $\sigma \in \mathfrak{S}_n$. Moreover, the dimension of rotation-invariant distributions in $(E^{\otimes n})^*$ is $(n-1)!!$.*

Lemma 5.5.7 *Let α, β, γ be non-negative integers and put $l = 2\alpha + \beta$, $m = 2\gamma + \beta$. Then,*

$$\Xi_{l,m}(\tau^{\otimes\alpha} \otimes \tau_\beta \otimes \tau^{\otimes\gamma}) = (\Delta_G^\alpha)^* \Xi_{\beta,\beta}(\tau_\beta)\Delta_G^\gamma, \tag{5.66}$$

where $\tau_\beta \in (E^{\otimes 2\beta})^*$ is defined in (4.56).

PROOF. Since both sides of (5.66) are continuous operator from (E) into $(E)^*$, it is sufficient to check that they coincide on the polynomials. But this is proved by a straightforward computation. qed

PROOF OF THEOREM 5.5.3. Suppose that $\Xi_{l,m}(\kappa)$ is rotation-invariant. Without loss of generality we may assume that $\kappa \in (E_{\mathbb{C}}^{\otimes(l+m)})^*_{\mathrm{sym}(l,m)}$. Then, κ is rotation-invariant by Lemma 5.5.5. If $l+m$ is odd, it follows from Proposition 5.5.6 that $\kappa = 0$ and hence $\Xi_{l,m}(\kappa) = 0$.

We next consider the case when $l+m$ is even. It follows again from Proposition 5.5.6 that κ is a linear combination of $(\tau^{\otimes(l+m)/2})^\sigma$, $\sigma \in \mathfrak{S}_{l+m}$. For each $\sigma \in \mathfrak{S}_{l+m}$ we may find $\sigma' \in \mathfrak{S}_l \times \mathfrak{S}_m$ such that

$$\begin{aligned} &(\tau^{\otimes(l+m)/2})^{\sigma\sigma'} \\ &= \sum e_{i_1}^{\otimes 2} \otimes \cdots \otimes e_{i_\alpha}^{\otimes 2} \otimes e_{j_1} \otimes \cdots \otimes e_{j_\beta} \otimes e_{j_1} \otimes \cdots \otimes e_{j_\beta} \otimes e_{k_1}^{\otimes 2} \otimes \cdots \otimes e_{k_\gamma}^{\otimes 2} \\ &= \tau^{\otimes\alpha} \otimes \tau_\beta \otimes \tau^{\otimes\gamma} \end{aligned}$$

for some non-negative integers α, β, γ with $2\alpha + \beta = l$ and $2\gamma + \beta = m$. Then, in view of Lemma 5.5.7 we have

$$\Xi_{l,m}((\tau^{\otimes(l+m)/2})^\sigma) = \Xi_{l,m}((\tau^{\otimes(l+m)/2})^{\sigma\sigma'}) = (\Delta_G^\alpha)^* \Xi_{\beta,\beta}(\tau_\beta)\Delta_G^\gamma.$$

On the other hand, it follows from Proposition 5.3.7 that $\Xi_{\beta,\beta}(\tau_\beta)$ is a polynomial in the number operator N of degree β. Consequently, $\Xi_{l,m}((\tau^{\otimes(l+m)/2})^\sigma)$ is a linear combination of $(\Delta_G^\alpha)^* N^\beta \Delta_G^\gamma$ with $\alpha+\beta+\gamma \le (l+m)/2$. Therefore so is $\Xi_{l,m}(\kappa)$.

qed

PROOF OF THEOREM 5.5.4. Since $\Xi \equiv \Xi_{l,m}(\kappa)$ is rotation-invariant as an operator in $\mathcal{L}((E),(E)^*)$ at any rate, it follows from Theorem 5.5.3 that Ξ is a finite linear combination of $(\Delta_G^\alpha)^* N^\beta \Delta_G^\gamma$, say,

$$\Xi = \sum_{\alpha,\beta,\gamma} C_{\alpha,\beta,\gamma} (\Delta_G^\alpha)^* N^\beta \Delta_G^\gamma, \tag{5.67}$$

where $C_{\alpha,\beta,\gamma} = 0$ except finitely many α, β, γ. For $\xi \in E_{\mathbb{C}}$ a straightforwad computation (e.g., with Lemma 5.3.4) yields

$$(\Delta_G^\alpha)^* N^\beta \Delta_G^\gamma \phi_\xi(x) = \sum_{n=0}^\infty \frac{n^\beta \, |\xi|_0^{2\gamma}}{n!} \left\langle :x^{\otimes(n+2\alpha)}:, \tau^{\otimes\alpha} \otimes \xi^{\otimes n} \right\rangle$$

and therefore,

$$\Xi \phi_\xi(x) = \sum_{m=0}^\infty \sum_{n+2\alpha=m} \sum_{\beta,\gamma} C_{\alpha,\beta,\gamma} \frac{n^\beta \, |\xi|_0^{2\gamma}}{n!} \left\langle :x^{\otimes m}:, \tau^{\otimes\alpha} \otimes \xi^{\otimes n} \right\rangle.$$

Since $\kappa \in (E_{\mathbb{C}}^{\otimes l}) \otimes (E_{\mathbb{C}}^{\otimes m})^*$ by assumption, $\Xi = \Xi_{l,m}(\kappa) \in \mathcal{L}((E),(E))$ by Theorem 4.3.9. Hence, in particular, $\Xi\phi_\xi \in (E)$ and therefore

$$\sum_{n+2\alpha=m} \left(\sum_{\beta,\gamma} C_{\alpha,\beta,\gamma} \frac{n^\beta \, |\xi|_0^{2\gamma}}{n!} \right) \tau^{\otimes\alpha} \otimes \xi^{\otimes n} \in E_{\mathbb{C}}^{\otimes m} \qquad \text{for all} \quad m = 0, 1, 2, \cdots. \tag{5.68}$$

It is easily verified that $\{\tau^{\otimes\alpha} \otimes \xi^{\otimes n};\ 2\alpha + n = m\} \subset (E_{\mathbb{C}}^{\otimes m})^*$ is linearly independent modulo $E_{\mathbb{C}}^{\otimes m}$ whenever $\xi \ne 0$. Then (5.68) means

$$\sum_{\beta,\gamma} C_{\alpha,\beta,\gamma} \frac{n^\beta \, |\xi|_0^{2\gamma}}{n!} = 0 \qquad \text{whenever} \quad \alpha > 0 \quad \text{and} \quad n + 2\alpha = m.$$

Hence for a fixed $\alpha > 0$,

$$\sum_{\beta,\gamma} C_{\alpha,\beta,\gamma} (m - 2\alpha)^\beta \, |\xi|_0^{2\gamma} = 0$$

for any $m \ge 2\alpha$ and $\xi \in E_{\mathbb{C}}$ with $\xi \ne 0$. Consequently, $C_{\alpha,\beta,\gamma} = 0$ for $\alpha > 0$, and (5.67) becomes

$$\Xi = \sum_{\beta,\gamma} C_{0,\beta,\gamma} N^\beta \Delta_G^\gamma,$$

as desired. qed

In the finite dimensional case where $\mathcal{S}(\mathbb{R}^n) \subset L^2(\mathbb{R}^n) \subset \mathcal{S}'(\mathbb{R}^n)$ the rotation-invariant operators are generated by the Laplacian

$$\Delta_n = \sum_{j=1}^n \frac{\partial^2}{\partial x_j^2} = -\sum_{j=1}^n \left(\frac{\partial}{\partial x_j}\right)^* \left(\frac{\partial}{\partial x_j}\right)$$

and by multiplication with the Euclidean norm:

$$R(x) = \sum_{j=1}^{n} x_j^2.$$

On the other hand, we have a white noise analogue of the Euclidean norm:

$$R(x) = \left\langle :x^{\otimes 2}:, \tau \right\rangle,$$

which is, as multiplication operator, expressed by Laplacians:

$$R = 2N + \Delta_G + \Delta_G^*,$$

see Corollary 5.3.8. We have thus observed an interesting contrast between rotation-invariant operators on white noise functionals and those on a finite dimensional Euclidean space.

The rest of the section is entirely devoted to the proof of Proposition 5.5.6. Let $F \in (E_{\mathbb{C}}^{\otimes n})^*$ and consider the Fourier series expansion:

$$F = \sum_{i_1,\cdots,i_n=0}^{\infty} \langle F, e_{i_1} \otimes \cdots \otimes e_{i_n} \rangle \, e_{i_1} \otimes \cdots \otimes e_{i_n}.$$

Obviously, F is rotation-invariant if and only if

$$\langle F, e_{i_1} \otimes \cdots \otimes e_{i_n} \rangle = \langle F, ge_{i_1} \otimes \cdots \otimes ge_{i_n} \rangle \tag{5.69}$$

for every choice of $i_1, \cdots, i_n$ and $g \in O(E; H)$.

PROOF OF PROPOSITION 5.5.6 (CASE OF n BEING ODD). Suppose that $i_1, \cdots, i_n$ are arbitrarily given. Since n is odd, there is some j appearing odd times in that sequence. Take k which is different from $i_1, \cdots, i_n$ and let g be a (two dimensional) rotation defined by $ge_j = -e_j$, $ge_k = -e_k$ and $ge_i = e_i$ for all $i \neq j, k$. It then follows from (5.69) that

$$\langle F, e_{i_1} \otimes \cdots \otimes e_{i_n} \rangle = - \langle F, e_{i_1} \otimes \cdots \otimes e_{i_n} \rangle = 0.$$

Hence $F = 0$ as desired. qed

From now on we assume that $n = 2m$ is even and that $F \in (E_{\mathbb{C}}^{\otimes n})^*$ is rotation-invariant. By the same argument as in the above proof, we see that $\langle F, e_{i_1} \otimes \cdots \otimes e_{i_n} \rangle$ can be non-zero only when $i_1, \cdots, i_n$ consist of pairwise identical numbers, namely, only when there exists a permutation $\sigma \in \mathfrak{S}_n$ such that $i_{\sigma(1)} = i_{\sigma(2)}, \cdots, i_{\sigma(n-1)} = i_{\sigma(n)}$. Let Ω_n be the collection of such sequences $(i_1, \cdots, i_n)$. Let $\mathfrak{S}_\infty$ denote the group of all finite permutations of the non-negative integers. Then $\mathfrak{S}_\infty$ acts on Ω_n by means of the maps:

$$(i_1, \cdots, i_n) \mapsto (\rho(i_1), \cdots, \rho(i_n)),$$

where $(i_1, \cdots, i_n) \in \Omega_n$ and $\rho \in \mathfrak{S}_\infty$. Let $\mathfrak{T}_n = \mathfrak{S}_\infty \backslash \Omega_n$ be the set of $\mathfrak{S}_\infty$-orbits in Ω_n. With these notations, we have

Lemma 5.5.8 *If $F \in (E_{\mathbb{C}}^{\otimes n})^*$, n being even, is rotation-invariant, then*

$$F = \sum_{\alpha \in \mathfrak{T}_n} C(\alpha) \sum_{(i_1, \cdots, i_n) \in \alpha} e_{i_1} \otimes \cdots \otimes e_{i_n}$$

with some constants $C(\alpha) \in \mathbb{C}$.

PROOF. We have already seen that such an F is expressed as

$$\begin{aligned} F &= \sum_{(i_1, \cdots, i_n) \in \Omega_n} \langle F, e_{i_1} \otimes \cdots \otimes e_{i_n} \rangle e_{i_1} \otimes \cdots \otimes e_{i_n} \\ &= \sum_{\alpha \in \mathfrak{T}_n} \sum_{(i_1, \cdots, i_n) \in \alpha} \langle F, e_{i_1} \otimes \cdots \otimes e_{i_n} \rangle e_{i_1} \otimes \cdots \otimes e_{i_n}. \end{aligned}$$

For any pair $(i_1, \cdots, i_n), (i'_1, \cdots, i'_n) \in \alpha$ there exists a (finite dimensional) rotation $g \in O(E; H)$ such that $ge_{i_1} = e_{i'_1}, \cdots, ge_{i_n} = e_{i'_n}$. Then, in view of (5.69) we see that

$$\langle F, e_{i_1} \otimes \cdots \otimes e_{i_n} \rangle = \left\langle F, e_{i'_1} \otimes \cdots \otimes e_{i'_n} \right\rangle.$$

This proves the assertion. qed

We need some notation. A sequence of non-negative integers $\pi = (p_1, p_2, \cdots)$ is called a *partition* of a natural number m if

(i) $m \ge p_1 \ge p_2 \ge \cdots \ge 0$;

(ii) $p_1 + p_2 + \cdots = m$.

Let $\mathfrak{P}_m$ be the set of partitions of m. We introduce a lexicographic order into $\mathfrak{P}_m$. For $\pi = (p_1, p_2, \cdots)$ and $\pi' = (p'_1, p'_2, \cdots)$ we write $\pi < \pi'$ if

$$p_1 = p'_1, \quad \cdots, \quad p_{l-1} = p'_{l-1}, \quad p_l > p'_l$$

for some $l = 1, 2, \cdots$. For simplicity we arrange all elements of $\mathfrak{P}_m$ as $\pi_1 > \pi_2 > \cdots$. For $\pi = (p_1, p_2, \cdots) \in \mathfrak{P}_m$ we define

$$\omega(\pi) = \sum_{j_1, j_2, \cdots = 0}^{\infty} e_{j_1}^{\otimes 2p_1} \otimes e_{j_2}^{\otimes 2p_2} \otimes \cdots \in (E^{\otimes n})^*. \tag{5.70}$$

For example, $\pi_1 = (1, 1, \cdots, 1, 0, \cdots)$ and $\omega(\pi_1) = \tau^{\otimes m}$. For $\pi \in \mathfrak{P}_m$ we define

$$\mathfrak{S}(\pi) = \{\sigma \in \mathfrak{S}_n; \omega(\pi)^\sigma = \omega(\pi)\}.$$

With these notations we can state

Lemma 5.5.9 *If $F \in (E_{\mathbb{C}}^{\otimes n})^*$, n being even, is rotation-invariant, then*

$$F = \sum_{\pi \in \mathfrak{P}_m} \sum_{\sigma \in \mathfrak{S}(\pi) \backslash \mathfrak{S}_n} c(\pi; \sigma) \omega(\pi)^\sigma \tag{5.71}$$

with some constants $c(\pi; \sigma) \in \mathbb{C}$. Moreover, the constants $c(\pi; \sigma)$ are unique.

PROOF. First note that for $\pi = (p_1, p_2, \cdots) \in \mathfrak{P}_m$

$$\langle\!\langle \omega(\pi), e_{i_1} \otimes \cdots \otimes e_{i_n} \rangle\!\rangle = \\ = \begin{cases} 1, & \text{if } \; i_1 = \cdots = i_{2p_1}, i_{2p_1+1} = \cdots = i_{2p_1+2p_2}, \cdots, \\ 0, & \text{otherwise.} \end{cases} \tag{5.72}$$

Therefore,

$$\langle\!\langle \omega(\pi), e_{i_1} \otimes \cdots \otimes e_{i_n} \rangle\!\rangle = 0, \qquad \text{if} \quad (i_1, \cdots, i_n) \notin \Omega_n. \tag{5.73}$$

If $(i_1, \cdots, i_n) \in \Omega_n$, we may find $\sigma \in \mathfrak{S}_n$ and $\pi' = (p'_1, p'_2, \cdots) \in \mathfrak{P}_m$ such that

$$i_{\sigma(1)} = \cdots = i_{\sigma(2p'_1)} \equiv j_1, i_{\sigma(2p'_1+1)} = \cdots = i_{\sigma(2p'_1+2p'_2)} \equiv j_2, \cdots$$

where $j_1, j_2, \cdots$ are mutually distinct. Then from (5.72) we see that

$$\langle\!\langle \omega(\pi), e_{i_1} \otimes \cdots \otimes e_{i_n} \rangle\!\rangle = 0, \qquad \text{if} \quad \pi < \pi', \tag{5.74}$$

where $\pi' \in \mathfrak{P}_m$ is the (unique) partition corresponding to $(i_1, \cdots, i_n) \in \Omega_n$.

For $\pi = (p_1, p_2, \cdots) \in \mathfrak{P}_m$ and $\theta \in \mathbb{R}$ we put

$$\eta(\pi; \theta) = ((\sin\theta)e_0 + (\cos\theta)e_1)^{\otimes 2p_1} \otimes e_2^{\otimes 2p_2} \otimes e_3^{\otimes 2p_3} \otimes \cdots \in E^{\otimes n}.$$

In view of (5.73) and (5.74) we obtain

$$\begin{aligned} \langle \omega(\pi), \eta(\pi'; \theta)^\sigma \rangle &= 0, & &\text{if} \quad \pi < \pi', \sigma \in \mathfrak{S}_n, \\ \langle \omega(\pi), \eta(\pi; \theta)^\sigma \rangle &= \begin{cases} (\sin\theta)^{2p_1} + (\cos\theta)^{2p_1}, \\ 0, \end{cases} & &\begin{matrix} \text{if } \sigma \in \mathfrak{S}(\pi) \\ \text{otherwise.} \end{matrix} \end{aligned} \tag{5.75}$$

We now go back to the proof of the original assertion. We first prove that $\{\omega(\pi)^\sigma; \pi \in \mathfrak{P}_m, \sigma \in \mathfrak{S}(\pi)\backslash\mathfrak{S}_n\}$ is linearly independent. In fact, suppose that

$$\sum_{\pi \in \mathfrak{P}_m} \sum_{\sigma \in \mathfrak{S}(\pi)\backslash\mathfrak{S}_n} c(\pi; \sigma)\omega(\pi)^\sigma = 0$$

with some $c(\pi; \sigma) \in \mathbb{C}$. It then follows from (5.75) that, for any $\sigma' \in \mathfrak{S}_n$,

$$0 = \sum_{\pi \in \mathfrak{P}_m} \sum_{\sigma \in \mathfrak{S}(\pi)\backslash\mathfrak{S}_n} c(\pi; \sigma) \left\langle \omega(\pi)^\sigma, \eta(\pi_1; 0)^{\sigma'} \right\rangle = c(\pi_1; \sigma').$$

Using the same argument successively according to the arrangement of partitions in $\mathfrak{P}_m$, we conclude that $c(\pi_2; \sigma') = \cdots = 0$. Hence $\{\omega(\pi)^\sigma; \pi \in \mathfrak{P}_m, \sigma \in \mathfrak{S}(\pi)\backslash\mathfrak{S}_n\}$ is linearly independent.

Let W be the subspace of $(E_{\mathbb{C}}^{\otimes n})^*$ spanned by $\omega(\pi)^\sigma$, where $\pi \in \mathfrak{P}_m$ and $\sigma \in \mathfrak{S}(\pi)\backslash\mathfrak{S}_n$. We then note that $\dim W = \sum_{\pi \in \mathfrak{P}_m} |\mathfrak{S}_n : \mathfrak{S}(\pi)|$. Let V be the subspace of $(E_{\mathbb{C}}^{\otimes n})^*$ spanned by $\sum_{(i_1, \cdots, i_n) \in \alpha} e_{i_1} \otimes \cdots \otimes e_{i_n}$, where α runs over $\mathfrak{T}_n$. Let $\pi = (p_1, p_2, \cdots) \in \mathfrak{P}_m$. Since $\omega(\pi)$ is decomposed into a finite sum:

$$\begin{aligned} \omega(\pi) &= \sum_{j_1, j_2, \cdots} e_{j_1}^{\otimes 2p_1} \otimes e_{j_2}^{\otimes 2p_2} \otimes \cdots \\ &= \sum_{\substack{j_1, j_2, \cdots \\ \text{distinct}}} + \sum_{\substack{j_1 = j_2, j_3, \cdots \\ \text{distinct}}} + \sum_{\substack{j_1, j_2 = j_3, \cdots \\ \text{distinct}}} + \cdots + \sum_{j_1 = j_2 = \cdots}, \end{aligned}$$

we see that $\omega(\pi) \in V$. Since $V^\sigma = V$ for any $\sigma \in \mathfrak{S}_n$, we obtain $W \subset V$. On the other hand, since

$$\dim V \le |\mathfrak{T}_n| = \sum_{\pi\in\mathfrak{P}_m} |\mathfrak{S}_n : \mathfrak{S}(\pi)| = \dim W,$$

we conclude that $W = V$.

We now suppose that $F \in (E_{\mathbb{C}}^{\otimes n})^*$ is rotation-invariant. It then follows from Lemma 5.5.8 that $F \in V$. As we have proved above, $V = W$ is spanned by $\{\omega(\pi)^\sigma; \pi \in \mathfrak{P}_m, \sigma \in \mathfrak{S}(\pi)\backslash\mathfrak{S}_n\}$ which is a linear basis of W. Therefore F is uniquely expressed in the form (5.71). qed

Proposition 5.5.10 *$(\tau^{\otimes m})^\sigma$ is rotation-invariant for any $\sigma \in \mathfrak{S}_n$.*

PROOF. It is sufficient to show that

$$\left\langle (\tau^{\otimes m})^{\sigma^{-1}}, \xi_1 \otimes \cdots \otimes \xi_n \right\rangle = \left\langle (\tau^{\otimes m})^{\sigma^{-1}}, g\xi_1 \otimes \cdots \otimes g\xi_n \right\rangle$$

for any $\xi, \cdots, \xi_n \in E_{\mathbb{C}}$ and for any $g \in O(E; H)$. We observe that

$$\begin{aligned}\left\langle (\tau^{\otimes m})^{\sigma^{-1}}, g\xi_1 \otimes \cdots \otimes g\xi_n \right\rangle &= \left\langle \tau^{\otimes m}, g\xi_{\sigma(1)} \otimes \cdots \otimes g\xi_{\sigma(n)} \right\rangle \\ &= \left\langle \tau, g\xi_{\sigma(1)} \otimes g\xi_{\sigma(2)} \right\rangle \cdots \left\langle \tau, g\xi_{\sigma(n-1)} \otimes g\xi_{\sigma(n)} \right\rangle \\ &= \left\langle g\xi_{\sigma(1)}, g\xi_{\sigma(2)} \right\rangle \cdots \left\langle g\xi_{\sigma(n-1)}, g\xi_{\sigma(n)} \right\rangle \\ &= \left\langle \xi_{\sigma(1)}, \xi_{\sigma(2)} \right\rangle \cdots \left\langle \xi_{\sigma(n-1)}, \xi_{\sigma(n)} \right\rangle.\end{aligned}$$

Similarly the left hand side is computed to get the same expression. qed

PROOF OF PROPOSITION 5.5.6 (CASE OF n BEING EVEN). Suppose F is given as in (5.71). First we prove

$$\begin{aligned}F' &\equiv F - \sum_{\sigma\in\mathfrak{S}(\pi_1)\backslash\mathfrak{S}_n} c(\pi_1;\sigma)\omega(\pi_1)^\sigma \\ &= \sum_{\pi\le\pi_2} \sum_{\sigma\in\mathfrak{S}(\pi)\backslash\mathfrak{S}_n} c(\pi;\sigma)\omega(\pi)^\sigma = 0.\end{aligned}$$

Take a (two dimensional) rotation $g_\theta \in O(E; H)$ defined by

$$\begin{cases} g_\theta e_0 &= (\cos\theta)e_0 - (\sin\theta)e_1 \\ g_\theta e_1 &= (\sin\theta)e_0 + (\cos\theta)e_1, \end{cases}$$

with $\theta \in \mathbb{R}$. It follows from Proposition 5.5.10 and the assumption that F' is rotation-invariant. Then,

$$\langle F', \eta(\pi_2;0)^\sigma\rangle = \left\langle (g_\theta^{\otimes n})^* F', \eta(\pi_2;0)^\sigma\right\rangle = \left\langle F', (g_\theta^{\otimes n}\eta(\pi_2;0))^\sigma\right\rangle, \qquad \theta \in \mathbb{R}.$$

Since $g_\theta^{\otimes n}\eta(\pi_2;0) = \eta(\pi_2;\theta)$, using (5.72) and (5.75), we obtain

$$c(\pi_2;\sigma) = \langle F', \eta(\pi_2;0)^\sigma\rangle = \langle F', \eta(\pi_2;\theta)^\sigma\rangle = \left((\sin\theta)^{2p_1} + (\cos\theta)^{2p_1}\right) c(\pi_2;\sigma)$$

for all $\theta \in \mathbb{R}$, and therefore, $c(\pi_2;\sigma) = 0$. Inductively, we may prove that $c(\pi_3;\sigma) = \cdots = 0$, namely, $F' = 0$ as desired. Since $\omega(\pi_1)^\sigma = (\tau^{\otimes m})^\sigma$, $\sigma \in \mathfrak{S}(\pi_1)\backslash\mathfrak{S}_n$, are independent, the dimension of the rotation-invariant distributions in $(E_{\mathbb{C}}^{\otimes n})^*$ is given by $|\mathfrak{S}_n : \mathfrak{S}(\pi_1)| = n!/(2^m \cdot m!) = (n-1)!!$. qed

During the above proof of Proposition 5.5.6, we have used only the subgroup of $O(E;H)$ consisting of rotations which act identically on the subspace spanned by $\{e_n, e_{n+1}, \cdots\}$ for a large $n \geq 0$. It would be interesting to investigate invariant-distributions under another subgroups of $O(E;H)$, for instance, certain transformation groups of T which are naturally imbedded in $O(E;H)$.

5.6 Fourier transform

We begin with the following

Lemma 5.6.1 *For each $\theta \in \mathbb{R}$ there exists an operator $\mathfrak{G}_\theta \in \mathcal{L}((E),(E))$ such that*

$$\mathfrak{G}_\theta \phi_\xi = \exp\left(\frac{i}{2} e^{i\theta} \sin\theta \langle \xi, \xi \rangle\right) \phi_{e^{i\theta}\xi}, \qquad \xi \in E_{\mathbb{C}}. \tag{5.76}$$

PROOF. It is sufficient to verify that

$$\Theta(\xi,\eta) = \langle\langle \mathfrak{G}_\theta \phi_\xi, \phi_\eta \rangle\rangle = \exp\left(\frac{i}{2} e^{i\theta} \sin\theta \langle \xi, \xi \rangle + e^{i\theta} \langle \xi, \eta \rangle\right)$$

satisfies conditions (i) and (ii) in Theorem 4.4.7. In fact, (i) is obvious. To see (ii) we need only to note that

$$|\Theta(\xi,\eta)| \leq \exp\left(\frac{1}{2} |\langle \xi, \xi \rangle| + |\langle \xi, \eta \rangle|\right), \qquad \xi, \eta \in E_{\mathbb{C}},$$

and to apply Lemma 4.6.6. As a result, we see that Θ is the symbol of an operator in $\mathcal{L}((E),(E))$ of which action on exponential vectors is given as in (5.76). qed

Lemma 5.6.2 *For $\theta, \theta_1, \theta_2 \in \mathbb{R}$,*

$$\mathfrak{G}_0 = I, \qquad \mathfrak{G}_{\theta+2\pi} = \mathfrak{G}_\theta, \qquad \mathfrak{G}_{\theta_1+\theta_2} = \mathfrak{G}_{\theta_1}\mathfrak{G}_{\theta_2}.$$

This is immediate from (5.76). It is now obvious that $\mathfrak{F}_\theta = \mathfrak{G}_\theta^*$ becomes a continuous linear operator on $(E)^*$ and forms a one-parameter group of linear transformations acting on $(E)^*$ satisfying

$$\mathfrak{F}_0 = I, \qquad \mathfrak{F}_{\theta+2\pi} = \mathfrak{F}_\theta, \qquad \mathfrak{F}_{\theta_1+\theta_2} = \mathfrak{F}_{\theta_1}\mathfrak{F}_{\theta_2}.$$

Definition 5.6.3 The operators $\mathfrak{F} = \mathfrak{F}_{-\pi/2}$ and $\mathfrak{F}_\theta$ are called *(Kuo's) Fourier transform* and *Fourier-Mehler transform* (with parameter $\theta \in \mathbb{R}$), respectively.

These operators are characterized as follows.

Proposition 5.6.4 *For each $\theta \in \mathbf{R}$ Fourier-Mehler transform $\mathfrak{F}_\theta$ is the unique operator satisfying*

$$S(\mathfrak{F}_\theta\Phi)(\xi) = S\Phi(e^{i\theta}\xi)\exp\left(\frac{i}{2}e^{i\theta}\sin\theta\,\langle\xi,\xi\rangle\right), \qquad \xi \in E_{\mathbf{C}}, \quad \Phi \in (E)^*. \tag{5.77}$$

In particular, the Fourier transform is characterized by

$$S(\mathfrak{F}\Phi)(\xi) = S\Phi(-i\xi)\exp\left(-\frac{1}{2}\langle\xi,\xi\rangle\right), \qquad \xi \in E_{\mathbf{C}}, \quad \Phi \in (E)^*. \tag{5.78}$$

PROOF. Let $\Phi \in (E)^*$ and $\xi \in E_{\mathbf{C}}$. Then by definition of $\mathfrak{F}_\theta$ we have

$$S(\mathfrak{F}_\theta\Phi)(\xi) = \langle\!\langle \mathfrak{F}_\theta\Phi, \phi_\xi\rangle\!\rangle = \langle\!\langle \mathfrak{G}_\theta^*\Phi, \phi_\xi\rangle\!\rangle = \langle\!\langle \Phi, \mathfrak{G}_\theta\phi_\xi\rangle\!\rangle.$$

Then, in view of (5.76), we obtain

$$\begin{aligned}\langle\!\langle \Phi, \mathfrak{G}_\theta\phi_\xi\rangle\!\rangle &= \langle\!\langle \Phi, \phi_{e^{i\theta}\xi}\rangle\!\rangle \exp\left(\frac{i}{2}e^{i\theta}\sin\theta\,\langle\xi,\xi\rangle\right)\\ &= S\Phi(e^{i\theta}\xi)\exp\left(\frac{i}{2}e^{i\theta}\sin\theta\,\langle\xi,\xi\rangle\right).\end{aligned}$$

This completes the proof of (5.77). For (5.78) we need only to put $\theta = -\pi/2$. qed

The explicit forms of $\mathfrak{F}_\theta\Phi$ and $\mathfrak{G}_\theta\phi$ are easily derived.

Proposition 5.6.5 *Let*

$$\Phi(x) = \sum_{n=0}^{\infty}\left\langle :x^{\otimes n}:, F_n\right\rangle, \qquad F_n \in (E_{\mathbf{C}}^{\otimes n})^*_{\mathrm{sym}},$$

be the Wiener-Itô expansion of $\Phi \in (E)^*$. Then the Wiener-Itô expansion of $\mathfrak{F}_\theta\Phi$ is given as

$$\mathfrak{F}_\theta\Phi(x) = \sum_{n=0}^{\infty}\left\langle :x^{\otimes n}:, F_n(\mathfrak{F}_\theta\Phi)\right\rangle,$$

where

$$F_n(\mathfrak{F}_\theta\Phi) = \sum_{l+2m=n}\frac{1}{m!}e^{i(l+m)\theta}\left(\frac{i}{2}\sin\theta\right)^m F_l\hat{\otimes}\tau^{\otimes m}.$$

PROOF. First recall that

$$S\Phi(\xi) = \sum_{n=0}^{\infty}\left\langle F_n, \xi^{\otimes n}\right\rangle, \qquad \xi \in E_{\mathbf{C}}.$$

Hence by (5.77) we obtain

$$\begin{aligned}S(\mathfrak{F}_\theta\Phi)(\xi) &= S\Phi(e^{i\theta}\xi)\exp\left(\frac{i}{2}e^{i\theta}\sin\theta\,\langle\xi,\xi\rangle\right)\\ &= \sum_{n=0}^{\infty}\left\langle F_n, (e^{i\theta}\xi)^{\otimes n}\right\rangle\sum_{m=0}^{\infty}\frac{1}{m!}\left(\frac{i}{2}e^{i\theta}\sin\theta\right)^m\langle\xi,\xi\rangle^m\\ &= \sum_{m,n=0}^{\infty}\frac{1}{m!}e^{i(m+n)\theta}\left(\frac{i}{2}\sin\theta\right)^m\langle F_n,\xi^{\otimes n}\rangle\langle\tau,\xi\otimes\xi\rangle^m\\ &= \sum_{m,n=0}^{\infty}\frac{1}{m!}e^{i(m+n)\theta}\left(\frac{i}{2}\sin\theta\right)^m\left\langle F_n\hat{\otimes}\tau^{\otimes m}, \xi^{\otimes(n+2m)}\right\rangle.\end{aligned}$$

Changing parameters, we obtain

$$S(\mathfrak{F}_\theta\Phi)(\xi) = \sum_{n=0}^{\infty} \sum_{l+2m=n} \frac{1}{m!} e^{i(l+m)\theta} \left(\frac{i}{2}\sin\theta\right)^m \left\langle F_l \widehat{\otimes} \tau^{\otimes m}, \xi^{\otimes n} \right\rangle.$$

The assertion then follows immediately. qed

By duality argument we have

Proposition 5.6.6 *Let*

$$\phi(x) = \sum_{n=0}^{\infty} \left\langle :x^{\otimes n}:, f_n \right\rangle, \qquad f_n \in E_{\mathbb{C}}^{\widehat{\otimes} n},$$

be the Fock expansion of $\phi \in (E)$. *Then the Wiener-Itô expansion of* $\mathfrak{G}_\theta\phi$ *is given as*

$$\mathfrak{G}_\theta\phi(x) = \sum_{n=0}^{\infty} \left\langle :x^{\otimes n}:, f_n(\mathfrak{G}_\theta\phi) \right\rangle,$$

where

$$f_n(\mathfrak{G}_\theta\phi) = \sum_{m=0}^{\infty} \frac{(2m+n)!}{n!m!} e^{i(m+n)\theta} \left(\frac{i}{2}\sin\theta\right)^m \tau^{\otimes m} \widehat{\otimes}_{2m} f_{2m+n}.$$

The symbol of $\mathfrak{G}_\theta$ appeared already during the proof of Lemma 5.6.1. The Fock expansion is then straightforward by Taylor expansion. Moreover by duality one may obtain easily the symbol and Fock expansion of the Fourier-Mehler transform.

Proposition 5.6.7 *Let* $\theta \in \mathbb{R}$. *For* $\xi, \eta \in E_{\mathbb{C}}$ *we have*

$$\begin{aligned} \widehat{\mathfrak{G}_\theta}(\xi, \eta) &= \exp\left(\frac{i}{2} e^{i\theta} \sin\theta \langle \xi, \xi \rangle + e^{i\theta} \langle \xi, \eta \rangle\right), \\ \widehat{\mathfrak{F}_\theta}(\xi, \eta) &= \widehat{\mathfrak{G}_\theta}(\eta, \xi) = \exp\left(\frac{i}{2} e^{i\theta} \sin\theta \langle \eta, \eta \rangle + e^{i\theta} \langle \xi, \eta \rangle\right). \end{aligned}$$

For the Fock expansions we have

$$\begin{aligned} \mathfrak{G}_\theta &= \sum_{l,m=0}^{\infty} \frac{1}{l!m!} \left(\frac{i}{2} e^{i\theta} \sin\theta\right)^m \left(e^{i\theta} - 1\right)^l \Xi_{l,l+2m}(\tau_l \otimes \tau^{\otimes m}), \\ \mathfrak{F}_\theta &= \sum_{l,m=0}^{\infty} \frac{1}{l!m!} \left(\frac{i}{2} e^{i\theta} \sin\theta\right)^l \left(e^{i\theta} - 1\right)^m \Xi_{2l+m,m}(\tau^{\otimes l} \otimes \tau_m), \end{aligned}$$

where $\tau_l \in (E^{\otimes 2l})^*$ *is defined in (4.62).*

Moreover, by similar argument as in the proof of Theorem 5.3.11 one may obtain the following interesting result.

Theorem 5.6.8 $\{\mathfrak{G}_\theta\}_{\theta\in\mathbb{R}}$ *is a regular one-parameter subgroup of* $GL((E))$ *with infinitesimal generator* $iN + \frac{i}{2}\Delta_G$.

There is a close relation between the Fourier transform and the S-transform. For simplicity we put

$$\epsilon_\xi(x) = e^{i\langle x, \xi\rangle}, \qquad \xi \in E_{\mathbf{C}}, \quad x \in E^*.$$

Then,

$$\epsilon_\xi = e^{-\langle \xi, \xi\rangle/2}\phi_{i\xi}, \qquad \xi \in E_{\mathbf{C}}, \tag{5.79}$$

and, in particular, $\epsilon_\xi \in (E)$.

Definition 5.6.9 The T-*transform* of $\Phi \in (E)^*$ is a function on $E_{\mathbf{C}}$ defined by

$$T\Phi(\xi) = \langle\langle \Phi, \epsilon_\xi \rangle\rangle, \qquad \xi \in E_{\mathbf{C}}.$$

Proposition 5.6.10 *For any $\Phi \in (E)^*$ it holds that*

$$T\Phi(\xi) = e^{-\langle \xi, \xi\rangle/2} S\Phi(i\xi), \qquad \xi \in E_{\mathbf{C}}.$$

PROOF. In view of (5.79) we have

$$S\Phi(i\xi) = \langle\langle \Phi, \phi_{i\xi} \rangle\rangle = e^{\langle \xi, \xi\rangle/2} \langle\langle \Phi, \epsilon_\xi \rangle\rangle = e^{\langle \xi, \xi\rangle/2} T\Phi(\xi), \qquad \xi \in E_{\mathbf{C}},$$

which proves the assertion. qed

Corollary 5.6.11 *Let $\Phi \in (E)^*$. If $T\Phi(\xi) = 0$ for all $\xi \in E$, then $\Phi = 0$.*

PROOF. By Proposition 5.6.10 we have $S\Phi(i\xi) = 0$ for all $\xi \in E$, and then from Corollary 3.3.9 we see that $\Phi = 0$. (Hence it follws that $T\Phi(\xi) = 0$ for all $\xi \in E_{\mathbf{C}}$.) qed

It is also interesting to ask how $T\Phi$ is characterized as a function on $E_{\mathbf{C}}$. But since the T- and S-transforms are connected directly as in Proposition 5.6.10, we obtain the following

Theorem 5.6.12 *Let F be a $\mathbf{C}$-valued function on $E_{\mathbf{C}}$. Then $F = T\Phi$ for some $\Phi \in (E)^*$ if and only if*

(i) *$z \mapsto F(z\xi + \eta)$ is an entire holomorphic function on $\mathbf{C}$ for any $\xi, \eta \in E_{\mathbf{C}}$;*

(ii) *there exist $C \geq 0$, $K \geq 0$ and $p \in \mathbf{R}$ such that*

$$|F(\xi)| \leq C \exp\left(K\, |\xi|_p^2\right), \qquad \xi \in E_{\mathbf{C}}.$$

In other words, the function spaces $\{S\Phi;\, \Phi \in (E)^*\}$ and $\{T\Phi;\, \Phi \in (E)^*\}$ are the same, see Theorem 3.6.1. Moreover, since T-transform is injective by Corollary 5.6.11, $T^{-1}S$ is a linear operator from $(E)^*$ into itself. (In fact, S-transform is also injective, see Corollary 2.3.9.)

Theorem 5.6.13 $\mathfrak{F} = T^{-1}S$.

PROOF. By Proposition 5.6.4 we observe

$$S(\mathfrak{F}_{\pi/2}\Phi)(\xi) = e^{-\langle \xi, \xi\rangle/2} S\Phi(i\xi) = T\Phi(\xi), \qquad \xi \in E_{\mathbf{C}}.$$

Hence $S\mathfrak{F}_{\pi/2} = T$. Since $\mathfrak{F} = \mathfrak{F}_{-\pi/2} = \mathfrak{F}_{\pi/2}^{-1}$, we see that $\mathfrak{F} = T^{-1}S$. qed

Finally we remark extension of T-transform. If $f \in \bigcup_{p>1} L^p(E^*, \mu; \mathbf{C})$, then $\Phi_f \in (E)^*$ is defined by

$$\langle\langle \Phi_f, \phi \rangle\rangle = \int_{E^*} f(x)\phi(x)\,\mu(dx), \qquad \phi \in (E), \tag{5.80}$$

see Theorem 3.5.12. Hence,

$$T\Phi_f(\xi) = \int_{E^*} f(x) e^{i\langle x, \xi\rangle}\,\mu(dx), \qquad \xi \in E_{\mathbf{C}}. \tag{5.81}$$

Apparently, this is more like the Fourier transform on a finite dimensional Euclidean space than the Fourier transform $\mathfrak{F}$. While, (5.81) suggests us to define an integral transform:

$$\mathcal{T}f(\xi) = \int_{E^*} f(x) e^{i\langle x, \xi\rangle}\,\mu(dx), \qquad \xi \in E, \quad f \in L^1(E^*, \mu; \mathbf{C}).$$

Note that $\mathcal{T}f$ is defined only on E (not on $E_{\mathbf{C}}$) instead f can be taken from a larger class of functions, namely, $f \in L^1(E^*, \mu; \mathbf{C})$. In this connection we only note the following fact.

Proposition 5.6.14 *Let $f \in L^1(E^*, \mu; \mathbf{C})$. If $\mathcal{T}f(\xi) = 0$ for all $\xi \in E$, then $f = 0$.*

PROOF. Let $\mathfrak{B}_n$ be the sub σ-field of $\mathfrak{B}$ generated by the functions: $x \mapsto \langle x, e_j\rangle$, $j = 1, 2, \cdots$, and let $E(f|\mathfrak{B}_n)$ denote the conditional expectation of f respect to $\mathfrak{B}_n$, see the proof of Proposition 2.3.2. Then, it is known that

$$\lim_{n\to\infty} E(f|\mathfrak{B}_n)(x) = f(x), \qquad \mu\text{-a.e. } x \in E^*. \tag{5.82}$$

Now suppose that $\mathcal{T}f(\xi) = 0$ for all $\xi \in E$. Then, in particular,

$$\begin{aligned} 0 &= \mathcal{T}f(t_1 e_1 + \cdots t_n e_n) \\ &= \int_{E^*} f(x) e^{i(t_1\langle x, e_1\rangle + \cdots + t_n\langle x, e_n\rangle)}\mu(dx) \\ &= \int_{E^*} E(f|\mathfrak{B}_n)(x) e^{i(t_1\langle x, e_1\rangle + \cdots + t_n\langle x, e_n\rangle)}\mu(dx). \end{aligned}$$

Then the uniqueness of the Fourier transform of $L^1(\mathbf{R}^n)$ implies that $E(f|\mathfrak{B}_n)(x) = 0$ for μ-a.e. $x \in E^*$. Consequently, we see from (5.82) that $f = 0$. qed

Corollary 5.6.15 *For $f \in \bigcup_{p>1} L^p(E^*, \mu; \mathbf{C})$ let $\Phi_f \in (E)^*$ be defined as in (5.80). Then the map $\Phi : \bigcup_{p>1} L^p(E^*, \mu; \mathbf{C}) \to (E)^*$ is a linear injection.*

PROOF. Suppose that $\Phi_f = 0$, $f \in L^p(E^*, \mu; \mathbf{C})$, $p > 1$. Then, $T\Phi_f = 0$ and therefore $\mathcal{T}f = 0$. (Here we note that $L^p \subset L^1$.) Then, applying Proposition 5.6.14, we see that $f = 0$. qed

5.7 Intertwining property of Fourier transform

We next consider characterization of the Fourier transform in terms of its intertwining properties. For that purpose with each $\xi \in E_{\mathbf{C}}$ we associate operators p_ξ and q_ξ by

$$\begin{aligned} p_\xi &= \frac{1}{2}\{\Xi_{0,1}(\xi) - \Xi_{1,0}(\xi)\} = \frac{1}{2}\int_T \xi(t)(\partial_t - \partial_t^*)\,dt, \\ q_\xi &= i\{\Xi_{0,1}(\xi) + \Xi_{1,0}(\xi)\} = i\int_T \xi(t)(\partial_t + \partial_t^*)dt. \end{aligned}$$

That $p_\xi, q_\xi \in \mathcal{L}((E),(E))$ follows from Theorem 4.3.9. By a simple computation we obtain

$$p_\xi^* = -p_\xi, \qquad q_\xi^* = q_\xi, \qquad \xi \in E_{\mathbf{C}},$$

and the canonical commutation relation:

$$[p_\xi, p_\eta] = [q_\xi, q_\eta] = 0, \qquad [p_\xi, q_\eta] = i\langle \xi, \eta\rangle I, \qquad \xi, \eta \in E.$$

Here we note that the symbol * means the adjoint with respect to the complex bilinear form $\langle\langle \cdot, \cdot \rangle\rangle$. For a real $\xi \in E$ we may write also

$$p_\xi = \frac{1}{2}(D_\xi - D_\xi^*) = D_\xi + \frac{i}{2}q_\xi, \qquad q_\xi = i(D_\xi + D_\xi^*).$$

Lemma 5.7.1 *For $\phi \in (E)$ and $\xi \in E_{\mathbf{C}}$, we have*

$$(q_\xi\phi)(x) = i\langle x, \xi\rangle\, \phi(x), \qquad x \in E^*.$$

PROOF. Since q_ξ depends linearly on $\xi \in E_{\mathbf{C}}$, it is sufficient to prove the relation for a real $\xi \in E$. Since D_ξ is a derivation, we have

$$D_\xi(\phi_1\phi_2) = (D_\xi\phi_1)\phi_2 + \phi_1(D_\xi\phi_2), \qquad \phi_1, \phi_2 \in (E),$$

and by duality

$$D_\xi^*(\phi_1\phi_2) = (D_\xi^*\phi_1)\phi_2 - \phi_1(D_\xi\phi_2), \qquad \phi_1, \phi_2 \in (E).$$

Therefore,

$$(D_\xi + D_\xi^*)(\phi_1\phi_2) = ((D_\xi + D_\xi^*)\phi_1)\phi_2, \qquad \phi_1, \phi_2 \in (E).$$

In other words,

$$q_\xi(\phi_1\phi_2) = (q_\xi\phi_1)\phi_2, \qquad \phi_1, \phi_2 \in (E),$$

and in particular, we obtain

$$q_\xi\phi = q_\xi 1 \cdot \phi, \qquad \phi \in (E).$$

Then the assertion follows from the relation: $(q_\xi 1)(x) = iD_\xi^* 1(x) = i\langle x, \xi\rangle$. qed

According to Theorem 4.3.12, for any $\xi \in E$ the operators D_ξ and q_ξ are extended to continuous operators in $\mathcal{L}((E)^*, (E)^*)$, which are denoted by $\widetilde{D}_\xi$ and $\widetilde{q}_\xi$, respectively.

Theorem 5.7.2 *The Fourier-Mehler transform $\mathfrak{F}_\theta$ has the following properties:*
(i) $\mathfrak{F}_\theta$ *is a continuous operator from* $(E)^*$ *into itself;*
(ii) $\mathfrak{F}_\theta \widetilde{D}_\xi = ((\cos\theta)\widetilde{D}_\xi - (\sin\theta)\widetilde{q}_\xi)\mathfrak{F}_\theta$;
(iii) $\mathfrak{F}_\theta \widetilde{q}_\xi = ((\sin\theta)\widetilde{D}_\xi + (\cos\theta)\widetilde{q}_\xi)\mathfrak{F}_\theta$.

Corollary 5.7.3 *The Fourier transform $\mathfrak{F}$ has the following properties:*
(i) $\mathfrak{F}$ *is a continuous operator from* $(E)^*$ *into itself;*
(ii) $\mathfrak{F}\widetilde{D}_\xi = \widetilde{q}_\xi \mathfrak{F}$;
(iii) $\mathfrak{F}\widetilde{q}_\xi = -\widetilde{D}_\xi \mathfrak{F}$.

Symbolically the last expressions are also written as

$$\mathfrak{F}\partial_t = ix(t)\mathfrak{F}, \qquad \mathfrak{F}x(t) = i\partial_t\mathfrak{F}, \qquad t \in T.$$

Thus $\mathfrak{F}$ inherits a typical property of the Fourier transfrom on a finite dimensional Euclidean space.

PROOF OF THEOREM 5.7.2. Property (i) is clear by definition. By duality (ii) and (iii) are respectively equivalent to the following

$$\begin{aligned} D_\xi^* \mathfrak{G}_\theta &= \mathfrak{G}_\theta((\cos\theta)D_\xi^* - (\sin\theta)q_\xi), \\ q_\xi \mathfrak{G}_\theta &= \mathfrak{G}_\theta((\cos\theta)q_\xi + (\sin\theta)D_\xi^*). \end{aligned}$$

The latter being obtained from the former by changing parameter θ, we shall prove the former relation. Since both sides are continuous operators on (E), it is sufficient to show that they coincide on exponential vectors. Let $\eta \in E_{\mathbb{C}}$. Then by definition,

$$\mathfrak{G}_\theta \phi_\eta(x) = \exp\left(\frac{i}{2} e^{i\theta} \sin\theta \,\langle \eta, \eta\rangle\right) \phi_{e^{i\theta}\eta},$$

and therefore,

$$D_\xi^* \mathfrak{G}_\theta \phi_\eta(x) = \exp\left(\frac{i}{2} e^{i\theta} \sin\theta \,\langle \eta, \eta\rangle\right) \sum_{n=0}^{\infty} \left\langle :x^{\otimes(n+1)}:, \xi \widehat{\otimes} \left(\frac{(e^{i\theta}\eta)^{\otimes n}}{n!}\right)\right\rangle.$$

Then applying Proposition 5.6.6, we obtain

$$\begin{aligned} &\mathfrak{G}_{-\theta} D_\xi^* \mathfrak{G}_\theta \phi_\eta(x) = \\ &= \exp\left(\frac{i}{2} e^{i\theta} \sin\theta \,\langle \eta, \eta\rangle\right) \sum_{2m+n\geq 1} \frac{(2m+n)!}{n!m!} e^{-i(m+n)\theta} \left(-\frac{i}{2}\sin\theta\right)^m \\ &\quad \times \left\langle :x^{\otimes n}:, \tau^{\otimes m} \widehat{\otimes}_{2m} \left(\xi \widehat{\otimes} \left(\frac{(e^{i\theta}\eta)^{\otimes(2m+n-1)}}{(2m+n-1)!}\right)\right)\right\rangle. \end{aligned} \tag{5.83}$$

In view of the obvious relation:

$$\begin{aligned} &(2m+n)\tau^{\otimes m} \widehat{\otimes}_{2m} \left(\xi \widehat{\otimes} \eta^{\otimes(2m+n-1)}\right) = \\ &= 2m \langle \xi, \eta\rangle \langle \eta, \eta\rangle^{m-1} \eta^{\otimes n} + n \langle \eta, \eta\rangle^m \xi \widehat{\otimes} \eta^{\otimes(n-1)}, \end{aligned}$$

we can proceed the computation of (5.83).

$$\sum_{2m+n\geq 1} \frac{(2m+n)!}{n!m!} e^{-i(m+n)\theta} \left(-\frac{i}{2}\sin\theta\right)^m$$
$$\times \left\langle :x^{\otimes n}:, \tau^{\otimes m}\widehat{\otimes}_{2m}\left(\xi\widehat{\otimes}\left(\frac{(e^{i\theta}\eta)^{\otimes(2m+n-1)}}{(2m+n-1)!}\right)\right)\right\rangle$$
$$= 2\langle \xi, \eta\rangle \sum_{m,n=0}^{\infty} \frac{m}{n!m!} e^{i(m-1)\theta}\left(-\frac{i}{2}\sin\theta\right)^m \langle \eta, \eta\rangle^{m-1} \left\langle :x^{\otimes n}:, \eta^{\otimes n}\right\rangle$$
$$+ \sum_{m,n=0}^{\infty} \frac{n}{n!m!} e^{i(m-1)\theta}\left(-\frac{i}{2}\sin\theta\right)^m \langle \eta, \eta\rangle^m \left\langle :x^{\otimes n}:, \xi\widehat{\otimes}\eta^{\otimes(n-1)}\right\rangle$$
$$= -i(\sin\theta)\langle \xi, \eta\rangle \exp\left(-\frac{i}{2}e^{i\theta}\sin\theta\langle \eta, \eta\rangle\right)\sum_{n=0}^{\infty}\frac{1}{n!}\left\langle :x^{\otimes n}:, \eta^{\otimes n}\right\rangle$$
$$+ e^{-i\theta}\exp\left(-\frac{i}{2}e^{i\theta}\sin\theta\langle \eta, \eta\rangle\right)\sum_{n=0}^{\infty}\frac{1}{n!}\left\langle :x^{\otimes(n+1)}:, \xi\widehat{\otimes}\eta^{\otimes n}\right\rangle.$$

Hence (5.83) becomes

$$\begin{aligned}\mathfrak{G}_{-\theta}D_\xi^*\mathfrak{G}_\theta\phi_\eta &= -i(\sin\theta)\langle \xi, \eta\rangle \phi_\eta + e^{-i\theta}D_\xi^*\phi_\eta \\ &= -i(\sin\theta)D_\xi\phi_\eta + e^{-i\theta}D_\xi^*\phi_\eta \\ &= ((\cos\theta)D_\xi^* - (\sin\theta)q_\xi)\phi_\eta.\end{aligned}$$

This completes the proof. qed

More interesting is that the converse of the above assertions are also true. Namely, the properties listed above are actually characteristic properties of the Fourier-Mehler and the Fourier transform. A precise assertion is the following

Theorem 5.7.4 *Suppose $\mathfrak{A}_\theta$ is a linear operator from $(E)^*$ into itself satisfying the properties (i), (ii) and (iii) in Theorem 5.7.2. Then $\mathfrak{A}_\theta$ is a constant multiple of $\mathfrak{F}_\theta$. In particular, if $\mathfrak{A}$ is a linear operator from $(E)^*$ into itself satisfying the properties (i), (ii) and (iii) in Corollary 5.7.3, then $\mathfrak{A}$ is a constant multiple of $\mathfrak{F}$.*

For the proof we need the following

Proposition 5.7.5 *Let $\Xi \in \mathcal{L}((E),(E))$ satisfy*
(i) $\Xi q_\xi = q_\xi \Xi$ *for every* $\xi \in E$;
(ii) $\Xi D_\xi = D_\xi \Xi$ *for every* $\xi \in E$.
Then Ξ is a scalar operator.

PROOF. Put $\lambda = \Xi 1 \in (E)$. We shall first prove that Ξ is a multiplication operator by λ. Since the multiplication is continuous bilinear map from $(E)\times(E)$ into (E), it is sufficient to show that $\Xi\phi = \lambda\phi$ for ϕ in a dense subspace of (E). Consider $\phi(x) = (i\langle x, \xi\rangle)^n$ with arbitrarily fixed $\xi \in E$ and $n = 0, 1, 2, \cdots$. Then, in view of assumption (i) we observe

$$\Xi\phi(x) = \Xi(q_\xi^n 1)(x) = q_\xi^n \Xi 1(x) = (i\langle x, \xi\rangle)^n \lambda(x) = \lambda(x)\phi(x).$$

This proves that $\Xi\phi = \lambda\phi$ for all $\phi \in (E)$.

For the assertion it is then sufficient to prove that λ is a constant function. Since D_ξ is a derivation,

$$D_\xi \Xi \phi = D_\xi(\lambda\phi) = D_\xi\lambda \cdot \phi + \lambda \cdot D_\xi\phi, \qquad \phi \in (E). \tag{5.84}$$

On the other hand, by assumption (ii) we have

$$D_\xi \Xi \phi = \Xi D_\xi \phi = \lambda \cdot D_\xi \phi. \tag{5.85}$$

Therefore, from (5.84) and (5.85) we see that $D_\xi\lambda \cdot \phi = 0$ for all $\phi \in (E)$, and hence $D_\xi\lambda = 0$ for all $\xi \in E$. Let

$$\lambda(x) = \sum_{n=0}^{\infty} \left\langle :x^{\otimes n}:, f_n \right\rangle, \qquad f_n \in E_{\mathbb{C}}^{\widehat{\otimes} n},$$

be the Wiener-Itô expansion of λ. Then,

$$0 = D_\xi\lambda(x) = \sum_{n=1}^{\infty} n \left\langle :x^{\otimes(n-1)}:, \xi \widehat{\otimes}_1 f_n \right\rangle.$$

Hence $\xi\widehat{\otimes}_1 f_n = 0$ for any $\xi \in E$ and $n = 1, 2, \cdots$, and therefore $f_n = 0$ for all $n = 1, 2, \cdots$. This completes the proof. qed

The above theorem yields easily a dual result for an operator acting on the space $(E)^*$. Since $D_\xi \in \mathcal{L}((E),(E))$, $D_\xi^* \in \mathcal{L}((E)^*,(E)^*)$.

Proposition 5.7.6 *Let Λ be a continuous linear operator on $(E)^*$ satisfying*
(i) $\Lambda\widetilde{q}_\xi = \widetilde{q}_\xi\Lambda$ *for any* $\xi \in E$;
(ii) $\Lambda D_\xi^* = D_\xi^*\Lambda$ *for any* $\xi \in E$.
Then Λ is a scalar operator.

PROOF OF THEOREM 5.7.4. Suppose $\mathfrak{A}_\theta$ is an operator on $(E)^*$ satisfying conditions (i)–(iii) in Theorem 5.7.2. We consider $\mathfrak{F}_\theta^{-1}\mathfrak{A}_\theta = \mathfrak{F}_{-\theta}\mathfrak{A}_\theta$. Then, a direct calculation implies that

$$\begin{aligned} (\mathfrak{F}_\theta^{-1}\mathfrak{A}_\theta)\widetilde{D}_\xi &= \widetilde{D}_\xi(\mathfrak{F}_\theta^{-1}\mathfrak{A}_\theta), \\ (\mathfrak{F}_\theta^{-1}\mathfrak{A}_\theta)\widetilde{q}_\xi &= \widetilde{q}_\xi(\mathfrak{F}_\theta^{-1}\mathfrak{A}_\theta). \end{aligned} \tag{5.86}$$

Since $\widetilde{q}_\xi = i(\widetilde{D}_\xi + D_\xi^*)$, from the above two relation we deduce

$$(\mathfrak{F}_\theta^{-1}\mathfrak{A}_\theta)D_\xi^* = D_\xi^*(\mathfrak{F}_\theta^{-1}\mathfrak{A}_\theta). \tag{5.87}$$

Thus (5.86) and (5.87) mean that $\mathfrak{F}_\theta^{-1}\mathfrak{A}_\theta$ satisfies conditions (i) and (ii) in Proposition 5.7.6, and therefore $\mathfrak{F}_\theta^{-1}\mathfrak{A}_\theta$ is a scalar operator. In other words, $\mathfrak{A}_\theta$ is a constant multiple of the Fourier-Mehler transform $\mathfrak{F}_\theta$. qed

Finally we discuss one-parameter subgroups of $GL((E))$ which generate p_ξ and q_ξ, $\xi \in E$. For $\phi \in (E)$ we put

$$\begin{aligned} P_\xi\phi(x) &= \left(\frac{\mu(dx+\xi)}{\mu(dx)}\right)^{1/2} \phi(x+\xi) = e^{\langle x,\xi\rangle/2 - \langle \xi,\xi\rangle/4}\phi(x+\xi) \\ Q_\xi\phi(x) &= e^{i\langle x,\xi\rangle}\phi(x). \end{aligned}$$

Lemma 5.7.7 *For any $\xi \in E$ both P_ξ and Q_ξ belong to $U((E);(L^2))$, and they give unitary representations of the additive group E, namely,*

$$P_{\xi+\eta} = P_\xi P_\eta, \qquad Q_{\xi+\eta} = Q_\xi Q_\eta, \qquad \xi, \eta \in E.$$

Moreover,

$$P_\xi Q_\eta P_{-\xi} = i\,\langle \xi\,,\eta\rangle\, Q_\eta, \qquad \xi, \eta \in E.$$

The proof is easy. We say that the above operators P_ξ, Q_ξ satisfy the *Weyl form* of the canonical commutation relation. By a straightforward computation we obtain

Proposition 5.7.8 *Let $\xi \in E$. Then for any $\eta, \zeta \in E_{\mathbf{C}}$,*

$$\begin{aligned} \widehat{P_\xi}(\eta,\zeta) &= e^{\langle \eta\,,\zeta\rangle - \langle \xi\,,\xi\rangle/8 + \langle \xi\,,\eta-\zeta\rangle/2}, \\ \widehat{Q_\xi}(\eta,\zeta) &= e^{\langle \eta\,,\zeta\rangle - \langle \xi\,,\xi\rangle/2 + i\langle \xi\,,\eta+\zeta\rangle}. \end{aligned}$$

Moreover,

$$\begin{aligned} P_\xi &= \exp\left(-\frac{\langle \xi\,,\xi\rangle}{8}\right) \sum_{l,m=0}^{\infty} \frac{(-1)^l}{l!m!} \left(\frac{1}{2}\right)^{l+m} \Xi_{l,m}(\xi^{\otimes(l+m)}), \\ Q_\xi &= \exp\left(-\frac{\langle \xi\,,\xi\rangle}{2}\right) \sum_{l,m=0}^{\infty} \frac{i^{l+m}}{l!m!} \Xi_{l,m}(\xi^{\otimes(l+m)}). \end{aligned}$$

By a similar argument as in the proof of Theorem 5.3.1 one may prove easily the following

Theorem 5.7.9 *Let $\xi \in E$. Then $\{P_{\theta\xi}\}_{\theta\in\mathbf{R}}$ and $\{Q_{\theta\xi}\}_{\theta\in\mathbf{R}}$ are regular one-parameter subgroups of $U((E);(L^2))$ with infinitesimal generators p_ξ and q_ξ, respectively.*

Bibliographical Notes

The algebraic characterization of first order differential operators in §5.1 was first discussed in Obata [11]. Note that each first order differential operator with smooth real coefficients gives rise to a vector field on E^*. In this connection it seems interesting to study relationship between one-parameter transformation groups and vector fields on E^*. A regular one-parameter subgroup introduced in §5.2 is expected to be one of the practical ideas to go further. Our argument seems to be related with the general theory of nuclear Fréchet Lie groups developed by Omori [1] and Kobayashi-Yoshioka-Maeda-Omori [1].

The infinite dimensional rotation group $O(E;H)$ was first introduced by Yoshizawa in 1960's and discussed with some interests from harmonic analysis, see Kôno [1], Orihara [1], Umemura-Kôno [1]. Then followed the investigation of symmetry of Brownian motion in terms of infinite dimensional rotation group by Hida-Kubo-Nomoto-Yoshizawa [1] and Yoshizawa [1], see also Yoshizawa [2]. This idea was developed by Hida-Lee-Lee [1] in case of Gaussian random fields, see also Hida [4], [6], [8]. As

Hida [1] has pointed out the importance of the infinite dimensional rotation group in white noise calculus, the direct approach to the infinite dimensional rotation group in terms of white noise calculus has been made only recently by Hida-Obata-Saitô [1] and Obata [6]. The discussion in §5.4 is based on those works.

As is seen from the proof of Proposition 5.4.2, the unitary representation $(\Gamma, (L^2))$ of $O(E; H)$ is decomposed according to the Wiener-Itô expansion and, in fact, this is an irreducible decomposition. In this connection see Matsushima-Okamoto-Sakurai [1] and Okamoto-Sakurai [1], [2].

The characterization theorem of the Gross Laplacian and the number operator was first obtained by Obata [6]. A similar question was discussed by Yamasaki [1], where only the number operator is characterized as rotation-invariant operator in Fock space by means of the ergodicity of the Gaussian measure under $O(E; H)$. Incidentally, the Gross Laplacian originated in Gross [4] where an infinite dimensional analog of Poisson's equation was discussed.

Answering a question raised by Hida [2], Kuo introduced the Fourier transform of white noise functionals in [2] and the Fourier-Mehler transform in [4], and found many interesting properties in [8]. However, mainly because of lack of the characterization theorem of generalized white noise functionals (Theorem 3.6.1), his discussion stayed at a somehow formal level. As soon as it appeared, he reformulated the Fourier and Fourier-Mehler transform in [10]. In §5.6 we introduced those transforms by means of our operator theory. The main part of §5.7 is based on the work of Hida-Kuo-Obata [1]. The Fourier transform has been discussed also by Ito-Kubo-Takenaka [1], Kubo-Kuo [1] and Kuo [11].

Finally we add some remarks on the Lévy Laplacian and certain permutation groups both of which are not discussed in these lecture notes. There is another infinite dimensional Laplacian acting on white noise functionals called the *Lévy Laplacian*. With the notation in §5.1 the Lévy Laplacian is defined by

$$\Delta_L = \lim_{n\to\infty} \frac{1}{n} \sum_{j=0}^{n-1} D_j^2.$$

It is noted that $\Delta_L \phi = 0$ for all $\phi \in (E)$ and Δ_L acts on a certain subspace of $(E)^*$. This operator possesses certain typical properties of a finite dimensional Laplacian, in particular, when it is discussed along with spherical means. In this connection see Hida-Saitô [1], Kuo [5], Kuo-Obata-Saitô [1], Obata [2], [4], [5], Saitô [1], [2] and Yan [5]. There have been published a number of papers on the Lévy Laplacian from various points of view, see e.g., Polishchuk [1]. In connection with invariance property of the Lévy Laplacian we have discussed some permutation groups contained in the infinite dimensional rotation group. The idea is originally due to Lévy [1] and has been developed by Blümlinger-Obata [1] and Obata [1], [3].

Chapter 6

Addendum

6.1 Integral-sum kernel operators

In order to develop quantum stochastic calculus Maassen [1] introduced a certain class of operators on Fock space and later on his idea has been considerably developed to the theory of integral-sum kernel operators by Meyer [2], Lindsay [2], Lindsay-Maassen [1], Belavkin [1], among others. From the viewpoint of operator theory on Fock space their approaches bear some common spirits and interests with our theory of Fock expansion and an interesting interplay is expected. The present section is intended for the first link in this direction. However, the discussion stays at a formal translation level and more careful and deep study should be made.

In this section we assume that the measure ν on T admits no atoms. That the diagonal of $T \times T$ is a $\nu \times \nu$-null set is important.

Let Ω_n be the collection of subsets $\sigma \subset T$ consisting of n points, $0 \leq n < \infty$. Since the measure is non-atomic, we may identify Ω with the factor space $T^n/\mathfrak{S}_n$ up to ν^n-null sets. Let λ_n be the measure on Ω such that $n!\lambda_n$ is the image of ν^n under the canonical map $T^n \to T^n/\mathfrak{S}_n$. We then put

$$\Omega = \bigcup_{n=0}^{\infty} \Omega_n, \qquad \lambda = \sum_{n=0}^{\infty} \lambda_n.$$

A straightforward verification implies the following

Lemma 6.1.1 *For $\phi \in (E)$ given with Wiener-Itô expansion:*

$$\phi(x) = \sum_{n=0}^{\infty} \left\langle :x^{\otimes n}:, f_n \right\rangle,$$

we define a function f on Ω by

$$f(\{t_1, \cdots, t_n\}) = n! f_n(t_1, \cdots, t_n), \qquad t_j \neq t_k, \quad n = 0, 1, 2, \cdots.$$

Then $f \in L^2(\Omega, \lambda)$ and

$$\|\phi\|_0^2 = \sum_{n=0}^{\infty} n! \, |f_n|_0^2 = \|f\|_{L^2(\Omega,\lambda)}^2.$$

Moreover, the correspondence $\phi \mapsto f$ extends to a unitary map from (L^2) onto $L^2(\Omega, \lambda)$.

Since $(L^2) = L^2(E^*, \mu)$ is canonically isomorphic to the Boson Fock space by Wiener-Itô-Segal isomorphism, $L^2(\Omega, \lambda)$ is also isomorphic to Fock space and is called *Guichardet's (realization of) Fock space.* An explicit form of exponential vectors in the new space $L^2(\Omega, \lambda)$ is easily derived.

Lemma 6.1.2 *Let $\xi \in E_{\mathbb{C}}$. Under the isomorphism between (L^2) and $L^2(\Omega, \lambda)$ the exponential vector ϕ_ξ corresponds to a function f_ξ defined by*

$$f_\xi(\sigma) = \prod_{t \in \sigma} \xi(t), \qquad \sigma \in \Omega.$$

Hence,

$$f_\xi(\sigma_1 \cup \sigma_2) = f_\xi(\sigma_1) f_\xi(\sigma_2), \qquad \text{whenever} \quad \sigma_1 \cap \sigma_2 = \emptyset.$$

Lemma 6.1.3 (Union-Partition Identity) *For an integrable function f on (Ω^d, λ^d) we have*

$$\int_{\Omega^d} f(\sigma_1, \cdots, \sigma_d) \lambda(d\sigma_1) \cdots \lambda(d\sigma_d) = \int_\Omega \sum_{\sigma_1 \cup \cdots \cup \sigma_d = \sigma} f(\sigma_1, \cdots, \sigma_d) \lambda(d\sigma),$$

where $\sigma_1 \cup \cdots \cup \sigma_d = \sigma$ is a partition of σ, i.e., $\sigma_j \cap \sigma_k = \emptyset$ for $j \neq k$.

This elementary result is useful, for the proof see e.g., Lindsay [2]. Maassen [1] introduced an operator of the form

$$\Xi f(\sigma) = \int_\Omega \sum_{\alpha_1 \cup \alpha_2 = \sigma} k(\alpha_1, \omega) f(\omega \cup \alpha_2) \lambda(d\omega), \qquad \sigma \in \Omega. \tag{6.1}$$

As is easily expected, under certain integrability condition Ξ becomes a densely defined operator in $L^2(\Omega, \lambda)$.

Proposition 6.1.4 *Let $\Xi = \sum_{l,m=0}^\infty \Xi_{l,m}(\kappa_{l,m})$ be the Fock expansion of (6.1). Then the kernel distributions are given as*

$$\kappa_{l,m}(s_1, \cdots, s_l, t_1, \cdots, t_m) = \frac{1}{l!m!} k(\{s_1, \cdots, s_l\}, \{t_1, \cdots, t_m\}).$$

Proof. Let us compute the operator symbol of (6.1). For $\xi, \eta \in E_{\mathbb{C}}$ we have

$$\begin{aligned}
\widehat{\Xi}(\xi, \eta) &= \langle\!\langle \Xi f_\xi, f_\eta \rangle\!\rangle \\
&= \int_\Omega \Xi f_\xi(\sigma) f_\eta(\sigma) \lambda(d\sigma) \\
&= \int_\Omega \int_\Omega \sum_{\alpha_1 \cup \alpha_2 = \sigma} k(\alpha_1, \omega) f_\xi(\omega \cup \alpha_2) f_\eta(\sigma) \lambda(d\omega) \lambda(d\sigma).
\end{aligned}$$

Noting that $f_\eta(\alpha_1 \cup \alpha_2) = f_\eta(\alpha_1)f_\eta(\alpha_2)$ and employing Lemma 6.1.3 we obtain

$$\int_\Omega \sum_{\alpha_1 \cup \alpha_2 = \sigma} k(\alpha_1, \omega) f_\xi(\omega \cup \alpha_2) f_\eta(\sigma) \lambda(d\sigma) =$$
$$= \int_{\Omega\times\Omega} k(\alpha_1, \omega) f_\xi(\omega \cup \alpha_2) f_\eta(\alpha_1) f_\eta(\alpha_2) \lambda(d\alpha_1)\lambda(d\alpha_2).$$

Hence

$$\hat{\Xi}(\xi,\eta) = \int_{\Omega^3} k(\alpha_1, \omega) f_\xi(\omega \cup \alpha_2) f_\eta(\alpha_1) f_\eta(\alpha_2) \lambda(d\alpha_1)\lambda(d\alpha_2)\lambda(d\omega).$$

Since $f_\xi(\omega \cup \alpha_2) = f_\xi(\omega) f_\xi(\alpha_2)$ for $\lambda \times \lambda$-a.e. (ω, α_2), we come to

$$\begin{aligned}\hat{\Xi}(\xi,\eta) &= \int_\Omega \left(f_\xi(\omega) \int_{\Omega\times\Omega} k(\alpha_1, \omega) f_\xi(\alpha_2) f_\eta(\alpha_1) f_\eta(\alpha_2) \lambda(d\alpha_1)\lambda(d\alpha_2)\right) \lambda(d\omega) \\ &= e^{\langle \xi, \eta \rangle} \int_{\Omega\times\Omega} f_\xi(\omega) k(\alpha_1, \omega) f_\eta(\alpha_1) \lambda(d\alpha_1)\lambda(d\omega).\end{aligned}$$

The assertion is then immediate by writing the last integral explicitly. qed

Using the characterization theorems for operator symbols, we may find explicit condition on the kernel function k in order that Ξ is a continuous operator on (E).

Note that smeared creation and annihilation operators are expressible in the form (6.1). For an annihilation operator $\Xi_{0,1}(\xi)$, $\xi \in E_{\mathbf{C}}$, we take a kernel function as

$$k(\emptyset, \{t\}) = \xi(t), \qquad k(\sigma, \omega) = 0 \quad \text{otherwise}.$$

On the other hand, for a creation operator $\Xi_{1,0}(\xi)$, $\xi \in E_{\mathbf{C}}$,

$$k(\{s\}, \emptyset) = \xi(s), \qquad k(\sigma, \omega) = 0 \quad \text{otherwise}.$$

However, pointwisely defined operators ∂_t and ∂_t^* are not expressible unless a distribution kernel is allowed for k.

As was pointed out by Meyer, the original class of Maassen's operators does not contain the number operator. In fact, the number operator is given as $\Xi_{1,1}(\tau)$ and therefore the kernel distributions in the Fock expansion are

$$\kappa_{1,1} = \tau, \qquad \kappa_{l,m} = 0 \quad \text{unless} \quad l = m = 1.$$

Then, in view of Proposition 6.1.4 we come to

$$k(\{s\}, \{t\}) = \tau(s,t), \qquad k(\sigma, \omega) = 0 \qquad \text{unless} \quad |\sigma| = |\omega| = 1.$$

Therefore, in order to express the number operator in the form (6.1) the kernel *function* k should be a distribution.

To overcome the difficulty Meyer [2] generalized Maassen's idea and introduced an integral-sum kernel operator with a three argument kernel function:

$$\Xi f(\sigma) = \int_\Omega \sum_{\alpha_1 \cup \alpha_2 \cup \alpha_3 = \sigma} \rho(\alpha_1, \alpha_2, \omega) f(\omega \cup \alpha_2 \cup \alpha_3) \lambda(d\omega), \qquad \sigma \in \Omega. \tag{6.2}$$

Certainly, this is a generalization of Maassen's operators given as in (6.1). By a straightforward computation as in Proposition 6.1.4 we obtain

Proposition 6.1.5 *Let $\Xi = \sum_{l,m=0}^{\infty} \Xi_{l,m}(\kappa_{l,m})$ be the Fock expansion of (6.2). Then the kernel distributions are given as*

$$\kappa_{l,m}(s_1,\cdots,s_l,t_1,\cdots,t_m) = \frac{1}{l!m!}\sum_{j=0}^{l\wedge m} j!\binom{l}{j}\binom{m}{j}$$
$$\times\rho(\{s_{j+1},\cdots,s_l\},\{s_1,\cdots,s_j\},\{t_{j+1},\cdots,t_m\})\tau(s_1,t_1)\cdots\tau(s_j,t_j).$$

Therefore, the number operator is expressed in the form (6.2) with the kernel function defined as

$$\rho(\emptyset,\{s\},\emptyset) = 1, \qquad \rho(\alpha,\omega,\beta) = 0 \quad \text{otherwise.}$$

6.2 Reduction to finite degree of freedom

Although a simple trick it is noteworthy that T can be a discrete space or even a finite set under the standard construction developed in §3.1. If we take a finite set $T = \{1,2,\cdots,D\}$, the corresponding white noise calculus, which is justifiably called *white noise calculus with finite degree of freedom*, yields a finite dimensional calculus based on a particular Gelfand triple $\mathcal{D} \subset L^2(\mathbb{R}^D,dx) \subset \mathcal{D}^*$. On the other hand, the formal similarity between $(E) \subset (L^2) = L^2(E^*,\mu) \subset (E)^*$ and $\mathcal{S}(\mathbb{R}^D) \subset L^2(\mathbb{R}^D,dx) \subset \mathcal{S}'(\mathbb{R}^D)$ has often offered us good motivations to introduce new concepts into white noise calculus and to study various problems by analogy. Although these Gelfand triples are constructed by the same method, it is noted that $\mathcal{D} \neq \mathcal{S}(\mathbb{R})$. The purpose of this section is to characterize the space $\mathcal{D}$ and to illustrate how our general framework unifies concepts and results in finite dimensional and white noise calculi.

The present discussion would be known to some extent, though no written literature is found except the old work of Takenaka [1] who attempted to explain white noise calculus by observing its one-dimensional version. His discussion reduces to the case of $D = 1$ in our terminology.

From now on let $T = \{1,2,\cdots,D\}$ be a finite set with discrete topology and counting measure ν. Then $H = L^2(T,\nu;\mathbb{R}) \cong \mathbb{R}^D$ under the natural identification. The L^2-norm and the Euclidean norm coincide:

$$|\xi|^2 = \sum_{j=1}^{D} |\xi_j|^2, \qquad \xi = (\xi_1,\cdots,\xi_D) \in H. \tag{6.3}$$

In this context the operator A needed to construct Gaussian space is merely a symmetric matrix with eigenvalues $1 < \lambda_1 \le \cdots \le \lambda_D$. Then, all the norms $|\xi|_p = |A^p\xi|$, $p \in \mathbb{R}$, being equivalent, we use only the Euclidean norm (6.3). Moreover, the corresponding Gelfand triple becomes $E = H = E^* = \mathbb{R}^D$.

The verification of hypotheses (H1)–(H3) is very simple by observing that T is a discrete space with a counting measure ν. The evaluation map $\delta_j : \xi = (\xi_1,\cdots,\xi_D) \mapsto \xi_j \in \mathbb{R}$ is merely a coordinate projection. Hence $\delta_j \in E^* = (\mathbb{R}^D)^*$, and is identified with a coordinate vector in $E = \mathbb{R}^D$ as

$$\delta_j = (0,\cdots,0,\overset{j\text{-th}}{1},0,\cdots,0), \qquad j = 1,2,\cdots,D, \tag{6.4}$$

through the canonical bilinear form $\langle \cdot , \cdot \rangle$ on $(\mathbb{R}^D)^* \times \mathbb{R}^D$.

The Gaussian measure μ on $E^* = \mathbb{R}^D$ is nothing but the product of one dimensional standard Gaussian measures:

$$\mu(dx) = \left(\frac{1}{\sqrt{2\pi}}\right)^D e^{-|x|^2/2} dx,$$

where $dx = dx_1 \cdots dx_D$, $x = (x_1, \cdots, x_D) \in \mathbb{R}^D$. Then, by means of $\Gamma(A)$ we obtain the Gelfand triple of white noise functionals with finite degree of freedom:

$$(E) \subset (L^2) = L^2(\mathbb{R}^D, \mu; \mathbb{C}) \subset (E)^*.$$

By the continuous version theorem (Theorem 3.2.1) we may regard (E) as a space of continuous functions on $\mathbb{R}^D$. It is noted that every polynomial belongs to (E). In fact, it follows from Lemma 2.2.7 that

$$\left\langle :x^{\otimes n}:, \xi^{\otimes n} \right\rangle = \frac{|\xi|^n}{2^{n/2}} H_n\left(\frac{\langle x , \xi \rangle}{\sqrt{2}|\xi|}\right), \qquad \xi \in E, \quad \xi \neq 0,$$

where H_n is the Hermite polynomial of degree n. Putting $\xi = \delta_j$, we obtain

$$\left\langle :x^{\otimes n}:, \delta_j^{\otimes n} \right\rangle = \frac{1}{2^{n/2}} H_n\left(\frac{x_j}{\sqrt{2}}\right) = x_j^n + \cdots .$$

Hence (E) contains every polynomial in x_j and therefore in $x_1, \cdots, x_D$ since (E) is closed under pointwise multiplication.

As is easily verified, if $\phi \cdot e^{-\epsilon|x|^2} \in L^1(\mathbb{R}^D, dx)$ for any $\epsilon > 0$, the Fourier transform

$$(\phi \cdot e^{-\epsilon|x|^2})\widehat{\ }(\xi) = \left(\frac{1}{\sqrt{2\pi}}\right)^D \int_{\mathbb{R}^D} \phi(x) e^{-\epsilon|x|^2} e^{i\langle x , \xi \rangle} dx,$$

converges absolutely at any $\xi \in \mathbb{C}^D$ and becomes an entire holomorphic function on $\mathbb{C}^D$. Using this fact and the characterization theorem (Theorem 3.6.2 or Theorem 5.6.12), we may prove

Theorem 6.2.1 *A continuous function $\phi : \mathbb{R}^D \to \mathbb{C}$ belongs to (E) if and only if*
(i) $\phi \cdot e^{-\epsilon|x|^2} \in L^1(\mathbb{R}^D, dx)$ *for any* $\epsilon > 0$;
(ii) *for any $\epsilon > 0$ there exists $C \geq 0$ such that*

$$\left| e^{\langle \xi , \xi \rangle/2} (\phi \cdot e^{-|x|^2/2})\widehat{\ }(\xi) \right| \leq C\, e^{\epsilon|\xi|^2}, \qquad \xi \in \mathbb{C}^D.$$

Recall that (E) is obtained from $L^2(\mathbb{R}^D, \mu)$ where μ is the Gaussian measure. On the other hand, there is a natural unitary isomorphism from $L^2(\mathbb{R}^D, \mu)$ onto $L^2(\mathbb{R}^D, dx)$ given by

$$U\phi(x) = \left(\frac{1}{\sqrt{2\pi}}\right)^{D/2} e^{-|x|^2/4} \phi(x), \qquad \phi \in L^2(\mathbb{R}^D, \mu). \tag{6.5}$$

Let $\mathcal{D}$ denote the image of (E) under the unitary map U. Then, the Gelfand triple $(E) \subset L^2(\mathbb{R}^D, \mu) \subset (E)^*$ is translated into a new Gelfand triple

$$\mathcal{D} \subset L^2(\mathbb{R}^D, dx) \subset \mathcal{D}^*.$$

All the results obtained within the general framework of white noise calculus can be restated in terms of the above finite dimensional calculus. We shall discuss a few interesting cases.

We first mention a characterization of the space $\mathcal{D}$, which is an immediate consequence of Theorem 6.2.1.

Theorem 6.2.2 *A continuous function $\psi : \mathbb{R}^D \to \mathbb{C}$ belongs to $\mathcal{D}$ if and only if*
(i) $\psi \cdot e^{\left(\frac{1}{4}-\epsilon\right)|x|^2} \in L^1(\mathbb{R}^D, dx)$ *for any* $\epsilon > 0$;
(ii) *for any $\epsilon > 0$ there exists $C \geq 0$ such that*

$$\left| e^{\langle \xi, \xi \rangle /2} (\psi \cdot e^{-|x|^2/4})\hat{\ } (\xi) \right| \leq C\, e^{\epsilon |\xi|^2}, \qquad \xi \in \mathbb{C}^D.$$

Using the fact that $\mathcal{S}(\mathbb{R}^D)$ is invariant under the Fourier transform, one may prove the following assertion easily.

Lemma 6.2.3 *If $\phi \in (E)$, then $\phi \cdot e^{-\epsilon|x|^2} \in \mathcal{S}(\mathbb{R}^D)$ for any $\epsilon > 0$.*

As an immediate consequence, we see that $\mathcal{D} \subset \mathcal{S}(\mathbb{R}^D)$. In fact, the inclusion is proper. To see that it is sufficient to consider $\psi(x) = e^{-|x|^2/8}$ with Theorem 6.2.2.

In the theory of operators on white noise functionals a principal role is played by annihilation (Hida's differential) and creation operators. In the present context Hida's differential operator is defined by

$$\partial_j \phi(x) = \lim_{\theta \to 0} \frac{\phi(x + \theta \delta_j) - \phi(x)}{\theta}, \qquad \phi \in (E), \quad x \in \mathbb{R}^D.$$

In other words,

$$\partial_j = \frac{\partial}{\partial x_j}, \qquad j = 1, 2, \cdots, D.$$

The adjoint with respect to the Gaussian measure μ is the creation operator ∂_j^* by definition. Since δ_j is not a distribution but belongs to $E = \mathbb{R}^D$, the creation operator ∂_j^* belongs to $\mathcal{L}((E), (E))$ as well as ∂_j. The relation

$$\partial_j^* = x_j - \partial_j$$

follows from gerenal theory though a direct proof is also possible and easy using the fact that $\psi(x)\phi(x)e^{-|x|^2/2} \in \mathcal{S}(\mathbb{R}^D)$ for any $\phi, \psi \in (E)$. Moreover,

$$[\partial_j, \partial_k^*] = \delta_{jk}.$$

This is the well known canonical commutation relation (CCR) in case of D degree of freedom.

Using the unitary operator $U : L^2(\mathbb{R}^D, \mu) \to L^2(\mathbb{R}^D, dx)$ introduced in (6.5), we study a few interesting operators in $\mathcal{L}((E),(E)^*)$. Note that if $\Xi \in \mathcal{L}((E),(E)^*)$ then $U\Xi U^{-1} \in \mathcal{L}(\mathcal{D}, \mathcal{D}^*)$. A simple computation yields the following

$$U\partial_j U^{-1} = \frac{x_j}{2} + \frac{\partial}{\partial x_j}, \qquad U\partial_j^* U^{-1} = \frac{x_j}{2} - \frac{\partial}{\partial x_j}, \qquad Ux_jU^{-1} = x_j.$$

In particular,

$$P_j = \frac{1}{2i}(U\partial_j U^{-1} - U\partial_j^* U^{-1}) = \frac{1}{i}\frac{\partial}{\partial x_j}, \qquad Q_j = U\partial_j U^{-1} + U\partial_j^* U^{-1} = x_j$$

are the Schrödinger representation of CCR on $L^2(\mathbb{R}^D, dx)$ with common domain $\mathcal{D}$.

An integral kernel operator §4.3 is merely a finite linear combination of compositions of creation and annihilation operators with normal ordering:

$$\Xi_{l,m}(\kappa) = \sum \kappa(i_1, \cdots, i_l, j_1, \cdots, j_m)\partial_{i_1}^* \cdots \partial_{i_l}^* \partial_{j_1} \cdots \partial_{j_m}, \tag{6.6}$$

where $i_1, \cdots, i_l, j_1, \cdots, j_m$ run over $T = \{1, 2, \cdots, D\}$. This is a differential operator with polynomial coefficients. It follows from Proposition 5.2 that $U\Xi_{l,m}(\kappa)U^{-1}$ is again a differential operator with polynomial coefficients:

$$U\Xi_{l,m}(\kappa)U^{-1} = \sum_{|\alpha|,|\beta| \le l+m} C(\alpha, \beta)x^\alpha \left(\frac{\partial}{\partial x}\right)^\beta. \tag{6.7}$$

It is also expressed in terms of P_j and Q_j introduced in Proposition 5.2:

$$U\Xi_{l,m}(\kappa)U^{-1} = \sum_{|\alpha|,|\beta| \le l+m} C(\alpha, \beta)Q^\alpha P^\beta. \tag{6.8}$$

Applying the theory of Fock expansion (see §4.5), we see that every operator $\Xi \in \mathcal{L}(\mathcal{D}, \mathcal{D}^*)$ is expressed in an infinite linear combination of operators of the form (6.7) or equivalently (6.8). Namely,

$$U\Xi U^{-1} = \sum_{l,m=0}^{\infty} \sum_{|\alpha|,|\beta| \le l+m} C_{l,m}(\alpha, \beta)Q^\alpha P^\beta. \tag{6.9}$$

Formally we may rearrange the above infinite series according to the usual order of multi-index notation:

$$U\Xi U^{-1} = \sum_{\alpha,\beta} C(\alpha, \beta)Q^\alpha P^\beta,$$

though the meaning of the convergence becomes unclear. Incidentally we note that (6.9) leads us to a statement of "irreducibility" of the Schrödinger representation of CCR on $L^2(\mathbb{R}^D, dx)$, where the common domain of P_j and Q_j is taken to be $\mathcal{D}$.

The Gross Laplacian and the number operator are defined respectively by

$$\Delta_G = \sum_{j=1}^{D} \partial_j^2, \qquad N = \sum_{j=1}^{D} \partial_j^* \partial_j.$$

Since

$$x_j^2 = (\partial_j^* + \partial_j)^2 = \partial_j^{*2} + \partial_j^2 + \partial_j^*\partial_j + \partial_j\partial_j^* = \partial_j^{*2} + \partial_j^2 + 2\partial_j^*\partial_j + 1,$$

we have

$$\sum_{j=1}^{D}(x_j^2 - 1) = \Delta_G^* + \Delta_G + 2N.$$

The left hand side is "renormalized" Euclidean norm. The white noise analogue was discussed in §§5.3 and 5.5. By a straightforward computation we obtain

$$\begin{aligned} U\Delta_G U^{-1} &= \sum_{j=1}^{D}\left(\frac{\partial^2}{\partial x_j^2} + x_j\frac{\partial}{\partial x_j} + \frac{x_j^2}{4} + \frac{1}{2}\right), \\ U\Delta_G^* U^{-1} &= \sum_{j=1}^{D}\left(\frac{\partial^2}{\partial x_j^2} - x_j\frac{\partial}{\partial x_j} + \frac{x_j^2}{4} - \frac{1}{2}\right), \\ UNU^{-1} &= \sum_{j=1}^{D}\left(-\frac{\partial^2}{\partial x_j^2} + \frac{x_j^2}{4} - \frac{1}{2}\right). \end{aligned}$$

On the other hand, for the usual Laplacian $\Delta = \sum_{j=1}^{D}\frac{\partial^2}{\partial x_j^2}$ on $L^2(\mathbb{R}^D, dx)$ we have

$$U^{-1}\Delta U = \sum_{j=1}^{D}\left(\partial_j^2 - x_j\partial_j + \frac{x_j^2}{4} - \frac{1}{2}\right) = \sum_{j=1}^{D}\left(-\partial_j^*\partial_j + \frac{x_j^2}{4} - \frac{1}{2}\right).$$

This expression motivated Yamasaki [1] to introduce an infinite dimensional Laplacian (in our terminology $-N$) by omitting the divergent terms $x_j^2/4 - 1/2$.

In §5.4 we discussed an operator $x(s)\partial_t - x(t)\partial_s = \partial_s^*\partial_t - \partial_t^*\partial_s$ in connection with the infinite dimensional rotation group. The translation into our finite dimensional calculus is easy.

$$U(x_j\partial_k - x_k\partial_j)U^{-1} = x_j\frac{\partial}{\partial x_k} - x_k\frac{\partial}{\partial x_j},$$

which coincides with an infinitesimal generator of rotations on $\mathbb{R}^D$.

Finally we discuss Kuo's Fourier transform $\mathfrak{F}$ and Fourier-Mehler transforms $\mathfrak{F}_\theta$, $\theta \in \mathbb{R}$, which is a one-parameter group of transformations on $(E)^*$ involving Kuo's Fourier transform as $\mathfrak{F} = \mathfrak{F}_{-\pi/2}$, for details see §5.6. By a step-by-step computation one obtains an explicit form of $\mathfrak{F}_\theta f$ as follows: for $\theta \not\equiv 0 \pmod{\pi}$,

$$\mathfrak{F}_\theta f(x) = \left(-2\pi i e^{i\theta}\sin\theta\right)^{-D/2}\int_{\mathbb{R}^D} f(y)\exp\left(\frac{-i(|x|^2 + |y|^2)\cos\theta + 2i\langle x, y\rangle}{2\sin\theta}\right)dy,$$

and for $\theta \equiv 0 \pmod{\pi}$ we have

$$\mathfrak{F}_\theta f(x) = \begin{cases} f(x), & \theta \equiv 0 \pmod{2\pi}, \\ f(-x), & \theta \equiv \pi \pmod{2\pi}. \end{cases}$$

In particular,

$$\mathfrak{F}f(x) = \mathfrak{F}_{-\pi/2}f(x) = \left(\frac{1}{\sqrt{2\pi}}\right)^D\int_{\mathbb{R}^D} f(y)e^{-i\langle x, y\rangle}dy.$$

These operators are defined (in the sense that the integral is absolutely convergent) on $L^1(\mathbb{R}^D, dx)$. Moreover, by the above explicit expression we see that $\mathfrak{F}_\theta$ is nothing but the Fourier-Mehler transform on $\mathbb{R}^D$ discussed in Hida [2: Chap. 7] and Wiener [1]. However, in fact, this is Kuo's original idea of finding the white noise version of Fourier-Mehler transform. Incidentally we note that $\{\mathfrak{F}_\theta\}_{\theta\in\mathbb{R}}$ becomes a one-parameter group of automorphisms of $\mathcal{S}(\mathbb{R}^D)$.

We then easily see that

$$U\mathfrak{F}_\theta U^{-1} = e^{-|x|^2/4} \circ \mathfrak{F}_\theta \circ e^{|x|^2/4},$$

and in particular,

$$U\mathfrak{F}U^{-1} = e^{-|x|^2/4} \circ \mathfrak{F} \circ e^{|x|^2/4}.$$

It follows from the characterization of Kuo's Fourier transform (Theorem 5.7.4) that the operator $\tilde{\mathfrak{F}} = U\mathfrak{F}U^{-1}$ is charactreized by the following intertwining properties:

$$\tilde{\mathfrak{F}}\left(\frac{x_j}{2} + \frac{\partial}{\partial x_j}\right) = ix_j\tilde{\mathfrak{F}}, \qquad \tilde{\mathfrak{F}}x_j = i\left(\frac{x_j}{2} + \frac{\partial}{\partial x_j}\right)\tilde{\mathfrak{F}}.$$

6.3 Vector-valued white noise functionals

Our study of white noise functionals has been so far restricted to the case of scalar-valued functionals, however, the extension of the theory to vector-valued functionals is non-trivial and important from various aspects. Such generalization is indispensable to discuss, for example, quantum interacting systems such as "System + Reservoir" models (see Accardi-Lu [1] and references cited therein), infinite dimensional Dirac operators defined on Boson-Fermion Fock space toward supersymmetric quantum field theory (see Arai [1], Arai-Mitoma [1]), and so on.

In the recent work Obata [10] has been proposed a theory of vector-valued white noise functionals where the values lie in a standard CH-space. The choice of a standard CH-space is based on the following reasons. First of all, taking applications into account, we must not exclude Hilbert spaces. This means that our theory will cover the case of Hilbert space-valued distributions on Gaussian space. Second, a standard countably Hilbert space possesses nice properties from the viewpoint of topological vector spaces, in particular, the theory of topological tensor products can be applied effectively, see §1. Finally, notation and results established so far for scalar-valued functionals help the study of the vector-valued case very much. The present section is then devoted to a quick review of the results in Obata [10] with no proofs.

Let $\mathfrak{H}$ be another complex Hilbert space whose norm is denoted by $|\cdot|_0$ again. This is often called an *initial space*. Let B be a positive selfadjoint operator on $\mathfrak{H}$ with $\inf \mathrm{Spec}(B) > 0$ and let $\mathfrak{E}$ be the standard CH-space constructed from $(\mathfrak{H}, B)$. Then $(E) \otimes \mathfrak{E}$ is the space of *$\mathfrak{E}$-valued test white noise functionals* and its dual space $((E) \otimes \mathfrak{E})^* = (E)^* \otimes \mathfrak{E}^*$ consists of *$\mathfrak{E}^*$-valued generalized white noise functionals*. We must keep it in mind that (E) is a nuclear Fréchet space. Set

$$\sigma = (\inf \mathrm{Spec}(B))^{-1} = \|B^{-1}\|_{\mathrm{op}} > 0.$$

If we take the identity operator on $\mathfrak{H}$ for B, identifying $\mathfrak{H}^*$ with $\mathfrak{H}$, we obtain $\mathfrak{H}$-valued test and generalized white noise functionals.

By Proposition 1.3.8 the standard CH-space construcd from $((L^2)\otimes\mathfrak{H}, \Gamma(A)\otimes B)$ is isomorphic to $(E)\otimes\mathfrak{E}$. This fact enables us to employ basic notations used so far for scalar-valued functionals. The canonical bilinear form on $((E)\otimes\mathfrak{E})^*\times((E)\otimes\mathfrak{E})$ is denoted by $\langle\!\langle\cdot,\cdot\rangle\!\rangle$ again. Similarly, the norms of $(E)\otimes\mathfrak{E}$ are denoted again by $\|\cdot\|_p$, i.e.,

$$\|\phi\|_p = \|(\Gamma(A)\otimes B)^p\phi\|_0, \qquad \phi\in(E)\otimes\mathfrak{E}, \quad p\in\mathbf{R},$$

where $\|\cdot\|_0$ is the Hilbertian norm of $(L^2)\otimes\mathfrak{H}$, the space of $\mathfrak{H}$-valued L^2-functions on (E^*,μ). While, the canonical bilinear form $\langle\cdot,\cdot\rangle$ on $(E_{\mathbf{C}}^{\otimes n})^*\times E_{\mathbf{C}}^{\otimes n}$ is extended to a separately continuous bilinear map from $(E_{\mathbf{C}}^{\otimes n})^*\times(E_{\mathbf{C}}^{\otimes n}\otimes\mathfrak{E})$ into $\mathfrak{E}$ in an obvious way.

With these notation each $\phi\in(L^2)\otimes\mathfrak{H}$ is expressed explicitly. The vector-valued version of Wiener-Itô decomposition takes form as

$$(L^2)\otimes\mathfrak{H} = \sum_{n=0}^{\infty}\oplus(\mathcal{H}_n(\mathbf{C})\otimes\mathfrak{H}),$$

and each $\phi\in(L^2)\otimes\mathfrak{H}$ admits a Wiener-Itô expansion

$$\phi(x) = \sum_{n=0}^{\infty}\left\langle :x^{\otimes n}:, f_n\right\rangle, \qquad f_n\in H_{\mathbf{C}}^{\widehat{\otimes} n}\otimes\mathfrak{H}, \quad x\in E^*, \tag{6.10}$$

where $\langle :x^{\otimes n}:, f_n\rangle$ and the convergence of the series are understood in L^2-sense. Moreover, it holds that

$$\|\phi\|_p^2 = \sum_{n=0}^{\infty} n!\,|f_n|_p^2, \qquad p\in\mathbf{R},$$

where

$$|f_n|_p = |(A^{\otimes n}\otimes B)^p f_n|_0, \qquad f_n\in E_{\mathbf{C}}^{\otimes n}\otimes\mathfrak{E}.$$

Therefore, $\phi\in(L^2)\otimes\mathfrak{H}$ belongs to $\phi\in(E)\otimes\mathfrak{E}$ if and only if $f_n\in E_{\mathbf{C}}^{\widehat{\otimes} n}\otimes\mathfrak{E}$ for all $n\geq 0$ and $\sum_{n=0}^{\infty} n!\,|f_n|_p^2<\infty$ for all $p\geq 0$.

A description of $\mathfrak{E}^*$-valued generalized functionals follows immediately by duality. We put $(E_{\mathbf{C}}^{\otimes n}\otimes\mathfrak{E})^*_{\text{sym}} = (E_{\mathbf{C}}^{\otimes n})^*_{\text{sym}}\otimes\mathfrak{E}^*$.

Proposition 6.3.1 *For each $\Phi\in((E)\otimes\mathfrak{E})^*$ there exists a unique sequence $(F_n)_{n=0}^{\infty}$ with the property*

$$F_n\in(E_{\mathbf{C}}^{\otimes n}\otimes\mathfrak{E})^*_{\text{sym}} \quad \textit{and} \quad \sum_{n=0}^{\infty} n!\,|F_n|_{-p}^2<\infty \quad \textit{for some} \quad p\geq 0, \tag{6.11}$$

such that

$$\langle\!\langle\Phi,\phi\rangle\!\rangle = \sum_{n=0}^{\infty} n!\,\langle F_n, f_n\rangle, \tag{6.12}$$

for any $\phi\in(E)\otimes\mathfrak{E}$ given as in (6.10). Conversely, if a sequence $(F_n)_{n=0}^{\infty}$ satisfies (6.11), a linear functional Φ defined by (6.12) belongs to $((E)\otimes\mathfrak{E})^*$. In that case,

$$\|\Phi\|_{-p}^2 = \sum_{n=0}^{\infty} n!\,|F_n|_{-p}^2.$$

If $\Phi \in ((E)\otimes\mathfrak{E})^*$ and $(F_n)_{n=0}^{\infty}$ are related as in Proposition 6.3.1, we adopt a formal expression:

$$\Phi(x) = \sum_{n=0}^{\infty} \left\langle :x^{\otimes n}: , F_n \right\rangle ,$$

which is also called Wiener-Itô expansion. Thus the widely accepted notations for scalar-valued functionals are now available for vector-valued case as well.

By construction each $\phi \in (E) \otimes \mathfrak{E}$ is an $\mathfrak{E}$-valued function on E^* determined only up to μ-null functions. It is therefore important to establish a vector-valued version of Kubo-Yokoi's continuous version theorem (Theorem 3.2.1).

Theorem 6.3.2 *For each $\phi \in (E) \otimes \mathfrak{E}$ there exists a unique continuous function $\tilde{\phi} : E^* \to \mathfrak{E}$ such that $\phi(x) = \tilde{\phi}(x)$ for μ-a.e. $x \in E^*$. Moreover, $\tilde{\phi}(x)$ is given by an absolutely convergent series:*

$$\tilde{\phi}(x) \equiv \sum_{n=0}^{\infty} \left\langle :x^{\otimes n}: , f_n \right\rangle , \qquad x \in E^*,$$

where $f_n \in E_{\mathbb{C}}^{\widehat{\otimes} n} \otimes \mathfrak{E}$ are determined through the Wiener-Itô expansion of ϕ.

In other words, $(E) \otimes \mathfrak{E}$ satisfies hypothesis (H1) in a vector-valued sense. Accordingly, when $\phi \in (E) \otimes \mathfrak{E}$ is expressed as Wiener-Itô expansion (6.10), we understand that the infinite series converges absolutely at each $x \in E^*$. By the continuous version theorem the evaluation $\phi \mapsto \phi(x) \in \mathfrak{E}$, $x \in E^*$, is defined and becomes a linear map from $(E) \otimes \mathfrak{E}$ into $\mathfrak{E}$. Moreover, it is proved that for $p, q \geq 0$,

$$|\phi(x)|_p \leq \sigma^{q+1} \|\Gamma(A)^{-1}\|_{\mathrm{HS}} \, \|\delta_x\|_{-q} \, \|\phi\|_{p+q+1} , \qquad \phi \in (E) \otimes \mathfrak{E}, \quad x \in E^*.$$

For $\|\delta_x\|_{-q}$ see Theorem 3.2.12. Thus we have observed a property corresponding to (H2), i.e., the white noise delta function $\delta_x \in (E)^*$ is regarded as continuous linear operator in $\mathcal{L}((E) \otimes \mathfrak{E}, \mathfrak{E})$. Hence the convergence in $(E) \otimes \mathfrak{E}$ implies the pointwise convergence as $\mathfrak{E}$-valued functions on E^*. As for (H3), we see immediately that

$$x \mapsto \delta_x \in \mathcal{L}((E) \otimes \mathfrak{E}, \mathfrak{E}), \qquad x \in E^*,$$

is continuous when $\mathcal{L}((E) \otimes \mathfrak{E}, \mathfrak{E})$ is equipped with the weak operator topology. However, we do not know the continuity with respect to the bounded convergence topology of $\mathcal{L}((E) \otimes \mathfrak{E}, \mathfrak{E})$.

It is known that the exponential vectors $\{\phi_\xi \,;\, \xi \in E_{\mathbb{C}}\}$ span a dense subspace of (E). The same is true for $\{\phi_\xi \otimes u \,;\, \xi \in E_{\mathbb{C}},\, u \in \mathfrak{E}\} \subset (E) \otimes \mathfrak{E}$ and therefore, it is important to investigate the behavior of $\Phi \in ((E) \otimes \mathfrak{E})^*$ on such special vectors. The *S-transform* of $\Phi \in ((E) \otimes \mathfrak{E})^*$ is an $\mathfrak{E}^*$-valued function on $E_{\mathbb{C}}$ defined by

$$\langle S\Phi(\xi) , u \rangle = \langle\langle \Phi , \phi_\xi \otimes u \rangle\rangle , \qquad u \in \mathfrak{E}, \quad \xi \in E_{\mathbb{C}}.$$

For a function $F : E_{\mathbb{C}} \to \mathfrak{E}^*$ we consider

(F1) $z \mapsto \langle F(z\xi + \eta) , u \rangle$ is an entire holomorphic function on $\mathbb{C}$ for $\xi, \eta \in E_{\mathbb{C}}$ and $u \in \mathfrak{E}$;

(F2) there exist $C \geq 0$, $K \geq 0$ and $p \in \mathbb{R}$ such that

$$|\langle F(\xi), u\rangle| \leq C\,|u|_p \exp\left(K\,|\xi|_p^2\right), \qquad u \in \mathfrak{E}, \quad \xi \in E_{\mathbb{C}}.$$

(F2′) for $\epsilon > 0$ and $p \geq 0$ there exists $C \geq 0$ such that

$$|\langle F(\xi), u\rangle| \leq C\,|u|_{-p} \exp\left(\epsilon\,|\xi|_{-p}^2\right), \qquad u \in \mathfrak{E}, \quad \xi \in E_{\mathbb{C}}.$$

With these notation we have a characterization of S-transform.

Theorem 6.3.3 *Let $F : E_{\mathbb{C}} \to \mathfrak{E}^*$ be a function. Then there exists some $\Phi \in ((E) \otimes \mathfrak{E})^*$ such that $F = S\Phi$ if and only if F satisfies (F1) and (F2). Moreover, $\Phi = \phi \in ((E) \otimes \mathfrak{E})$ if and only if F satisfies (F1) and (F2′).*

We now introduce an integral kernel operator acting on $\mathfrak{E}$-valued white noise functionals. For a linear map $\kappa : E_{\mathbb{C}}^{\otimes(l+m)} \to \mathcal{L}(\mathfrak{E}, \mathfrak{E}^*)$ and $p, q, r, s \in \mathbb{R}$ we put

$$\|\kappa\|_{l,m;p,q;r,s} = \sup\left\{\sum_{\mathbf{i},\mathbf{j}} |\langle \kappa(e(\mathbf{i}) \otimes e(\mathbf{j}))u\,, v\rangle|^2\,|e(\mathbf{i})|_p^2\,|e(\mathbf{j})|_q^2\,;\ \begin{matrix} u, v \in \mathfrak{E} \\ |u|_{-s} \leq 1,\ |v|_{-r} \leq 1 \end{matrix}\right\}^{1/2},$$

and for brevity

$$\|\kappa\|_{l,m;p,q} = \|\kappa\|_{l,m;p,q;p,q}\,, \qquad \|\kappa\|_p = \|\kappa\|_{l,m;p,p;p,p}\,.$$

The next result is useful.

Lemma 6.3.4 *For a linear map $\kappa : E_{\mathbb{C}}^{\otimes(l+m)} \to \mathcal{L}(\mathfrak{E}, \mathfrak{E}^*)$ the following four conditions are equivalent:*
(i) $\kappa \in \mathcal{L}(E_{\mathbb{C}}^{\otimes(l+m)}, \mathcal{L}(\mathfrak{E}, \mathfrak{E}^*))$;
(ii) $\sup\left\{|\langle \kappa(\eta)u\,, v\rangle|\,;\ \begin{matrix} \eta \in E_{\mathbb{C}}^{\otimes(l+m)}, & |\eta|_p \leq 1 \\ u, v \in \mathfrak{E}, & |u|_p \leq 1, |v|_p \leq 1 \end{matrix}\right\} < \infty$ *for some $p \geq 0$;*
(iii) $\|\kappa\|_{-p} < \infty$ *for some $p \geq 0$;*
(iv) $\|\kappa\|_{l,m;p,q;r,s} < \infty$ *for some $p, q, r, s \in \mathbb{R}$.*

Each $\kappa \in \mathcal{L}(E_{\mathbb{C}}^{\otimes(l+m)}, \mathcal{L}(\mathfrak{E}, \mathfrak{E}^*))$ might be called an $\mathcal{L}(\mathfrak{E}, \mathfrak{E}^*)$-valued distribution on T^{l+m}. If $\mathcal{L}(\mathfrak{E}, \mathfrak{E}^*)$ is a Fréchet space, we have a canonical isomorphism

$$\mathcal{L}(E_{\mathbb{C}}^{\otimes(l+m)}, \mathcal{L}(\mathfrak{E}, \mathfrak{E}^*)) \cong (E_{\mathbb{C}}^{\otimes(l+m)})^* \otimes \mathcal{L}(\mathfrak{E}, \mathfrak{E}^*)$$

by the kernel theorem. However, it is noted that $\mathcal{L}(\mathfrak{E}, \mathfrak{E}^*)$ is Fréchet if and only if $\mathfrak{E}$ is a Hilbert space.

We need to generalize the contraction of tensor product introduced in §3.4. For $\kappa \in \mathcal{L}(E_{\mathbb{C}}^{\otimes(l+m)}, \mathcal{L}(\mathfrak{E}, \mathfrak{E}^*))$ and $f \in E_{\mathbb{C}}^{\otimes(m+n)} \otimes \mathfrak{E}$ there exists an element $\kappa \otimes_m f \in (E_{\mathbb{C}}^{\otimes(l+m)} \otimes \mathfrak{E})^*$ uniquely determined by

$$\langle \kappa \otimes_m (f_0 \otimes u)\,, g_0 \otimes v\rangle = \langle \kappa(g_0 \otimes_n f_0)u\,, v\rangle\,, \tag{6.13}$$

$$f_0 \in E_{\mathbf{C}}^{\otimes(m+n)}, \quad g_0 \in E_{\mathbf{C}}^{\otimes(l+n)}, \quad u, v \in \mathfrak{E}.$$

Then, $(\kappa, f) \mapsto \kappa \otimes_m f$ becomes a separately continuous bilinear map:

$$\otimes_m : \mathcal{L}(E_{\mathbf{C}}^{\otimes(l+m)}, \mathcal{L}(\mathfrak{E}, \mathfrak{E}^*)) \times (E_{\mathbf{C}}^{\otimes(m+n)} \otimes \mathfrak{E}) \to (E_{\mathbf{C}}^{\otimes(l+m)} \otimes \mathfrak{E})^*.$$

We have thus generalized usual contraction of tensor products, however, the above defined $\kappa \otimes_m f$ is slightly different from the original one. More precisely, in case of $\mathfrak{E} = \mathbf{C}$ the definition (6.13) gives

$$\kappa \otimes_m f = \sum_{\mathbf{i},\mathbf{k}} \left(\sum_{\mathbf{j}} \langle \kappa, e(\mathbf{i}) \otimes e(\mathbf{j}) \rangle \langle f, e(\mathbf{j}) \otimes e(\mathbf{k}) \rangle \right) e(\mathbf{i}) \otimes e(\mathbf{k}),$$

which should be carefully compared with the original definitions in §3.4. Nevertheless, in actual application to white noise calculus $f \in E_{\mathbf{C}}^{\otimes(l+m)} \otimes \mathfrak{E}$ appears only as a symmetric element, i.e., $f \in E_{\mathbf{C}}^{\widehat{\otimes}(l+m)} \otimes \mathfrak{E}$ and there is no difference between the two definitions.

As for norm estimate we record the following

Proposition 6.3.5 *Let $\kappa \in \mathcal{L}(E_{\mathbf{C}}^{\otimes(l+m)}, \mathcal{L}(\mathfrak{E}, \mathfrak{E}^*))$ and $f \in E_{\mathbf{C}}^{\otimes(m+n)} \otimes \mathfrak{E}$. Then, for any $p \in \mathbf{R}$ and $q \geq 0$ it holds that*

$$|\kappa \otimes_m f|_p \leq \delta^{l+m+2n} \sigma^2 \rho^{qn} \|\kappa\|_{l,m;p+1,-(p+q+1)} |f|_{p+q+2} \,. \tag{6.14}$$

With each $\kappa \in \mathcal{L}(E_{\mathbf{C}}^{\otimes(l+m)}, \mathcal{L}(\mathfrak{E}, \mathfrak{E}^*))$ we shall associate a continuous operator $\Xi_{l,m}(\kappa)$ from $(E) \otimes \mathfrak{E}$ into $((E) \otimes \mathfrak{E})^*$. Let $\phi \in (E) \otimes \mathfrak{E}$ be given with Wiener-Itô expansion:

$$\phi(x) = \sum_{n=0}^{\infty} \left\langle :x^{\otimes n}:, f_n \right\rangle, \qquad f_n \in E_{\mathbf{C}}^{\widehat{\otimes} n} \otimes \mathfrak{E}.$$

We then put

$$\Xi_{l,m}(\kappa)\phi(x) = \sum_{n=0}^{\infty} \frac{(n+m)!}{n!} \left\langle :x^{\otimes(l+n)}:, \kappa \otimes_m f_{n+m} \right\rangle. \tag{6.15}$$

By a direct estimate of the norm of the right hand side we may prove that $\Xi_{l,m}(\kappa) \in \mathcal{L}((E) \otimes \mathfrak{E}, ((E) \otimes \mathfrak{E})^*)$. In order to mention the estimate we need notation. We put

$$\Delta_q = \frac{\delta}{-e\rho^{q/2} \log(\delta^2 \rho^q)}, \qquad q > q_0 \equiv \inf\{q > 0;\, \delta^2 \rho^q \leq 1\}.$$

Remind that q_0 and the function Δ_q are determined only by the constant numbers ρ and δ.

Proposition 6.3.6 *Let $\kappa \in \mathcal{L}(E_{\mathbf{C}}^{\otimes(l+m)}, \mathcal{L}(\mathfrak{E}, \mathfrak{E}^*))$. Then for any $p \in \mathbf{R}$ and $q_0 < q_1 \leq q$ we have*

$$\|\Xi_{l,m}(\kappa)\phi\|_p \leq \rho^{-q/2} \delta^{-1} \sigma^2 \left(l^l m^m\right)^{1/2} \Delta_{q_1}^{(l+m)/2} \|\kappa\|_{l,m;p+1,-(p+q+1)} \|\phi\|_{p+q+2} \,.$$

Theorem 6.3.7 *The operator $\Xi_{l,m}(\kappa)$ belongs to $\mathcal{L}((E) \otimes \mathfrak{E}, ((E) \otimes \mathfrak{E})^*)$ for any $\kappa \in \mathcal{L}(E_{\mathbb{C}}^{\otimes(l+m)}, \mathcal{L}(\mathfrak{E}, \mathfrak{E}^*))$. In that case, for any $p > q_0/2$ and $\phi \in (E) \otimes \mathfrak{E}$ we have*

$$\|\Xi_{l,m}(\kappa)\phi\|_{-(p+1)} \le \rho^{-p}\delta^{-1}\sigma^2 \left(l^l m^m\right)^{1/2} \Delta_{2p}^{(l+m)/2} \|\kappa\|_{-p} \|\phi\|_{p+1}. \tag{6.16}$$

The above defined operator $\Xi_{l,m}(\kappa)$ is called an *integral kernel operator* with *kernel distribution* κ. As is easily seen, $\Xi = \Xi_{l,m}(\kappa)$ is uniquely determined by

$$\langle\langle \Xi(\phi \otimes u), \psi \otimes v \rangle\rangle = \langle \kappa(\eta_{\phi,\psi})u, v \rangle, \qquad \phi, \psi \in (E), \quad u, v \in \mathfrak{E}, \tag{6.17}$$

where

$$\eta_{\phi,\psi}(s_1, \cdots, s_l, t_1, \cdots, t_m) = \left\langle\left\langle \partial_{s_1}^* \cdots \partial_{s_l}^* \partial_{t_1} \cdots \partial_{t_m}\phi, \psi \right\rangle\right\rangle.$$

Recall that $\eta_{\phi,\psi} \in E_{\mathbb{C}}^{\otimes(l+m)}$, see §4.3. Hence, modelled after the case of scalar valued functionals, we adopt an intuitive expression

$$\Xi_{l,m}(\kappa) = \int_{T^{l+m}} \kappa(s_1, \cdots, s_l, t_1, \cdots t_m)\partial_{s_1}^* \cdots \partial_{s_l}^* \partial_{t_1} \cdots \partial_{t_m} ds_1 \cdots ds_l dt_1 \cdots dt_m,$$

on the understanding that ∂_s^* and ∂_t are respectively shortened notation for $(\partial_s \otimes I)^*$ and $\partial_t \otimes I$, I being the identity operator on $\mathfrak{E}$, and κ for $I \otimes \kappa$ with I being the identity operator on (E).

For the uniqueness of kernel distributions we need only to introduce the symmetrized distributions:

$$\mathcal{L}(E_{\mathbb{C}}^{\otimes(l+m)}, \mathcal{L}(\mathfrak{E}, \mathfrak{E}^*))_{\mathrm{sym}(l,m)} = \left\{\kappa \in \mathcal{L}(E_{\mathbb{C}}^{\otimes(l+m)}, \mathcal{L}(\mathfrak{E}, \mathfrak{E}^*))\,;\, s_{l,m}(\kappa) = \kappa\right\},$$

where

$$s_{l,m}(\kappa) = \frac{1}{l!m!} \sum_{\sigma \in \mathfrak{S}_l \times \mathfrak{S}_m} \kappa^\sigma, \qquad \kappa \in \mathcal{L}(E_{\mathbb{C}}^{\otimes(l+m)}, \mathcal{L}(\mathfrak{E}, \mathfrak{E}^*)).$$

The result is parallel to the case of scalar-valued functionals.

In order to characterize an integral kernel operator in $\mathcal{L}((E) \otimes \mathfrak{E}, (E) \otimes \mathfrak{E})$. We need notation.

Lemma 6.3.8 *For $\kappa \in \mathcal{L}(E_{\mathbb{C}}^{\otimes(l+m)}, \mathcal{L}(\mathfrak{E}, \mathfrak{E}^*))$ the following four conditions are equivalent:*

(i) *$f \mapsto \kappa \otimes_m f$ is a continuous linear map from $E_{\mathbb{C}}^{\otimes(m+n)} \otimes \mathfrak{E}$ into $E_{\mathbb{C}}^{\otimes(l+n)} \otimes \mathfrak{E}$ for all $n \ge 0$;*

(ii) *the bilinear map $(\xi, \eta) \mapsto \kappa(\xi \otimes \eta)$, $\xi \in E_{\mathbb{C}}^{\otimes l}$, $\eta \in E_{\mathbb{C}}^{\otimes m}$, is extended to a separately continuous bilinear map from $(E_{\mathbb{C}}^{\otimes l})^* \times (E_{\mathbb{C}}^{\otimes m})$ into $\mathcal{L}(\mathfrak{E}, \mathfrak{E})$;*

(iii) *for any $p \ge 0$ there exists $q \ge 0$ such that $|\kappa|_{l,m;p,-(p+q)} < \infty$;*

(iv) *for any $p \ge 0$ there exist $r, s \in \mathbb{R}$ such that $|\kappa|_{l,m;p,r;p,s} < \infty$.*

Let $\mathcal{B}_{\mathrm{sep}}((E_{\mathbb{C}}^{\otimes l})^*, E_{\mathbb{C}}^{\otimes m}; \mathcal{L}(\mathfrak{E}, \mathfrak{E}))$ denote the space of all separately continuous bilinear maps from $(E_{\mathbb{C}}^{\otimes l})^* \times (E_{\mathbb{C}}^{\otimes m})$ into $\mathcal{L}(\mathfrak{E}, \mathfrak{E})$. Each element in $\mathcal{B}_{\mathrm{sep}}((E_{\mathbb{C}}^{\otimes l})^*, E_{\mathbb{C}}^{\otimes m}; \mathcal{L}(\mathfrak{E}, \mathfrak{E}))$ is identified with $\kappa \in \mathcal{L}(E_{\mathbb{C}}^{\otimes(l+m)}, \mathcal{L}(\mathfrak{E}, \mathfrak{E}^*))$ satisfying one (therefore all) of the conditions in Lemma 6.3.8. In general, $\mathcal{B}_{\mathrm{sep}}((E_{\mathbb{C}}^{\otimes l})^*, E_{\mathbb{C}}^{\otimes m}; \mathcal{L}(\mathfrak{E}, \mathfrak{E}))$ does not coinside with $\mathcal{L}((E_{\mathbb{C}}^{\otimes l})^* \otimes E_{\mathbb{C}}^{\otimes m}, \mathcal{L}(\mathfrak{E}, \mathfrak{E}))$, see §1.3.

Theorem 6.3.9 *Let $\kappa \in \mathcal{L}(E_{\mathbb{C}}^{\otimes(l+m)}, \mathcal{L}(\mathfrak{E}, \mathfrak{E}^*))$. Then $\Xi_{l,m}(\kappa) \in \mathcal{L}((E) \otimes \mathfrak{E}, (E) \otimes \mathfrak{E})$ if and only if $\kappa \in \mathcal{B}_{\mathrm{sep}}((E_{\mathbb{C}}^{\otimes l})^*, E_{\mathbb{C}}^{\otimes m}; \mathcal{L}(\mathfrak{E}, \mathfrak{E}))$. In that case, for any $p \in \mathbb{R}$ and $q_0 < q_1 \le q$ we have*

$$\|\Xi_{l,m}(\kappa)\phi\|_p \le \rho^{-q/2}\delta^{-1}\sigma^2 \left(l^l m^m\right)^{1/2} \Delta_{q_1}^{(l+m)/2} \|\kappa\|_{l,m;p+1,-(p+q+1)} \|\phi\|_{p+q+2},$$

for any $\phi \in (E) \otimes \mathfrak{E}$.

The *symbol* of $\Xi \in \mathcal{L}((E)\otimes\mathfrak{E}, ((E)\otimes\mathfrak{E})^*)$ is an $\mathcal{L}(\mathfrak{E}, \mathfrak{E}^*)$-valued function on $E_{\mathbb{C}} \times E_{\mathbb{C}}$ defined by

$$\left\langle \widehat{\Xi}(\xi,\eta)u\,,v\right\rangle = \langle\!\langle \Xi(\phi_\xi \otimes u)\,, \phi_\eta \otimes v\rangle\!\rangle, \qquad \xi,\eta \in E_{\mathbb{C}}. \tag{6.18}$$

For example, for $\kappa \in \mathcal{L}(E_{\mathbb{C}}^{\otimes(l+m)}, \mathcal{L}(\mathfrak{E}, \mathfrak{E}^*))$, we have

$$\widehat{\Xi_{l,m}(\kappa)}(\xi,\eta) = e^{\langle \xi,\eta\rangle}\kappa(\eta^{\otimes l} \otimes \xi^{\otimes m}), \qquad \xi,\eta \in E_{\mathbb{C}}. \tag{6.19}$$

In particular, for $\xi, \eta \in E_{\mathbb{C}}$ we have

$$(\partial_t \otimes I)\widehat{\ }(\xi,\eta) = e^{\langle \xi,\eta\rangle}\xi(t)I, \quad (\partial_t^* \otimes I)\widehat{\ }(\xi,\eta) = e^{\langle \xi,\eta\rangle}\eta(t)I,$$

where I is the identity operator on $\mathfrak{E}$.

For a function $\Theta : E_{\mathbb{C}} \times E_{\mathbb{C}} \to \mathcal{L}(\mathfrak{E}, \mathfrak{E}^*)$ we consider the following properties:

(O1) for any $\xi, \xi_1, \eta, \eta_1 \in E_{\mathbb{C}}$ and $u, v \in \mathfrak{E}$ the function

$$(z,w) \mapsto \langle \Theta(z\xi + \xi_1, w\eta + \eta_1)u\,, v\rangle$$

is entire holomorphic on $\mathbb{C} \times \mathbb{C}$;

(O2) there exist constant numbers $C \ge 0$, $K \ge 0$ and $p \in \mathbb{R}$ such that

$$|\langle \Theta(\xi,\eta)u\,,v\rangle| \le C\,|u|_p\,|v|_p \exp K\left(|\xi|_p^2 + |\eta|_p^2\right), \qquad \xi,\eta \in E_{\mathbb{C}}, \quad u,v \in \mathfrak{E};$$

(O2′) for any $p \ge 0$ and $\epsilon > 0$ there exist $C \ge 0$ and $q \ge 0$ such that

$$|\langle \Theta(\xi,\eta)u\,,v\rangle| \le C\,|u|_{p+q}\,|v|_{-p} \exp \epsilon\left(|\xi|_{p+q}^2 + |\eta|_{-p}^2\right), \qquad \xi,\eta \in E_{\mathbb{C}}, \quad u,v \in \mathfrak{E}.$$

Note that if Θ satisfies (O2′), the values lie in $\mathcal{L}(\mathfrak{E}, \mathfrak{E})$.

It is straightforward to see that $\Theta = \widehat{\Xi}$ for $\Xi \in \mathcal{L}((E)\otimes\mathfrak{E}, ((E)\otimes\mathfrak{E})^*)$ satisfies (O1) and (O2). While, if $\Xi \in \mathcal{L}((E) \otimes \mathfrak{E}, (E) \otimes \mathfrak{E})$, then Θ satisfies (O1) and (O2′). More important is that the above listed properties reproduce an operator on vector-valued white noise functionals.

Theorem 6.3.10 *Let $\Theta : E_{\mathbb{C}} \times E_{\mathbb{C}} \to \mathcal{L}(\mathfrak{E}, \mathfrak{E}^*)$ be a function satisfying (O1) and (O2). Then, there exists a unique family of kernel distributions $(\kappa_{l,m})_{l,m=0}^\infty$, $\kappa_{l,m} \in \mathcal{L}(E_{\mathbb{C}}^{\otimes(l+m)}, \mathcal{L}(\mathfrak{E}, \mathfrak{E}^*))_{\mathrm{sym}(l,m)}$, such that*

$$\langle \Theta(\xi,\eta)u\,,v\rangle = \sum_{l,m=0}^\infty \langle\!\langle \Xi_{l,m}(\kappa_{l,m})(\phi_\xi \otimes u)\,, \phi_\eta \otimes v\rangle\!\rangle, \qquad \xi,\eta \in E_{\mathbb{C}}, \quad u,v \in \mathfrak{E}. \tag{6.20}$$

In that case, the series

$$\Xi\phi = \sum_{l,m=0}^{\infty} \Xi_{l,m}(\kappa_{l,m})\phi, \qquad \phi \in (E) \otimes \mathfrak{E}, \tag{6.21}$$

converges in $((E) \otimes \mathfrak{E})^*$. *Moreover,* $\Xi \in \mathcal{L}((E) \otimes \mathfrak{E}, ((E) \otimes \mathfrak{E})^*)$ *and* $\widehat{\Xi} = \Theta$. *If in addition* Θ *satisfies (O2'), each* $\kappa_{l,m}$ *belongs to* $\mathcal{B}_{\text{sep}}((E_{\mathbb{C}}^{\otimes l})^*, E_{\mathbb{C}}^{\otimes m}; \mathcal{L}(\mathfrak{E}, \mathfrak{E}))_{\text{sym}(l,m)}$ *for* $l, m = 0, 1, 2, \cdots$, *the series (6.21) converges in* $(E) \otimes \mathfrak{E}$ *and* $\Xi \in \mathcal{L}((E) \otimes \mathfrak{E}, (E) \otimes \mathfrak{E})$.

As immediate consequences we establish

Theorem 6.3.11 *Let* Θ *be an* $\mathcal{L}(\mathfrak{E}, \mathfrak{E}^*)$*-valued function on* $E_{\mathbb{C}} \times E_{\mathbb{C}}$. *Then, there exists* $\Xi \in \mathcal{L}((E) \otimes \mathfrak{E}, ((E) \otimes \mathfrak{E})^*)$ *with* $\Theta = \widehat{\Xi}$ *if and only if* Θ *satisfies (O1) and (O2). Moreover,* $\Xi \in \mathcal{L}((E) \otimes \mathfrak{E}, (E) \otimes \mathfrak{E})$ *if and only if* Θ *satisfies (O1) and (O2').*

Theorem 6.3.12 *For any* $\Xi \in \mathcal{L}((E) \otimes \mathfrak{E}, ((E) \otimes \mathfrak{E})^*)$ *there exists a unique family of kernel distributions* $\kappa_{l,m} \in \mathcal{L}((E_{\mathbb{C}}^{\otimes(l+m)})^*, \mathcal{L}(\mathfrak{E}, \mathfrak{E}^*))_{\text{sym}(l,m)}$ *such that (6.21) holds where the series converges in* $((E) \otimes \mathfrak{E})^*$. *If* $\Xi \in \mathcal{L}((E) \otimes \mathfrak{E}, (E) \otimes \mathfrak{E})$, *then every kernel distribution* $\kappa_{l,m}$ *belongs to* $\mathcal{B}_{\text{sep}}((E_{\mathbb{C}}^{\otimes l})^*, E_{\mathbb{C}}^{\otimes m}; \mathcal{L}(\mathfrak{E}, \mathfrak{E}))_{\text{sym}(l,m)}$ *and the right hand side of (6.21) converges in* $(E) \otimes \mathfrak{E}$.

The unique expression of $\Xi \in \mathcal{L}((E) \otimes \mathfrak{E}, ((E) \otimes \mathfrak{E})^*)$ given as in (6.21) is called the *Fock expansion* of Ξ. In that case we have

$$e^{-\langle \xi, \eta \rangle} \left\langle \widehat{\Xi}(\xi, \eta)u, v \right\rangle = \sum_{l,m=0}^{\infty} \left\langle \kappa_{l,m}(\eta^{\otimes l} \otimes \xi^{\otimes m})u, v \right\rangle, \qquad \xi, \eta \in E_{\mathbb{C}}, \quad u, v \in \mathfrak{E}.$$

Hence, in order to find kernel distributions $\kappa_{l,m}$ one need only to calculate the Taylor expansion of $e^{-\langle \xi, \eta \rangle} \left\langle \widehat{\Xi}(\xi, \eta)u, v \right\rangle$.

Note also that every bounded operator Ξ on $(L^2) \otimes \mathfrak{H}$ belongs to $\mathcal{L}((E) \otimes \mathfrak{E}, ((E) \otimes \mathfrak{E})^*)$ and therefore admits a Fock expansion. However, the convergence of the Fock expansion can not be discussed within the framework of Hilbert space in general.

Appendices

A Polarization formula

Let $\mathfrak{X}$ and $\mathfrak{Y}$ be vector spaces. Let F be a symmetric n-linear map from $\mathfrak{X} \times \cdots \times \mathfrak{X}$ (n-times) to $\mathfrak{Y}$ and put

$$A(\xi) = F(\underbrace{\xi, \cdots, \xi}_{n \text{ times}}), \qquad \xi \in \mathfrak{X}.$$

Then a simple calculation leads us to the polarization formula:

$$F(\xi_1, \cdots, \xi_n) = \frac{1}{2^n n!} \sum_{\epsilon} \epsilon_1 \cdots \epsilon_n A(\epsilon_1 \xi_1 + \cdots + \epsilon_n \xi_n), \tag{A.1}$$

where $\sum_{\epsilon}$ means the summation over $\epsilon_1 = \pm 1, \cdots, \epsilon_n = \pm 1$. As an immediate consequence, we obtain

$$\xi_1 \hat{\otimes} \cdots \hat{\otimes} \xi_n = \frac{1}{2^n n!} \sum_{\epsilon} \epsilon_1 \cdots \epsilon_n \left(\epsilon_1 \xi_1 + \cdots + \epsilon_n \xi_n\right)^{\otimes n}, \tag{A.2}$$

for any $\xi_1, \cdots, \xi_n \in \mathfrak{X}$.

Suppose that $\mathfrak{X}$ and $\mathfrak{Y}$ are equipped with seminorms $|\cdot|$. Let F and A be the same as above. Put

$$\begin{aligned} \|F\| &= \sup\left\{|F(\xi_1, \cdots, \xi_n)| \,;\, |\xi_1| \le 1, \cdots, |\xi_n| \le 1\right\}, \\ \|A\| &= \sup\{|A(\xi)| \,;\, |\xi| \le 1\}. \end{aligned}$$

Then

$$\|A\| \le \|F\| \le \frac{n^n}{n!} \|A\|. \tag{A.3}$$

Although the coefficient $n^n/n!$ is best possible,

$$\|F\| \le e^n \|A\| \tag{A.4}$$

is sometimes more useful.

B Hermite polynomials

The *Hermite polynomials* $H_n(x)$ are defined by the generating function:

$$\exp(2xt - t^2) = \sum_{n=0}^{\infty} \frac{t^n}{n!} H_n(x) \tag{B.1}$$

The explicit form is easily derived:

$$H_n(x) = n! \sum_{k=0}^{[n/2]} \frac{(-1)^k}{k!} \frac{(2x)^{n-2k}}{(n-2k)!} \tag{B.2}$$

$$x^n = \frac{n!}{2^n} \sum_{k=0}^{[n/2]} \frac{H_{n-2k}(x)}{k!(n-2k)!} \tag{B.3}$$

A recursion formula:

$$\begin{cases} H_0(x) &= 1 \\ H_1(x) &= 2x \\ H_n(x) &= 2xH_{n-1}(x) - 2(n-1)H_{n-2}(x), \qquad n \geq 2 \end{cases} \tag{B.4}$$

Formulae including derivatives:

$$H_n''(x) - 2xH_n'(x) + 2nH_n(x) = 0 \tag{B.5}$$

$$H_n'(x) = 2nH_{n-1}(x) \tag{B.6}$$

An expansion of binomial type: for $a, b \in \mathbb{C}$ with $a^2 + b^2 = 1$,

$$H_n(ax + by) = \sum_{k=0}^{n} \binom{n}{k} a^k b^{n-k} H_k(x) H_{n-k}(y) \tag{B.7}$$

The orthogonal relation:

$$\frac{1}{\sqrt{2\pi}} \int_{-\infty}^{+\infty} H_m\left(\frac{x}{\sqrt{2}}\right) H_n\left(\frac{x}{\sqrt{2}}\right) e^{-x^2/2} dx = 2^n n! \delta_{mn} \tag{B.8}$$

An integral of Fourier type:

$$\frac{1}{\sqrt{2\pi}} \int_{-\infty}^{+\infty} H_n\left(\frac{x}{\sqrt{2}}\right) e^{xy} e^{-x^2/2} dx = \left(\sqrt{2}\, y\right)^n e^{y^2/2}, \qquad y \in \mathbb{C} \tag{B.9}$$

C Norm estimates of contractions

Here is a list of norm estimates of contraction of tensor products. For definition and further details see §3.4.

A general inequality: for $p, q, r \in \mathbb{R}$,

$$|f \otimes^l g|_{m,n;p,q} \le |f|_{l,m;r,p}\, |g|_{l,n;-r,q}\,, \qquad f \in E_{\mathbb{C}}^{\otimes(l+m)}, \quad g \in E_{\mathbb{C}}^{\otimes(l+n)} \tag{C.1}$$

By specializing the parameters we obtain: for $p \ge 0$,

$$|f \otimes^l g|_p \le \rho^{2pl}\, |f|_p\, |g|_p\,, \qquad f \in E_{\mathbb{C}}^{\otimes(l+m)}, \quad g \in E_{\mathbb{C}}^{\otimes(l+n)} \tag{C.2}$$

$$|F \otimes^l g|_{-p} \le \rho^{2pn}\, |F|_{-p}\, |g|_p\,, \qquad F \in (E_{\mathbb{C}}^{\otimes(l+m)})^*, \quad g \in E_{\mathbb{C}}^{\otimes(l+n)} \tag{C.3}$$

For $p \in \mathbb{R}$ and $q \ge 0$,

$$|F \otimes^l g|_p \le \rho^{qn}\, |F|_{l,m;-(p+q),p}\, |g|_{p+q}\,, \qquad F \in (E_{\mathbb{C}}^{\otimes(l+m)})^*, \quad g \in E_{\mathbb{C}}^{\otimes(l+n)} \tag{C.4}$$

$$|F \otimes^l g|_p \le \rho^{qn}\, |F|_{-(p+q)}\, |g|_{p+q}\,, \qquad F \in (E_{\mathbb{C}}^{\otimes l})^*, \quad g \in E_{\mathbb{C}}^{\otimes(l+n)} \tag{C.5}$$

Symmetrized contraction: for $p \ge 0$,

$$|f \widehat{\otimes}_l g|_p \le \rho^{2pl}\, |f|_p\, |g|_p\,, \qquad f \in E_{\mathbb{C}}^{\widehat{\otimes}(l+m)}, \quad g \in E_{\mathbb{C}}^{\widehat{\otimes}(l+n)} \tag{C.6}$$

$$|F \widehat{\otimes}_l g|_{-p} \le \rho^{2pn}\, |F|_{-p}\, |g|_p\,, \qquad F \in (E_{\mathbb{C}}^{\otimes(l+m)})^*_{\mathrm{sym}}, \quad g \in E_{\mathbb{C}}^{\widehat{\otimes}(l+n)} \tag{C.7}$$

For $p \in \mathbb{R}$ and $q \ge 0$,

$$|F \widehat{\otimes}_l g|_p \le \rho^{qn}\, |F|_{-(p+q)}\, |g|_{p+q}\,, \qquad F \in (E_{\mathbb{C}}^{\otimes l})^*, \quad g \in E_{\mathbb{C}}^{\widehat{\otimes}(l+n)} \tag{C.8}$$

References

ACCARDI, L. AND LU, Y. G.

[1] *The low density limit in finite temperature case*, Nagoya Math. J. **126** (1992), 25–87.

ALBEVERIO, S., HIDA, T., POTTHOFF, J., RÖCKNER, M. AND STREIT, L.

[1] *Dirichlet forms in terms of white noise analysis I - Construction and QFT examples*, Rev. Math. Phys. **1** (1990), 291-312.

[2] *Dirichlet forms in terms of white noise analysis II - Closability and diffusion processes*, Rev. Math. Phys. **1** (1990), 313–323.

ALBEVERIO, S., HIDA, T., POTTHOFF, J. AND STREIT, L.

[1] *The vacuum of the Hoegh-Krohn model as a generalized white noise functional*, Phys. Lett. **B217** (1989), 511–514.

ARAI, A.

[1] *A general class of infinite dimensional Dirac operators and path integral representation of their index*, J. Funct. Anal. **105** (1992), 342–408.

ARAI, A. AND MITOMA, I.

[1] *De Rham-Hodge-Kodaira decomposition in ∞-dimensions*, Math. Ann. **291** (1991), 51–73.

[2] *Comparison and nuclearity of spaces of differential forms on topological vector spaces*, J. Funct. Anal. **111** (1993), 278–294.

BELAVKIN, V. P.

[1] *A quantum nonadapted Ito formula and stochastic analysis in Fock scale*, J. Funct. Anal. **102** (1991), 414–447.

BEREZANSKY, YU. M. AND KONDRAT'EV YU. G.

[1] "Spectral Methods in Infinite Dimensional Analysis," Naukova Dumka, Kiev, 1988 (Russian).

BEREZIN, F. A.

[1] "The Method of Second Quantization," Academic Press, 1966.

[2] *Some notes on representations of the commutation relations*, Russian Math. Surveys **24** (1969), 65–88.

[3] *Wick and anti-Wick operator symbols*, Math. USSR Sbornik **15** (1971), 577–606.

[4] *Quantization*, Math. USSR Izv. **8** (1974), 1109–1165.

BLÜMLINGER, M. AND OBATA, N.

[1] *Permutations preserving the Cesàro mean, densities of natural numbers and uniform distribution of sequences*, Ann. Inst. Fourier, Grenoble **41** (1991), 665–678.

BOGOLUBOV, N. N., LOGUNOV, A. A. AND TODOROV, I. T.
[1] "Introduction to Axiomatic Quantum Field Theory," Benjamin, Massachusetts, 1975.

BORCHERS, H. J.
[1] Algebraic aspects of Wightman field theory, in "Statistical Mechanics and Field Theory (Sen, R. N. and Weil, C. Eds.)," Halsted Press, New York 1973, pp. 31–79.
[2] Algebras of unbounded operators in quantum field theory, Physica **124A** (1984), 127–144.

COLOMBEAU, J. F.
[1] *Some aspects of infinite-dimensional holomorphy in mathematical physics*, in "Aspects of Mathematics and its Application (Barroso, J. A. Ed.)," North-Holland, 1986, pp.253–263.

DE FARIA, M. AND KUO, H.-H.
[1] *A delta white noise functional*, Acta Appl. Math. **17** (1989), 287–298.

DE FARIA, M., POTTHOFF, J. AND STREIT, L.
[1] *The Feynman integrand as a Hida distribution*, J. Math. Phys. **32** (1991), 2123-2127.

DIRAC, P. A. M.
[1] *Theory of the emission and absorption of radiation*, Proc. Royal Soc. London **A144** (1927), 243–262.

FOCK, V.
[1] *Konfigurationsraum und zweite Quantelung*, Z. Phys. **75** (1932), 622–647.

GELFAND, I. M. AND VILENKIN, N. YA
[1] "Generalized Functions, Vol.4," Academic Press, 1964.

GLIMM, J. AND JAFFE, A.
[1] "Quantum Physics," 2nd ed., Springer-Verlag, 1987.

GROSS, L.
[1] *Integration and non-linear transformations in Hilbert space*, Trans. Amer. Math. Soc. **94** (1960), 404–440.
[2] "Harmonic Analysis on Hilbert Space," Mem. Amer. Math. Soc. No. 46, Providence, 1963.
[3] *Abstract Wiener space*, in "Proc. Fifth Berkeley Symp. Math. Stat. Prob. Vol. II, Part 1 (Le Cam, L. M. and Neyman, J. Eds.)," Univ. of California Press, 1967, pp. 31–42.
[4] *Potential theory on Hilbert space*, J. Funct. Anal. **1** (1967), 123–181.

GROSSMANN, A.
[1] *Elementary properties of nested Hilbert spaces*, Commun. Math. Phys. **2** (1966), 1-30.
[2] *Homomorphisms and direct sums of nested Hilbert spaces*, Commun. Math. Phys. **4** (1967), 190–202.
[3] *Fields at a point*, Commun. Math. Phys. **4** (1967), 203–216.

GUICHARDET, A.
[1] "Symmetric Hilbert Spaces and Related Topics," Lect. Notes in Math. Vol. 261, Springer-Verlag, 1972.

HIDA, T.

[1] "Analysis of Brownian Functionals," Carleton Math. Lect. Notes Vol. 13, Carleton University, Ottawa, 1975.

[2] "Brownian Motion," Springer-Verlag, 1980.

[3] *Generalized Brownian functionals*, in "Theory and Application of Random Fields (Kallianpur, G. Ed.)," Lect. Notes in Control and Information Sciences Vol. 49, Springer-Verlag, 1983, pp. 89–95.

[4] *Brownian functionals and the rotation group*, in "Mathematics + Physics, Vol.1 (Streit, L. Ed.)," World Scientific, 1985, pp. 167–194.

[5] *A note on generalized Gaussian random fields*, J. Multivariate Anal. **27** (1988), 255–260.

[6] *Infinite-dimensional rotation group and unitary group*, in "Probability Measures on Groups IX (Heyer, H. Ed.)," Lect. Notes in Math. Vol. 1379, Springer-Verlag, 1989, pp. 125–134.

[7] *White noise and Gaussian random fields*, in "Probability Theory (Chen, L. Y. et al. Eds.)," Walter de Gruyter, Berlin/New York, 1992, pp. 83–90.

[8] *Stochastic variational calculus*, preprint, 1991.

HIDA, T. AND IKEDA, N.

[1] *Analysis on Hilbert space with reproducing kernel arising from multiple Wiener integral*, in "Proc. Fifth Berkeley Symp. Math. Stat. Prob. Vol. II, Part 1 (Le Cam, L. M. and Neyman, J. Eds.)," Univ. of California Press, 1967, pp. 117–143.

HIDA, T., KUBO, I., NOMOTO, H. AND YOSHIZAWA, H.

[1] *On projective invariance of Brownian motion*, Publ. RIMS, Kyoto Univ. Ser. A **4** (1969), 595–609.

HIDA, T., KUO, H.-H. AND OBATA, N.

[1] *Transformations for white noise functionals*, J. Funct. Anal. **111** (1993), 259–277.

HIDA, T., KUO, H.-H., POTTHOFF, J. AND STREIT, L.

[1] "White Noise: An Infinite Dimensional Calculus," Kluwer Academic, 1993.

HIDA, T., LEE, K.-S. AND LEE, S.-S.

[1] *Conformal invariance of white noise*, Nagoya Math. J. **98** (1985), 87–98.

HIDA, T., OBATA, N. AND SAITÔ, K.

[1] *Infinite dimensional rotations and Laplacians in terms of white noise calculus*, Nagoya Math. J. **128** (1992), 65–93.

HIDA, T. AND POTTHOFF, J.

[1] *White noise analysis – An overview*, in "White Noise Analysis (Hida, T. et al. Eds.)," World Scientific, 1990, pp. 140–165.

HIDA, T., POTTHOFF, J. AND STREIT, L.

[1] *Dirichlet forms and white noise analysis*, Commun. Math. Phys. **116** (1988), 235–245.

[2] *White noise analysis and applications*, in "Mathematics + Physics, Vol. 3 (Streit, L. Ed.)," World Scientific, 1988, pp. 143–178.

HIDA, T. AND SAITÔ, K.

[1] *White noise analysis and the Lévy Laplacian*, in "Stochastic Processes in Physics and Engineering (Albeverio, S. et al. Eds.)," D. Reidel Pub., Dordrecht, 1988, pp. 177–184.

HIDA, T. AND STREIT, L.
[1] *On quantum theory in terms of white noise*, Nagoya Math. J. **68** (1977), 21–34.
[2] *Generalized Brownian functionals and the Feynman integral*, Stoch. Proc. Appl. **16** (1983), 55–69.

HOLDEN, H., LINDSTRØM, T., ØKSENDAL, B., UBØE, J. AND ZHANG, T.-S.
[1] *Stochastic boundary value problems: a white noise functional approach*, Probab. Theory Relat. Fields **95** (1993), 391–419.

HUANG, Z.-Y.
[1] *Quantum white noises – White noise approach to quantum stochastic calculus*, Nagoya Math. J. **129** (1993), 23–42.

HUANG, Z.-Y. AND REN J.-G.
[1] *Analytic functionals and a new distribution theory over infinite dimensional spaces*, preprint, 1992.

HUDSON, R. L. AND PARTHASARATHY, K. R.
[1] *Quantum Itô's formula and stochastic evolution*, Commun. Math. Phys. **93** (1984), 301–323.

ITÔ, K.
[1] *Multiple Wiener integral*, J. Math. Soc. Japan **3** (1951), 157–169.
[2] "Foundations of Stochastic Differential Equations in Infinite Dimensional Spaces," CBMS-NSF Regional Conference Series in Applied Mathematics, SIAM, Philadelphia, 1984.

ITO, Y., KUBO, I. AND TAKENAKA, S.
[1] *Calculus on Gaussian white noise and Kuo's Fourier transformation*, in "White Noise Analysis – Mathematics and Applications (Hida, T., et al. Eds.)," World Scientific, 1990, pp. 180–207.

JORDAN, P. AND WIGNER, E. P.
[1] *Über das Paulische Äquivalenzverbot*, Z. Phys. **47** (1928), 631–658.

KHANDEKAR, D. C. AND STREIT, L.
[1] *Constructing the Feynman integrand*, Ann. Physik **1** (1992), 49–55.

KOBAYASHI, O., YOSHIOKA, A., MAEDA, Y. AND OMORI, H.
[1] *The theory of infinite-dimensional Lie groups and its applications*, Acta Appl. Math. **3** (1985), 71–106.

KONDRAT'EV, YU. G.
[1] *Nuclear spaces of entire functions in problems of infinite-dimensional analysis*, Soviet Math. Dokl. **22** (1980), 588–592.

KONDRAT'EV, YU. G. AND SAMOYLENKO, YU. S.
[1] *An integral representation for generalized positive definite kernels in infinitely many variables*, Soviet Math. Dokl. **17** (1976), 517–521.
[2] *The spaces of trial and generalized functions of infinite number of variables*, Rep. Math. Phys. **14** (1978), 325–350.

KONDRAT'EV, YU. G. AND STREIT, L.
[1] *A remark about a norm estimate for white noise distributions*, Uklainean Math. J. **44** (1992), 832–835.
[2] *Positive definite white noise functionals*, Madeira University preprint, 1991.

KÔNO, N.
[1] *Special functions connected with representations of the infinite dimensional motion group*, J. Math. Kyoto Univ. **6** (1966), 61–83.

KRÉE, M.
[1] *Propriété de trace en dimension infinie d'éspaces du type Sobolev*, CR Acad. Sc. Paris **279A** (1974), 157–160.

KRÉE, P.
[1] *Calcul symbolique et seconde quantification des fonctions sesquiholomorphes*, CR Acad. Sc. Paris **284A** (1977), 25–28.
[2] *Triplets nucléaires et théorème des noyaux en analyse en dimension infinie*, CR Acad. Sc. Paris **287A** (1978), 635–637.
[3] *La théorie des distributions en dimension quelconque et l'intégration stochastique*, in "Stochastic Analysis and Related Topics (Korezlioglu, H. and Ustunel, A. S. Eds.)," Lect. Notes in Math. Vol. 1316, Springer-Verlag, 1988, pp. 170–233.

KRÉE, P. AND RĄCZKA, R.
[1] *Kernels and symbols of operators in quantum field theory*, Ann. Inst. Henri Poincaré Sect. A **28** (1978), 41–73.

KRISTENSEN, P., MEJLBO, L. AND THUE POULSEN, E.
[1] *Tempered distributions in infinitely many dimensions*, Commun. Math. Phys. **1** (1965), 175–214.

KUBO, I.
[1] *Ito formula for generalized Brownian functionals*, in "Theory and Application of Random Fields (Kallianpur, G. Ed.)," Lect. Notes in Control and Information Sciences Vol. 49, Springer-Verlag, 1983, pp. 156–166.

KUBO, I. AND KUO, H.-H.
[1] *Fourier transform and cylindrical Hida distributions*, preprint, 1991.
[2] *Finite dimensional Hida distributions*, preprint, 1992.

KUBO, I. AND TAKENAKA, S.
[1] *Calculus on Gaussian white noise I*, Proc. Japan Acad. **56A** (1980), 376–380.
[2] *Calculus on Gaussian white noise II*, Proc. Japan Acad. **56A** (1980), 411–416.
[3] *Calculus on Gaussian white noise III*, Proc. Japan Acad. **57A** (1981), 433–437.
[4] *Calculus on Gaussian white noise IV*, Proc. Japan Acad. **58A** (1982), 186–189.

KUBO, I. AND YOKOI, Y.
[1] *A remark on the space of testing random variables in the white noise calculus*, Nagoya Math. J. **115** (1989), 139–149.

KUO, H.-H.
[1] "Gaussian Measures in Banach Spaces," Lect. Notes in Math. Vol. 463, Springer-Verlag, 1975.
[2] *On Fourier transform of generalized Brownian functionals*, J. Multivariate Anal. **12** (1982), 415–431.
[3] *Brownian functionals and applications*, Acta Appl. Math. **1** (1983), 175–188.
[4] *Fourier-Mehler transforms of generalized Brownian functionals*, Proc. Japan Acad. **59A** (1983), 312–314.

[5] *Donsker's delta function as a generalized Brownian functional and its application*, in "Theory and Application of Random Fields (Kallianpur, G. Ed.)," Lect. Notes in Control and Information Sciences Vol. 49, Springer-Verlag, 1983, pp. 167–178.
[6] *On Laplacian operators of generalized Brownian functionals*, in "Stochastic Processes and Their Applications (Itô, K. and Hida, T. Eds.)," Lect. Notes in Math. Vol. 1203, Springer-Verlag, 1986, pp. 119–128.
[7] *White noise calculus*, in "Algebra, Analysis and Geometry (Kang, M.-C. and Lih, K.-W. Eds.)," World Scientific, 1989, pp. 77–119.
[8] *The Fourier transform in white noise calculus*, J. Multivariate Anal. **31** (1989), 311–327.
[9] *Lectures on white noise analysis*, Soochow J. Math. **18** (1992), 229–300.
[10] *Fourier-Mehler transforms in white noise analysis*, in "Gaussian Random Fields (Itô, K. and Hida, T. Eds.)," World Scientific, 1991, pp. 257–271.
[11] *Convolution and Fourier transform of Hida distributions*, preprint, 1991.

KUO, H.-H., OBATA, N. AND SAITÔ, K.
[1] *Lévy Laplacian of generalized functions on a nuclear space*, J. Funct. Anal. **94** (1990), 74–92.

KUO, H.-H., POTTHOFF, J. AND STREIT, L.
[1] *A characterization of white noise test functionals*, Nagoya Math. J. **121** (1991), 185–194.

KUO, H.-H. AND RUSSEK, A.
[1] *White noise approach to stochastic integration*, J. Multivariate Anal. **24** (1988), 218–236.

LEE, K.-S.
[1] *White noise approach to Gaussian random fields*, Nagoya Math. J. **119** (1990), 93–106.

LEE, Y.-J.
[1] *On the convergence of Wiener-Itô decomposition*, Bull. Inst. Math. Acad. Sinica **17** (1989), 305–312.
[2] *Generalized functions on infinite dimensional spaces and its application to white noise calculus*, J. Funct. Anal. **82** (1989), 429–464.
[3] *Calculus of generalized white noise functionals – An abstract Wiener space approach*, in "Proc. Preseminar for International Conference on Gaussian Random Fields (Hida, T. and Saitô, K. Eds.)," Nagoya, 1991, pp. 66–125.
[4] *A characterization of generalized functions on infinite dimensional spaces and Bargman-Segal analytic functions*, in "Gaussian Random Fields (Itô, K. and Hida, T. Eds.)," World Scientific, 1991, pp. 272–284.
[5] *Analytic version of test functionals, Fourier transform and a characterization of measures in white noise calculus*, J. Funct. Anal. **100** (1991), 359–380.

LÉVY, P.
[1] "Problèmes Concrets d'Analyse Fonctionnelle," Gauthier-Villars, Paris, 1951.

LINDSAY, J. M.
[1] *Gaussian hypercontractivity revisited*, J. Funct. Anal. **92** (1990), 313–324.
[2] *On set convolutions and integral-sum kernel operators*, in "Probability Theory and Mathematical Statistics, Vol. 2 (Grigelionis, B. et al. Eds.)," Mokslas, Vilnius, 1990, pp. 105–123.

LINDSAY, J. M. AND MAASSEN, H.

[1] *An integral kernel approach to noise*, in "Quantum Probability and Applications III (Accardi, L. and von Waldenfels, W. Eds.)," Lect. Notes in Math. Vol. 1303, Springer-Verlag, 1988, pp. 192–208.

LIU, K. AND YAN, J. A.

[1] *Euler operator and homogeneous Hida distributions*, Acta Math. Sinica, to appear.

MAASSEN, H.

[1] *Quantum Markov processes on Fock space described by integral kernels*, in "Quantum Probability and Applications II (Accardi, L. and von Waldenfels, W. Eds.)," Lect. Notes in Math. Vol. 1136, Springer-Verlag, 1985, pp. 361–374.

MATSUSHIMA, H., OKAMOTO, K. AND SAKURAI, T.

[1] *On a certain class of irreducible unitary representations of the infinite dimensional rotation group I*, Hiroshima Math. J. **11** (1981), 181–193.

MEYER, P.-A.

[1] *Distributions, noyaux, symboles d'après Krée*, in "Séminaire de Probabilités XXII (Azéma, J. et al. Eds.)," Lect. Notes in Math. Vol. 1321, Springer-Verlag, 1988, pp. 467–476.

[2] "Quantum Probability for Probabilists," Lect. Notes in Math. Vol. 1538, Springer-Verlag, 1993.

OBATA, N.

[1] *A note on certain permutation groups in the infinite dimensional rotation group*, Nagoya Math. J. **109** (1988), 91–107.

[2] *Analysis of the Lévy Laplacian*, Soochow J. Math. **14** (1988), 105–109.

[3] *Density of natural numbers and the Lévy group*, J. Number Theory **30** (1988), 288–297.

[4] *The Lévy Laplacian and the mean value theorem*, in "Probability Measures on Groups IX (Heyer, H. Ed.)," Lect. Notes in Math. Vol. 1379, Springer-Verlag, 1989, pp. 242–253.

[5] *A characterization of the Lévy Laplacian in terms of infinite dimensional rotation groups*, Nagoya Math. J. **118** (1990), 111–132.

[6] *Rotation-invariant operators on white noise functionals*, Math. Z. **210** (1992), 69–89.

[7] *Fock expansion of operators on white noise functionals*, in "Proc. 3rd International Conference on Stochastic Processes, Physics and Geometry (Albeverio, S. et al. Eds.)," to appear.

[8] *An analytic characterization of symbols of operators on white noise functionals*, J. Math. Soc. Japan **45** (1993), 421–445.

[9] *White noise delta functions and continuous version theorem*, Nagoya Math. J. **129** (1993), 1–22.

[10] *Operator calculus on vector-valued white noise functionals*, J. Funct. Anal. **119** (1994), to appear.

[11] *Derivations on white noise functionals*, preprint, 1993.

OKAMOTO, K. AND SAKURAI, T.

[1] *On a certain class of irreducible unitary representations of the infinite dimensional rotation group II*, Hiroshima Math. J. **12** (1982), 385–397.

[2] *An analogue of Peter-Weyl theorem for the infinite dimensional unitary group*, Hiroshima Math. J. **12** (1982), 529–541.

OMORI, H.

[1] "Infinite Dimensional Lie Transformation Groups," Lect. Notes in Math. Vol. 427, Springer-Verlag, 1974.

ORIHARA, A.

[1] *Hermite polynomials and infinite dimensional motion group*, J. Math. Kyoto Univ. **6** (1966), 1–12.

OUERDIANE, H.

[1] *Application des methodes d'holomorphie et de distributions en dimension quelconque a l'anlyse sur les espaces gaussiens*, BiBoS (Universität Bielefeld) preprint **491**, 1991.

PARTHASARATHY, K. R.

[1] "An Introduction to Quantum Stochastic Calculus," Birkhäuser, 1992.

POLISHCHUK, E. M.

[1] "Continual Means and Boundary Value Problems in Function Spaces," Birkhäuser, Basel/Boston/Berlin, 1988.

POTTHOFF, J.

[1] *On positive generalized functionals*, J. Funct. Anal. **74** (1987), 81–95.

POTTHOFF, J. AND STREIT, L.

[1] *A characterization of Hida distributions*, J. Funct. Anal. **101** (1991), 212–229.

[2] *Invariant states on random and quantum fields: ϕ-bounds and white noise analysis*, J. Funct. Anal. **111** (1993), 295–311.

POTTHOFF, J. AND YAN J.-A.

[1] *Some results about test and generalized functionals of white noise*, in "Probability Theory (Chen, L.Y. et al. Eds.)," Walter de Gruyter, Berlin/New York, 1992, pp. 121–145.

RAZAFIMANANTENA, E. A.

[1] *A general class of Dirichlet forms in terms of white noise analysis: Markovian property and construction of the diffusion process*, Soochow J. Math. **19** (1993), 173–197.

REED, M. AND SIMON, B.

[1] "Method of Modern Mathematical Physics, Vol. 1: Functional Analysis," Academic Press, 1980.

SAITÔ, K.

[1] *Itô's formula and Lévy's Laplacian*, Nagoya Math. J. **108** (1987), 67–76.

[2] *Itô's formula and Lévy's Laplacian, II*, Nagoya Math. J. **123** (1991), 153–169.

SCHAEFER, H. H.

[1] "Topological Vector Spaces," 4th corrected printing, Springer-Verlag, 1980.

SEGAL, I. E.

[1] *Tensor algebras over Hilbert spaces I*, Trans. Amer. Math. Soc. **81** (1956), 106–134.

[2] *Tensor algebras over Hilbert spaces II*, Ann. of Math. **63** (1956), 160–175.

[3] *Distributions in Hilbert space and canonical systems of operators*, Trans. Amer. Math. Soc. **88** (1958), 12–41.

SIMON, B.

[1] "The $P(\phi)_2$ Euclidean (Quantum) Field Theory," Princeton Univ. Press, 1974.

STRĂTILĂ, S. AND ZSIDÓ, L.

[1] "Lectures on Von Neumann Algebras," Abacus Press, 1979.

STREIT, L.

[1] *White noise analysis – Theory and applications*, in "Quantum Probability and Related Topics, Vol. 7, (Accardi, L. Manag. Ed.)," World Scientific, 1992, pp. 337–347.

[2] *The characterization theorem for Hida distributions. Generalizations and applications*, Madeira University preprint, 1991.

STREIT, L. AND WESTERKAMP, W.

[1] *A generalization of the characterization theorem for generalized functionals of white noise*, BiBoS (Universität Bielefeld) preprint **480**, 1991.

TAKENAKA, S.

[1] *Invitation to white noise calculus*, in "Theory and Application of Random Fields (Kallianpur, G. Ed.)," Lect. Notes in Control and Information Sciences Vol. 49, Springer-Verlag, 1983, pp. 249–257.

TREVES, F.

[1] " Topological Vector Spaces, Distributions and Kernels," Academic Press, 1967.

UMEMURA, Y. AND KÔNO, N.

[1] *Infinite dimensional Laplacian and spherical harmonics*, Publ. RIMS, Kyoto Univ. **1** (1966), 163–186.

WICK, G. C.

[1] *The evaluation of the collision matrix*, Phys. Rev. **80** (1950), 268–272.

WIENER, N.

[1] *Hermitian polynomials and Fourier analysis*, J. Math. Phys. MIT **8** (1928/29), 70–73.

[2] *Generalized harmonic analysis*, Acta Math. **55** (1930), 117–258.

YAMASAKI (UMEMURA), Y.

[1] *On the infinite dimensional Laplacian operator*, J. Math. Kyoto Univ. **4** (1965), 477–492.

[2] "Measures on Infinite Dimensional Spaces," World Scientific, 1985.

YAN, J.-A.

[1] *Products and transforms of white noise functionals*, preprint, 1990.

[2] *Notes on the Wiener semigroup and renormalization*, in "Séminaire de Probabilités XXV (Azéma, J. et al. Eds.)," Lect. Notes in Math. Vol. 1485, Springer-Verlag, 1991, pp. 79–94.

[3] *An elementary proof of a theorem of Lee*, preprint, 1990.

[4] *Some recent developments in white noise analysis*, preprint, 1991.

[5] *A note on the Lévy Laplacian*, preprint, 1991.

[6] *Inequalities for products of white noise functionals*, preprint, 1991.

YOKOI, Y.

[1] *Positive generalized white noise functionals*, Hiroshima Math. J. **20** (1990), 137–157.

[2] *Positive generalized Brownian functionals*, in "White Noise Analysis (Hida, T. et al. Eds.)," World Scientific, 1990, pp. 407–422.

YOSHIZAWA, H.

[1] *Rotation group of Hilbert space and its application to Brownian motion*, in "Proc. Internat. Conf. on Functional Analysis and Related Topics," Univ. of Tokyo Press, 1970, pp. 414–423.

[2] *Infinite-dimensional rotation group and Brownian motion*, in "Gaussian Random Fields (Itô, K. and Hida, T. Eds.)," World Scientific, 1991, pp. 406–419.

ZHANG, T.-S.

[1] *Characterization of white noise test function space* (S), Oslo University preprint, 1991.

Index

Printing: Weihert-Druck GmbH, Darmstadt
Binding: Theo Gansert Buchbinderei GmbH, Weinheim

Vol. 1482: J. Chabrowski, The Dirichlet Problem with L^2-Boundary Data for Elliptic Linear Equations. VI, 173 pages. 1991.

Vol. 1483: E. Reithmeier, Periodic Solutions of Nonlinear Dynamical Systems. VI, 171 pages. 1991.

Vol. 1484: H. Delfs, Homology of Locally Semialgebraic Spaces. IX, 136 pages. 1991.

Vol. 1485: J. Azéma, P. A. Meyer, M. Yor (Eds.), Séminaire de Probabilités XXV. VIII, 440 pages. 1991.

Vol. 1486: L. Arnold, H. Crauel, J.-P. Eckmann (Eds.), Lyapunov Exponents. Proceedings, 1990. VIII, 365 pages. 1991.

Vol. 1487: E. Freitag, Singular Modular Forms and Theta Relations. VI, 172 pages. 1991.

Vol. 1488: A. Carboni, M. C. Pedicchio, G. Rosolini (Eds.), Category Theory. Proceedings, 1990. VII, 494 pages, 1991.

Vol. 1489: A. Mielke, Hamiltonian and Lagrangian Flows on Center Manifolds. X, 140 pages. 1991.

Vol. 1490: K. Metsch, Linear Spaces with Few Lines. XIII, 196 pages. 1991.

Vol. 1491: E. Lluis-Puebla, J.-L. Loday, H. Gillet, C. Soulé, V. Snaith, Higher Algebraic K-Theory: an overview. IX, 164 pages. 1992.

Vol. 1492: K. R. Wicks, Fractals and Hyperspaces. VIII, 168 pages. 1991.

Vol. 1493: E. Benoît (Ed.), Dynamic Bifurcations. Proceedings, Luminy 1990. VII, 219 pages. 1991.

Vol. 1494: M.-T. Cheng, X.-W. Zhou, D.-G. Deng (Eds.), Harmonic Analysis. Proceedings, 1988. IX, 226 pages. 1991.

Vol. 1495: J. M. Bony, G. Grubb, L. Hörmander, H. Komatsu, J. Sjöstrand, Microlocal Analysis and Applications. Montecatini Terme, 1989. Editors: L. Cattabriga, L. Rodino. VII, 349 pages. 1991.

Vol. 1496: C. Foias, B. Francis, J. W. Helton, H. Kwakernaak, J. B. Pearson, H_∞-Control Theory. Como, 1990. Editors: E. Mosca, L. Pandolfi. VII, 336 pages. 1991.

Vol. 1497: G. T. Herman, A. K. Louis, F. Natterer (Eds.), Mathematical Methods in Tomography. Proceedings 1990. X, 268 pages. 1991.

Vol. 1498: R. Lang, Spectral Theory of Random Schrödinger Operators. X, 125 pages. 1991.

Vol. 1499: K. Taira, Boundary Value Problems and Markov Processes. IX, 132 pages. 1991.

Vol. 1500: J.-P. Serre, Lie Algebras and Lie Groups. VII, 168 pages. 1992.

Vol. 1501: A. De Masi, E. Presutti, Mathematical Methods for Hydrodynamic Limits. IX, 196 pages. 1991.

Vol. 1502: C. Simpson, Asymptotic Behavior of Monodromy. V, 139 pages. 1991.

Vol. 1503: S. Shokranian, The Selberg-Arthur Trace Formula (Lectures by J. Arthur). VII, 97 pages. 1991.

Vol. 1504: J. Cheeger, M. Gromov, C. Okonek, P. Pansu, Geometric Topology: Recent Developments. Editors: P. de Bartolomeis, F. Tricerri. VII, 197 pages. 1991.

Vol. 1505: K. Kajitani, T. Nishitani, The Hyperbolic Cauchy Problem. VII, 168 pages. 1991.

Vol. 1506: A. Buium, Differential Algebraic Groups of Finite Dimension. XV, 145 pages. 1992.

Vol. 1507: K. Hulek, T. Peternell, M. Schneider, F.-O. Schreyer (Eds.), Complex Algebraic Varieties. Proceedings, 1990. VII, 179 pages. 1992.

Vol. 1508: M. Vuorinen (Ed.), Quasiconformal Space Mappings. A Collection of Surveys 1960-1990. IX, 148 pages. 1992.

Vol. 1509: J. Aguadé, M. Castellet, F. R. Cohen (Eds.), Algebraic Topology - Homotopy and Group Cohomology. Proceedings, 1990. X, 330 pages. 1992.

Vol. 1510: P. P. Kulish (Ed.), Quantum Groups. Proceedings, 1990. XII, 398 pages. 1992.

Vol. 1511: B. S. Yadav, D. Singh (Eds.), Functional Analysis and Operator Theory. Proceedings, 1990. VIII, 223 pages. 1992.

Vol. 1512: L. M. Adleman, M.-D. A. Huang, Primality Testing and Abelian Varieties Over Finite Fields. VII, 142 pages. 1992.

Vol. 1513: L. S. Block, W. A. Coppel, Dynamics in One Dimension. VIII, 249 pages. 1992.

Vol. 1514: U. Krengel, K. Richter, V. Warstat (Eds.), Ergodic Theory and Related Topics III, Proceedings, 1990. VIII, 236 pages. 1992.

Vol. 1515: E. Ballico, F. Catanese, C. Ciliberto (Eds.), Classification of Irregular Varieties. Proceedings, 1990. VII, 149 pages. 1992.

Vol. 1516: R. A. Lorentz, Multivariate Birkhoff Interpolation. IX, 192 pages. 1992.

Vol. 1517: K. Keimel, W. Roth, Ordered Cones and Approximation. VI, 134 pages. 1992.

Vol. 1518: H. Stichtenoth, M. A. Tsfasman (Eds.), Coding Theory and Algebraic Geometry. Proceedings, 1991. VIII, 223 pages. 1992.

Vol. 1519: M. W. Short, The Primitive Soluble Permutation Groups of Degree less than 256. IX, 145 pages. 1992.

Vol. 1520: Yu. G. Borisovich, Yu. E. Gliklikh (Eds.), Global Analysis – Studies and Applications V. VII, 284 pages. 1992.

Vol. 1521: S. Busenberg, B. Forte, H. K. Kuiken, Mathematical Modelling of Industrial Process. Bari, 1990. Editors: V. Capasso, A. Fasano. VII, 162 pages. 1992.

Vol. 1522: J.-M. Delort, F. B. I. Transformation. VII, 101 pages. 1992.

Vol. 1523: W. Xue, Rings with Morita Duality. X, 168 pages. 1992.

Vol. 1524: M. Coste, L. Mahé, M.-F. Roy (Eds.), Real Algebraic Geometry. Proceedings, 1991. VIII, 418 pages. 1992.

Vol. 1525: C. Casacuberta, M. Castellet (Eds.), Mathematical Research Today and Tomorrow. VII, 112 pages. 1992.

Vol. 1526: J. Azéma, P. A. Meyer, M. Yor (Eds.), Séminaire de Probabilités XXVI. X, 633 pages. 1992.

Vol. 1527: M. I. Freidlin, J.-F. Le Gall, Ecole d'Eté de Probabilités de Saint-Flour XX – 1990. Editor: P. L. Hennequin. VIII, 244 pages. 1992.

Vol. 1528: G. Isac, Complementarity Problems. VI, 297 pages. 1992.

Vol. 1529: J. van Neerven, The Adjoint of a Semigroup of Linear Operators. X, 195 pages. 1992.

Vol. 1530: J. G. Heywood, K. Masuda, R. Rautmann, S. A. Solonnikov (Eds.), The Navier-Stokes Equations II – Theory and Numerical Methods. IX, 322 pages. 1992.

Vol. 1531: M. Stoer, Design of Survivable Networks. IV, 206 pages. 1992.

Vol. 1532: J. F. Colombeau, Multiplication of Distributions. X, 184 pages. 1992.

Vol. 1533: P. Jipsen, H. Rose, Varieties of Lattices. X, 162 pages. 1992.

Vol. 1534: C. Greither, Cyclic Galois Extensions of Commutative Rings. X, 145 pages. 1992.

Vol. 1535: A. B. Evans, Orthomorphism Graphs of Groups. VIII, 114 pages. 1992.

Vol. 1536: M. K. Kwong, A. Zettl, Norm Inequalities for Derivatives and Differences. VII, 150 pages. 1992.

Vol. 1537: P. Fitzpatrick, M. Martelli, J. Mawhin, R. Nussbaum, Topological Methods for Ordinary Differential Equations. Montecatini Terme, 1991. Editors: M. Furi, P. Zecca. VII, 218 pages. 1993.

Vol. 1538: P.-A. Meyer, Quantum Probability for Probabilists. X, 287 pages. 1993.

Vol. 1539: M. Coornaert, A. Papadopoulos, Symbolic Dynamics and Hyperbolic Groups. VIII, 138 pages. 1993.

Vol. 1540: H. Komatsu (Ed.), Functional Analysis and Related Topics, 1991. Proceedings. XXI, 413 pages. 1993.

Vol. 1541: D. A. Dawson, B. Maisonneuve, J. Spencer, Ecole d' Eté de Probabilités de Saint-Flour XXI - 1991. Editor: P. L. Hennequin. VIII, 356 pages. 1993.

Vol. 1542: J.Fröhlich, Th.Kerler, Quantum Groups, Quantum Categories and Quantum Field Theory. VII, 431 pages. 1993.

Vol. 1543: A. L. Dontchev, T. Zolezzi, Well-Posed Optimization Problems. XII, 421 pages. 1993.

Vol. 1544: M.Schürmann, White Noise on Bialgebras. VII, 146 pages. 1993.

Vol. 1545: J. Morgan, K. O'Grady, Differential Topology of Complex Surfaces. VIII, 224 pages. 1993.

Vol. 1546: V. V. Kalashnikov, V. M. Zolotarev (Eds.), Stability Problems for Stochastic Models. Proceedings, 1991. VIII, 229 pages. 1993.

Vol. 1547: P. Harmand, D. Werner, W. Werner, M-ideals in Banach Spaces and Banach Algebras. VIII, 387 pages. 1993.

Vol. 1548: T. Urabe, Dynkin Graphs and Quadrilateral Singularities. VI, 233 pages. 1993.

Vol. 1549: G. Vainikko, Multidimensional Weakly Singular Integral Equations. XI, 159 pages. 1993.

Vol. 1550: A. A. Gonchar, E. B. Saff (Eds.), Methods of Approximation Theory in Complex Analysis and Mathematical Physics IV, 222 pages, 1993.

Vol. 1551: L. Arkeryd, P. L. Lions, P.A. Markowich, S.R. S. Varadhan. Nonequilibrium Problems in Many-Particle Systems. Montecatini, 1992. Editors: C. Cercignani, M. Pulvirenti. VII, 158 pages 1993.

Vol. 1552: J. Hilgert, K.-H. Neeb, Lie Semigroups and their Applications. XII, 315 pages. 1993.

Vol. 1553: J.-L- Colliot-Thélène, J. Kato, P. Vojta. Arithmetic Algebraic Geometry. Trento, 1991. Editor: E. Ballico. VII, 223 pages. 1993.

Vol. 1554: A. K. Lenstra, H. W. Lenstra, Jr. (Eds.), The Development of the Number Field Sieve. VIII, 131 pages. 1993.

Vol. 1555: O. Liess, Conical Refraction and Higher Microlocalization. X, 389 pages. 1993.

Vol. 1556: S. B. Kuksin, Nearly Integrable Infinite-Dimensional Hamiltonian Systems. XXVII, 101 pages. 1993.

Vol. 1557: J. Azéma, P. A. Meyer, M. Yor (Eds.), Séminaire de Probabilités XXVII. VI, 327 pages. 1993.

Vol. 1558: T. J. Bridges, J. E. Furter, Singularity Theory and Equivariant Symplectic Maps. VI, 226 pages. 1993.

Vol. 1559: V. G. Sprindžuk, Classical Diophantine Equations. XII, 228 pages. 1993.

Vol. 1560: T. Bartsch, Topological Methods for Variational Problems with Symmetries. X, 152 pages. 1993.

Vol. 1561: I. S. Molchanov, Limit Theorems for Unions of Random Closed Sets. X, 157 pages. 1993.

Vol. 1562: G. Harder, Eisensteinkohomologie und die Konstruktion gemischter Motive. XX, 184 pages. 1993.

Vol. 1563: E. Fabes, M. Fukushima, L. Gross, C. Kenig, M. Röckner, D. W. Stroock, Dirichlet Forms. Varenna, 1992. Editors: G. Dell'Antonio, U. Mosco. VII, 245 pages. 1993.

Vol. 1564: J. Jorgenson, S. Lang, Basic Analysis of Regularized Series and Products. IX, 122 pages. 1993.

Vol. 1565: L. Boutet de Monvel, C. De Concini, C. Procesi, P. Schapira, M. Vergne. D-modules, Representation Theory, and Quantum Groups. Venezia, 1992. Editors: G. Zampieri, A. D'Agnolo. VII, 217 pages. 1993.

Vol. 1566: B. Edixhoven, J.-H. Evertse (Eds.), Diophantine Approximation and Abelian Varieties. XIII, 127 pages. 1993.

Vol. 1567: R. L. Dobrushin, S. Kusuoka, Statistical Mechanics and Fractals. VII, 98 pages. 1993.

Vol. 1568: F. Weisz, Martingale Hardy Spaces and their Application in Fourier Analysis. VIII, 217 pages. 1994.

Vol. 1569: V. Totik, Weighted Approximation with Varying Weight. VI, 117 pages. 1994.

Vol. 1570: R. deLaubenfels, Existence Families, Functional Calculi and Evolution Equations. XV, 234 pages. 1994.

Vol. 1571: S. Yu. Pilyugin, The Space of Dynamical Systems with the C^0-Topology. X, 188 pages. 1994.

Vol. 1572: L. Göttsche, Hilbert Schemes of Zero-Dimensional Subschemes of Smooth Varieties. IX, 196 pages. 1994.

Vol. 1573: V. P. Havin, N. K. Nikolski (Eds.), Linear and Complex Analysis – Problem Book 3 – Part I. XXII, 489 pages. 1994.

Vol. 1574: V. P. Havin, N. K. Nikolski (Eds.), Linear and Complex Analysis – Problem Book 3 – Part II. XXII, 507 pages. 1994.

Vol. 1575: M. Mitrea, Clifford Wavelets, Singular Integrals, and Hardy Spaces. XI, 116 pages. 1994.

Vol. 1576: K. Kitahara, Spaces of Approximating Functions with Haar-Like Condition. X, 110 pages. 1994.

Vol. 1577: N. Obata, White Noise Calculus and Fock Space. X, 183 pages. 1994.

Vol. 1358: D. Mumford, The Red Book of Varieties and Schemes. 2nd Printing. VII, 310 pages. 1994.